Titles in This Series

For a complete list of titles in this series, visit the AMS Bookstore at **www.ams.org/bookstore/**.

Mathematical
Surveys
and
Monographs

Volume 95

Braid and Knot Theory in Dimension Four

Seiichi Kamada

American Mathematical Society

2000 *Mathematics Subject Classification.* Primary 57Q45;
Secondary 20F36, 57M05, 57M12, 57M25, 57Q35.

ABSTRACT. Braid theory and knot theory are related to each other via two famous results due to Alexander and Markov. Alexander's theorem states that any knot or link can be put into braid form. Markov's theorem gives necessary and sufficient conditions to conclude that two braids represent the same knot or link. Thus one can use braid theory to study knot theory, and vice versa. In this book we generalize braid theory to dimension four. We develop the theory of surface braids and apply it to study surface links. Especially, the generalized Alexander and Markov theorems in dimension four are given. This book is the first place that contains a complete proof of the generalized Markov theorem.

Surface links are also studied via the motion picture method, and some important techniques of this method are studied. For surface braids, various methods to describe them are introduced and developed: the motion picture method, the chart description, the braid monodromy, and the braid system. These tools are fundamental to understanding and computing invariants of surface braids and surface links.

A table of knotted surfaces is included with a computation of Alexander polynomials. The braid techniques are extended to represent link homotopy classes.

Library of Congress Cataloging-in-Publication Data

Kamada, Seiichi, 1964–
   Braid and knot theory in dimension four / Seiichi Kamada.
      p. cm. — (Mathematical surveys and monographs ; v. 95)
   Includes bibliographical references and index.
   ISBN 0-8218-2969-6 (alk. paper)
   1. Braid theory. 2. Knot theory. I. Title. II. Mathematical surveys and monographs ; no. 95.

QA612.23 .K36   2002
514′.224—dc21
                                                            2002018274

Contents

Preface

Knot theory is currently one of the most active research fields in topology. In the classical sense it is the study of circles (closed 1-manifold) in Euclidean 3-space or a 3-sphere S^3. This is generalized to higher dimensional knot theory and furthermore to the study of manifold pairs or topological space pairs up to homeomorphism (in the topological, PL, or smooth category). Since topology is the study of topological spaces up to homeomorphism, knot theory in this global sense is a quite wide area of topology. Two-dimensional knot theory or knot theory in dimension four deals with surfaces in 4-space. We will refer to them as "surface links" in this book. This is one branch of knot theory in the global sense. However it is very mysterious. Classical knots have been studied for a long time. Since they are objects in 3-space, one can watch them directly and handle them without difficulty. One can apply 3-manifold theory, which has also been studied for a long time, to knot theory and vice versa. Of course this does not mean that classical knot theory is easy. Since surface links are objects in 4-space, we cannot see and handle them directly. One method to compute with surface links is to use motion pictures which were introduced in 1962 in Fox's famous article "A quick trip through knot theory" [168]. Another method is to use projection images in 3-space called surface link diagrams. This method is quite convenient when we describe a local configuration of a surface link or try to generalize some notion in classical knot theory to 2-dimensional knot theory, because a lot of important notions for classical knots are defined or interpreted by use of classical knot diagrams. In this book we will mainly discuss motion pictures. One reason is that there is already a good book on surface link diagrams by Carter and Saito [89]. Another reason is that there are important techniques in the motion picture method that are not so familiar. One of the main goals of this book is to introduce 2-dimensional knot theory and the technique of the motion picture method.

Braid theory also has been studied for a long time. Pioneering and systematic studies of braids were introduced in Artin's "Theorie der Zopfe" [15] in 1925 and "Theory of braids" [17] in 1947. Braid theory and knot theory are related by two famous results due to Alexander [5] in 1923 and Markov [567] in 1935. Alexander's theorem states that any knot or link can be presented as a closed braid and Markov's theorem states that such a braid form is uniquely determined up to "braid isotopy", "conjugation" and "stabilization". Therefore one can use braid theory to study knot theory and vice versa. Birman's book "Braids, links, and mapping class groups" [42] in 1974 contains a lot of results that relate knots and braids. This book is also famous for a proof of Markov's theorem. A remarkable application of braids in knot theory is the polynomial invariant discovered by Jones [298, 299] in 1985. This is one of the most powerful tools in knot theory.

There are various generalizations of classical braids: Brieskorn and Saito [62] generalized Artin's braid group from a group theoretical point of view. Another generalization is the fundamental group of a certain space. For such a space, Dahm [124] and Goldsmith [201, 202] considered a space of n-links, and Manin and Schechtman [566] a space of hyperplane arrangements. Another generalization is a braided surface defined by Rudolph [778, 783]. This gave a lot of interesting applications to knot theory; especially, Seifert ribbons for closed braids, ribbon surfaces in the 4-disk and quasi-positive braids [778, 779, 781, 782, 783, 790, 791, 792]. Viro [928] (cf. [316]) introduced the notion of a 2-dimensional braid and established generalized Alexander theorem in dimension four: Any surface link can be described in a braid form (Theorem 23.6). In this book, 2-dimensional braids are referred to as surface braids. They are similar to the braided surfaces of Rudolph. In fact they are regarded as braided surfaces with trivial boundary. As an application, a characterization of 2-knot groups and surface link groups is obtained [316]. González-Acuña [206] defined another braid form for surface links and gave another characterization of surface link groups. In this book, we study surface braids.

This book is divided into five parts. **I.** The first part is introductory material in classical braid and knot theory. This part covers the material necessary for later use. One can read this part easily. **II.** The second part is an introduction to 2-dimensional knot theory (knot theory in dimension four). It is written mainly from the view point of the motion picture method. The first two parts are written at an elementary level and are almost self-contained so that beginners and undergraduates can easily read and understand. (Some parts of the second part are technical. The reader who has difficulty may skip through such parts.) **III.** The third part is the main introduction to surface braid theory (braid theory in dimension four). The goal of this part is to understand two important notions, "braid monodromy" and "chart description". These notions can be generalized and used not only for surface braids but also other materials related to braids. **IV.** The fourth part is devoted to establishing a relationship between surface braids and surface links. Generalized Alexander and Markov theorems are given in this part. The generalized Alexander theorem was proved by Viro [928] and the author [316]. The generalized Markov theorem was announced in [317]. This book is the first place that contains a complete proof of the generalized Markov theorem. The proof is based on a manuscript [315] and prepared here for specialists. The reader who wants to learn the basics is encouraged to skip over the proof part.) **V.** In the final part, surface links are studied from the view point of surface braid theory.

This book is written for a wide target audience from beginners (including graduates) to specialists. It can be used as a graduate textbook and also as a handbook for researchers.

I would like to thank J. Scott Carter for reading a draft, a lot of conversations and encouragement. I thank Daniel Silver, Susan Williams and John Dean for fruitful discussions, and Stephen Brick for his help with computers. I thank the Department of Mathematics and Statistics, University of South Alabama, for their hospitality. Part of this book was written while I was visiting the department. I also thank Akio Kawauchi for his advice and encouragement and Taizo Kanenobu for his help. I would like to express my personal thanks to Naoko, my wife, for her constant encouragement.

Seiichi Kamada

Basic Notions and Notation

0.1. Properness and Local Flatness

DEFINITION 0.1. A manifold K embedded in a manifold M is *proper* if $\partial K = K \cap \partial M$. (In the smooth category it is also assumed that the embedding of K is transversal to ∂M.) We also say that the embedding is *proper*.

Consider Fig. 0.1, in which 1-manifolds K in a 3-ball are illustrated. The first one is proper. The second one is not proper, because one of the boundary points of K is an interior point of M. The third one is not proper, because there is an interior point x of K which is a boundary point of M.

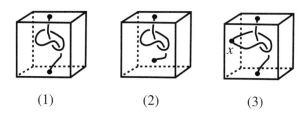

$$(1) \qquad\qquad (2) \qquad\qquad (3)$$

FIGURE 0.1. Properly and non-properly embedded manifolds

DEFINITION 0.2. Let K be a properly embedded k-manifold in an m-manifold M. (1) K is *locally flat at a point* $x \in K$ if there exists a regular neighborhood N of x in M such that the pair $(N, N \cap K)$ is homeomorphic to a standard ball pair (D^m, D^k). (2) K is *locally flat* if it is locally flat at every point of M.

EXAMPLE 0.3. Prepare an infinite series of copies of a knotted arc K_0 as in Fig. 0.1(1) whose sizes are getting smaller and arrange them in the 3-space $\mathbb{R}^3$ such that they converge into a point, say x, and the union of the infinite copies of K_0 forms an arc. Connect the endpoints of the arc by another arc in $\mathbb{R}^3$ trivially as in Fig. 0.2. Now we have a knotted circle in 3-space, say K. It is a topologically embedded 1-dimensional in $\mathbb{R}^3$. It is not locally flat at x, because any regular neighborhood N of x in $\mathbb{R}^3$ contains some copies of K_0 and the pair $(N, N \cap K)$ is never homeomorphic to a standard (3,1)-ball pair. K is locally flat at each point of K except x.

The previous example is excluded in the PL (piecewise-linear) or smooth category: K is not a PL or smoothly embedded circle in $\mathbb{R}^3$. In the PL category, any compact 1-manifold in $\mathbb{R}^3$ is regarded as the union of a finite number of 1-simplices in $\mathbb{R}^3$, although K is not. Any PL embedded circle (or a closed 1-manifold) in $\mathbb{R}^3$ is locally flat. In the smooth category, properly embedded manifolds are always locally flat.

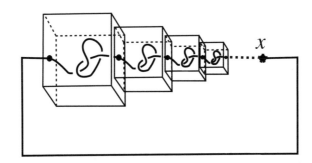

FIGURE 0.2. A topological knot with a non-locally flat point

EXAMPLE 0.4. Let K_0 be the PL embedded circle in a 3-ball M_0 illustrated on the left hand side of Fig. 0.3. Let M be a 4-ball that is the product of M_0 and the interval $[0,1]$. We identify M_0 with $M_0 \times \{0\}$. Take an interior point of M, say x, and consider a cone of K_0 from x in M, say K. The cone K is a PL properly embedded 2-manifold in M with $\partial K = K \cap \partial M = K_0 \subset M_0 \times \{0\}$. The right hand side of Fig. 0.3 is a schematic picture of K in M. K is not locally flat at x.

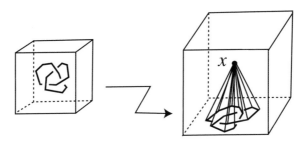

FIGURE 0.3. A disk in a 4-ball with a non-locally flat point

In this book, we work in the PL or smooth category and assume that embedded manifolds are proper and locally flat unless otherwise stated.

0.2. Isotopy

DEFINITION 0.5. Two (topological, PL or smooth) embeddings $f, f' : X \to Y$ are *isotopic* if there exists a one-parameter family of embeddings $f_t : X \to Y$ ($t \in I = [0,1]$) with $f_0 = f$ and $f_1 = f'$ such that the map

$$F : X \times I \to Y \times I \quad ; \quad F(x,t) = (f_t(x), t)$$

is a (topological, PL or smooth) embedding. The one-parameter family $\{f_t\}$ or F is called an *isotopy* connecting f and f'.

We do not require $F : X \times I \to Y \times I$ to be locally flat. When it is locally flat, we call it a *locally flat isotopy*. In the smooth category, by definition, we assume that F is extended to a map

$$F' : X \times I' \to Y \times I' \quad ; \quad F'(x,t) = \begin{cases} (f_0(x), t) & \text{for } t < 0, \\ (f_t(x), t) & \text{for } t \in I = [0,1], \\ (f_1(x), t) & \text{for } t > 1, \end{cases}$$

which is a smooth embedding, where I' is an open interval containing I. The one-parameter family $\{f_t\}_{t \in I'}$ or F' is also called an *isotopy* connecting f and f'.

0.3. Ambient Isotopy

DEFINITION 0.6. A (topological, PL, or smooth) *ambient isotopy* of Y is a one-parameter family $\{h_t\}_{t \in [0,1]}$ of (topological, PL, or smooth) homeomorphisms $h_t : Y \to Y$ such that $h_0 = \mathrm{id}$ and the map

$$H : Y \times I \to Y \times I \quad ; \quad H(y, t) = (h_t(y), t)$$

is a (topological, PL, or smooth) homeomorphism. The map H is also called an *ambient isotopy* of Y.

Let A be a subset of Y. An ambient isotopy of Y *relative to A* (or rel A) is an ambient isotopy $\{h_t\}_{t \in I}$ of Y such that $h_t|_A = \mathrm{id}_A$ for all t.

DEFINITION 0.7. Two subsets X and X' of Y are *ambient isotopic* (rel A) if there exists an ambient isotopy $\{h_t\}_{t \in I}$ of Y (rel A) such that $h_1(X) = X'$.

DEFINITION 0.8. Two embeddings $f, f' : X \to Y$ are *ambient isotopic* if there exists an ambient isotopy $\{h_t\}$ of Y such that

$$f' = h_1 \circ f.$$

The one-parameter family $\{h_t\}$ or H is called an *ambient isotopy* of Y connecting f and f'.

0.4. Notation

The following notations are used frequently.

∂X	the boundary of a manifold X
$\partial_Y X$	the boundary of X as a topological subset of Y (which is also called the *frontier* of X in Y)
$\mathrm{int}\, X$	the interior of a manifold X
$\mathrm{int}_Y X$	the interior of X as a topological subset of Y
$\mathrm{cl}(X)$, $\mathrm{cl}_Y(X)$ or $\overline{X}$	the closure of X as a topological subset of Y
$N(X)$ or $N(X; Y)$	a closed regular neighborhood of X in Y (or an open regular neighborhood when specified)

In the motion picture method, for subsets $A \subset \mathbb{R}^3$ and $J \subset \mathbb{R}^1$, we denote by AJ the subset $A \times J$ of $\mathbb{R}^4 = \mathbb{R}^3 \times \mathbb{R}^1$. In particular, $\mathbb{R}^3[a]$ means a hyperplane

$$\{(x, y, z, t) \in \mathbb{R}^4 \mid t = a\}.$$

Part 1

Classical Braids and Links

CHAPTER 1

Braids

1.1. Geometric Braids and the Braid Group

Let D^2 be a 2-disk and I an interval. We usually assume that I is the unit interval $[0, 1]$. Throughout this chapter, $pr_1 : D^2 \times I \to D^2$ and $pr_2 : D^2 \times I \to I$ will stand for the projections. Let Q_m be a set of m interior points of D^2 and fix the points.

A *geometric m-braid* is the union $b = a_1 \cup \cdots \cup a_m$ of m strings a_i $(i = 1, \ldots, m)$ in the cylinder $D^2 \times I$ such that for each $t \in I = [0, 1]$, b intersects the 2-disk $D^2 \times \{t\}$ transversely in m distinct interior points of $D^2 \times \{t\}$ and such that if $t \in \{0, 1\}$, then the intersection coincides with Q_m. The strings are assumed to be oriented in the same direction as the interval I.

The definition of a geometric m-braid is restated as follows:

DEFINITION 1.1. A *geometric m-braid* is a 1-manifold b embedded in the cylinder $D^2 \times I$ such that the restriction map $pr_2|_b : b \to I$ of pr_2 is a covering map of degree m and $\partial b = b \cap \partial(D^2 \times I) = Q_m \times \partial I$.

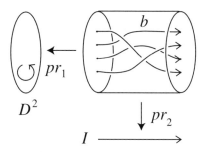

FIGURE 1.1. Geometric braid

Two geometric m-braids b and b' are *equivalent* if there exists a continuous sequence of geometric m-braids b_s $(s \in [0, 1])$ with $b_0 = b$ and $b_1 = b'$. This is an equivalence relation on the set of geometric m-braids. Each equivalence class is called an *m-braid*.

It is common to use the same symbol b for a geometric m-braid b and its equivalence class $[b]$, and this convention usually does not cause confusion. Furthermore, a geometric m-braid is often called an 'm-braid' for short.

Let $b_1 \subset D^2 \times I_1$ and $b_2 \subset D^2 \times I_2$ be geometric m-braids, where I_i $(i = 1, 2)$ is the unit interval $[0, 1]$. Identifying I_1 with $[0, 1/2]$ and I_2 with $[1/2, 1]$, we have a geometric m-braid $b_1 \cup b_2$ in $D^2 \times (I_1 \cup I_2) = D^2 \times I$. We call it the *product* or *concatenation product* of b_1 and b_2 and denote it by $b_1 \cdot b_2$ or simply by $b_1 b_2$. It is obvious that if $[b_1] = [b_1{}']$ and $[b_2] = [b_2{}']$, then $[b_1 b_2] = [b_1{}' b_2{}']$. The *product* $b_1 b_2$ of two m-braids $b_1 = [b_1]$ and $b_2 = [b_2]$ is defined by the equivalence class $[b_1 b_2]$ of the geometric m-braid $b_1 b_2$. The set of m-braids, together with this product, forms a group.

DEFINITION 1.2. The *m-braid group*, denoted by B_m, is the group of m-braids.

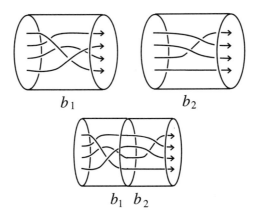

FIGURE 1.2. Product of braids

EXERCISE 1.3. Prove that the set B_m of m-braids forms a group. [*Hint*: A geometric m-braid $Q_m \times I$ is called a *trivial geometric m-braid* and its equivalence class is called the *trivial m-braid*, which is the trivial element of B_m. For a geometric m-braid b, consider the geometric m-braid b^{-1} such that $pr_1(b \cap (D^2 \times \{t\})) = pr_1(b^{-1} \cap (D^2 \times \{1 - t\}))$ for $t \in I = [0, 1]$. This is called the *inverse* of b. Its equivalence class $b^{-1} = [b^{-1}] \in B_m$ is the inverse of $b = [b]$ in B_m.]

We work in the PL or smooth category. However, until 1.6 we assume that braids are defined in the topological category.

1.2. Configuration Space

Let $P^m(W) = W \times \cdots \times W$ denote the product space of m copies of a topological space W. The *configuration space of ordered m distinct points* of W is

$$\widetilde{C}_m(W) = \{(w_1, \cdots, w_m) \in P^m(W) \mid w_i \neq w_j \text{ for } i \neq j\}.$$

The symmetric group on m letters $\{1, 2, \ldots, m\}$ acts naturally on this space by permuting the coordinates; namely,

$$\sigma(w_1, \cdots, w_m) = (w_{\sigma^{-1}(1)}, \cdots, w_{\sigma^{-1}(m)}).$$

The quotient space

$$C_m(W) = \widetilde{C}_m(W)/_\sim = \{(w_1, \cdots, w_m) \in P^m(W) \mid w_i \neq w_j \text{ for } i \neq j\}/_\sim$$

is called the *configuration space of unordered m distinct points* of W. It is naturally identified with the set

$$\{\{w_1, \cdots, w_m\} \subset W \mid w_i \neq w_j \text{ for } i \neq j\},$$

which is also called the *configuration space of unordered m distinct points* of W and is denoted by the same symbol $C_m(W)$.

In what follows, let C_m denote the configuration space $C_m(\text{int} D^2)$ of the interior of D^2.

Let $b \subset D^2 \times I$ be a geometric m-braid. For each $t \in I = [0, 1]$, $pr_1(b \cap pr_2^{-1}(t))$ is an element of C_m. Since $\partial b = Q_m \times \partial I$, we have a loop

$$g : (I, \partial I) \to (C_m, Q_m)$$

such that

$$g(t) = pr_1(b \cap pr_2^{-1}(t)) \quad \text{for } t \in I.$$

In other words, a geometric m-braid b determines a *motion* of m distinct points of $\text{int} D^2$ whose initial and terminal positions are the fixed set Q_m.

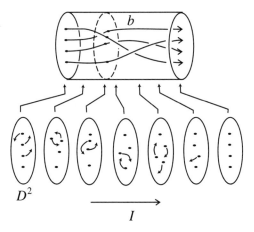

FIGURE 1.3. A motion of m points

Conversely, for any loop $g : (I, \partial I) \to (C_m, Q_m)$, there is a unique geometric m-braid b satisfying the above condition. So geometric m-braids are in one-to-one correspondence to loops in C_m with base point Q_m. This is an alternative definition of a geometric m-braid.

This correspondence preserves the product, i.e.,

$$b \cdot b' \longleftrightarrow g \cdot g'.$$

Moreover, if b and b' are equivalent by a one-parameter family of braids $\{b_s\}_{s \in [0,1]}$ with $b_0 = b$ and $b_1 = b'$, then g and g' are homotopic by the corresponding one-parameter family $\{g_s\}_{s \in [0,1]}$, and vice versa. Therefore we have

PROPOSITION 1.4. *The above bijection induces an isomorphism*

$$B_m \cong \pi_1(C_m, Q_m).$$

A geometric m-braid can also be considered as a collection of m paths $\alpha_i : I \to D^2 \times I$, $i = 1, \ldots, m$, such that (i) $pr_2 \circ \alpha_i = \text{id} : I \to I$, (ii) for each $t \in I$, $pr_1(\{\alpha_1(t), \ldots, \alpha_m(t)\})$ consists of m distinct interior points of D^2 and (iii) if $t \in \partial I$, then $pr_1(\{\alpha_1(t), \ldots, \alpha_m(t)\})$ is Q_m.

1.3. Pure Braids and Braid Permutation

Fix an order to Q_m (i.e., name the m points with integers $1, 2, \ldots, m$). The *ith string* of a geometric m-braid b is the string starting from the ith point of $Q_m = Q_m \times \{0\} \subset D^2 \times I$.

A geometric m-braid b induces a permutation, σ_b, of $\{1, 2, \ldots, m\}$ so that the ith string of b ends at the $\sigma_b(i)$th point of $Q_m = Q_m \times \{1\}$.

For example, the braids illustrated in Fig. 1.4 induce the permutations

$$\sigma_{b_1} = \begin{pmatrix} 1 & 2 & 3 & 4 \\ 3 & 2 & 4 & 1 \end{pmatrix}, \quad \sigma_{b_2} = \begin{pmatrix} 1 & 2 & 3 & 4 \\ 1 & 4 & 2 & 3 \end{pmatrix}, \quad \sigma_{b_1 b_2} = \begin{pmatrix} 1 & 2 & 3 & 4 \\ 2 & 4 & 3 & 1 \end{pmatrix}.$$

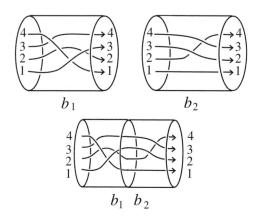

FIGURE 1.4. Braid permutation

Obviously, $\sigma_b = \sigma_{b'}$ if b is equivalent to b'. So the permutation σ_b is well-defined for an m-braid $b = [b]$.

DEFINITION 1.5. An m-braid $b = [b]$ is a *pure m-braid* if the permutation σ_b is trivial; i.e., the ith string of b ends at the ith point of $Q_m = Q_m \times \{1\}$ for every $i \in \{1, 2, \ldots, m\}$. The *pure m-braid group*, denoted by P_m, is the subgroup of B_m consisting of pure m-braids.

Let $\mathcal{S}_m$ be the *symmetric group*, which is the set of permutations of $\{1, 2, \ldots, m\}$ with the product defined by $\sigma\sigma'(i) = \sigma'(\sigma(i))$.

PROPOSITION 1.6. *The mapping $B_m \to \mathcal{S}_m$, $b \mapsto \sigma_b$ is a surjective homomorphism (epimorphism) and the kernel is P_m. So there is a short exact sequence*

$$1 \to P_m \to B_m \to \mathcal{S}_m \to 1.$$

Proof. The ith string of the product $b_1 b_2$ of two geometric m-braids is the concatenation of the ith string of b_1 and the $\sigma_{b_1}(i)$th string of b_2. Since the latter string ends at the $\sigma_{b_2}(\sigma_{b_1}(i))$th point of Q_m, we have $\sigma_{b_1 b_2}(i) = \sigma_{b_2}(\sigma_{b_1}(i))$. Thus $B_m \to \mathcal{S}_m$ is a homomorphism. The surjectivity is left to the reader as an easy exercise. By definition, an m-braid is a pure braid if and only if it is in the kernel of this homomorphism. ∎

EXERCISE 1.7. Verify the surjectivity of $B_m \to \mathcal{S}_m$.

1.4. **Brick Regular Neighborhood**

A *division* of $I = [0, 1]$ is a collection of intervals $(I_1, \ldots, I_s)$ such that $I_1 = [0, t_1], I_2 = [t_1, t_2], \ldots, I_s = [t_{s-1}, 1]$ for some real numbers $t_1, \ldots, t_{s-1}$ with $0 < t_1 < \cdots < t_{s-1} < 1$.

Let $b = a_1 \cup \cdots \cup a_m$ be a geometric m-braid in $D^2 \times I$. A closed regular neighborhood $N(b) = N(a_1) \cup \cdots \cup N(a_m)$ of b in $D^2 \times I$ is called a *brick regular neighborhood* of b if there exists a division $(I_1, \ldots, I_s)$ of I such that each $N(a_k)$ $(k = 1, \ldots, m)$ is the union of cylinders $D_{k,i} \times I_i$ $(i = 1, \ldots, s)$, where $D_{k,i}$ $(i = 1, \ldots, s)$ is a (PL or smooth) convex 2-disk in $\text{int} D^2$.

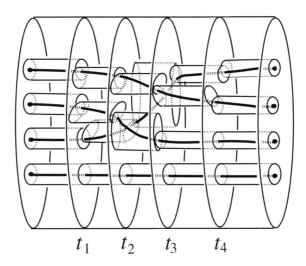

$$t_1 \qquad t_2 \qquad t_3 \qquad t_4$$

FIGURE 1.5. Brick neighborhood

LEMMA 1.8. *Any (topological) geometric m-braid b has a brick regular neighborhood $N(b)$.*

Proof. For $k = 1, \ldots, m$, let $\alpha_k : I \to D^2 \times I$ be a (topological) embedding such that $b = \alpha_1(I) \cup \cdots \cup \alpha_m(I)$ and that $pr_2 \circ \alpha_k(t) = t$ for $t \in I$. Take a value $t \in I$ and fix it. Since $\alpha_1(t), \ldots, \alpha_m(t)$ are distinct interior points of $pr_2^{-1}(t) = D^2 \times \{t\}$, there exist mutually disjoint (PL or smooth) convex 2-disks $D_1, \ldots, D_m$ in $\text{int} D^2$ such that $\alpha_k(t) \in (\text{int} D_k) \times \{t\}$ for each k $(k = 1, \ldots, m)$. There exists a positive number ϵ such that for every k $(k = 1, \ldots, m)$, the image $pr_1 \circ \alpha_k([t - \epsilon, t + \epsilon])$ is contained in $\text{int} D_k$. Then $\alpha_k([t - \epsilon, t + \epsilon]) \subset (\text{int} D_k) \times [t - \epsilon, t + \epsilon]$ for every k $(k = 1, \ldots, m)$. By compactness of I, we see the existence of a brick regular neighborhood. ∎

We call a homeomorphism $h : D^2 \times I \to D^2 \times I$ an *I-level-preserving homeomorphism* if $pr_2 \circ h = pr_2$, where $pr_2 : D^2 \times I \to I$ is the projection.

LEMMA 1.9. *Let $N(b)$ be a brick regular neighborhood of a geometric m-braid b in $D^2 \times I$, and $\{b_u\}_{u \in [0,1]}$ a continuous sequence of geometric m-braids in $D^2 \times I$ such that $b = b_0$ and $b_u \subset \text{int} N(b)$ for each $u \in [0, 1]$. There exists an ambient isotopy $\{h_u\}_{u \in [0,1]}$ of $D^2 \times I$ such that for each $u \in [0, 1]$,*

(1) $h_u(b) = b_u$,
(2) h_u is an I-level-preserving homeomorphism of $D^2 \times I$,
(3) $h_u|_{\partial(D^2 \times I)} = \mathrm{id}$ and
(4) $\mathrm{cl}\{x \in D^2 \times I | h_u(x) \neq x\} \subset N(b)$.

Proof. First we consider a special case where b is the trivial 1-braid $\{q_1\} \times I$ in $D^2 \times I$ and $N(b)$ is $D' \times I$ where D' is a 2-disk with $q_1 \in \mathrm{int}D'$ and $D' \subset \mathrm{int}D^2$. This elementary but essential case is left to the reader as an exercise. Now, consider a general case. Let $b = a_1 \cup \cdots \cup a_m$, $b_u = a_1^u \cup \cdots \cup a_m^u$ ($u \in [0,1]$), and $N(b) = N(a_1) \cup \cdots \cup N(a_m)$. Let $D_1 \times I_1, \ldots, D_s \times I_s$ be the cylinders which cover $N(a_1)$ as in the definition of a brick regular neighborhood, where $(I_1, \ldots, I_s)$ is a division of I. Let $I_1 = [0, t_1]$. Then $\mathrm{int}N(a_1) \cap (D^2 \times \{t_1\}) = (\mathrm{int}D_1 \cap \mathrm{int}D_2) \times \{t_1\}$, and $a_1 \cap (D^2 \times \{t_1\})$ and $a_1^u \cap (D^2 \times \{t_1\})$ ($u \in [0,1]$) are contained in this open disk. By an isotopy of $D^2 \times I$ satisfying (2)–(4), we may assume without loss of generality that $a_1 \cap (D^2 \times \{t_1\})$ and $a_1^u \cap (D^2 \times \{t_1\})$ ($u \in [0,1]$) are the same point. From the previous case, we see that the string $a_1 \cap (D_1 \times I_1)$ is deformed through $a_1^u \cap (D_1 \times I_1)$ by an isotopy of $D^2 \times I$ satisfying (2)–(4). Continuing this procedure, we have a desired isotopy of $D^2 \times I$. ∎

EXERCISE 1.10. Let $N(b)$ be a brick regular neighborhood of a geometric m-braid b. Prove that any geometric m-braid contained in $N(b)$ is equivalent to b.

1.5. Braid Isotopy Extension Theorem

Here we introduce Artin's braid isotopy extension theorem [**17**].

THEOREM 1.11 (Braid Isotopy Extension Theorem). *For a continuous sequence* $\{b_u\}_{u \in [0,1]}$ *of geometric m-braids in $D^2 \times I$, there exists an ambient isotopy* $\{h_u\}_{u \in [0,1]}$ *of $D^2 \times I$ such that for each $u \in [0,1]$,*

(1) $h_u(b_0) = b_u$,
(2) h_u *is an I-level-preserving homeomorphism of $D^2 \times I$, and*
(3) $h_u|_{\partial(D^2 \times I)} = \mathrm{id}$.

Proof. For each $u \in [0,1]$, let $N(b_u)$ be a brick regular neighborhood of b_u. There exists a positive number ϵ (depending on u) such that all braids b_s with $s \in [u - \epsilon, u + \epsilon]$ are contained in $\mathrm{int}_{D^2 \times I} N(b_u)$. By Lemma 1.9, the assertion of the theorem holds for the continuous sequence $\{b_s\}_{s \in [u-\epsilon, u+\epsilon]}$. Using compactness of $[0,1]$, we have the result. ∎

We regard $D^2 \times I$ as a trivial D^2-bundle over I and say that a homeomorphism $h : D^2 \times I \to D^2 \times I$ is a *fiber-preserving homeomorphism* if there exists a homeomorphism $\underline{h} : I \to I$ such that $pr_2 \circ h = \underline{h} \circ pr_2$. If $\underline{h}$ is the identity, then h is an I-level-preserving homeomorphism of $D^2 \times I$.

THEOREM 1.12. *For geometric m-braids b and b', the following conditions are mutually equivalent:*

(1) *There exists an ambient isotopy $\{h_u\}$ of $D^2 \times I$ such that $h_1(b) = b'$ and such that for each $u \in [0,1]$, h_u is an I-level-preserving homeomorphism of $D^2 \times I$ and $h_u|_{\partial(D^2 \times I)} = \mathrm{id}$.*
(2) *There exists an ambient isotopy $\{h_u\}$ of $D^2 \times I$ such that $h_1(b) = b'$ and such that for each $u \in [0,1]$, h_u is fiber-preserving and $h_u|_{D^2 \times \partial I} = \mathrm{id}$.*
(3) *There exists an ambient isotopy $\{h_u\}$ of $D^2 \times I$ such that $h_1(b) = b'$ and such that for each $u \in [0,1]$, $h_u(b)$ is a geometric m-braid and $h_u|_{D^2 \times \partial I} = \mathrm{id}$.*

(4) b and b' are equivalent; i.e., there is a continuous sequence of geometric m-braids from b to b'.

Proof. It is obvious that $(1) \Rightarrow (2) \Rightarrow (3) \Rightarrow (4)$. Theorem 1.11 implies that $(4) \Rightarrow (1)$. ∎

1.6. Polygonal Braids

Assume that the cylinder $D^2 \times I$ is in $\mathbb{R}^3$. For convenience, we assume that the braid direction is parallel to the z-axis (see Fig. 1.6).

By a *polygonal m-braid* we mean a geometric m-braid in $D^2 \times I \subset \mathbb{R}^3$ which is the union of m polygonal arcs in the cylinder.

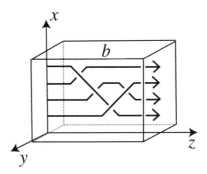

FIGURE 1.6. Polygonal braid

Let $\Delta = |p_0, p_1, p_2|$ be a linear 2-simplex in $D^2 \times I$ such that the z-coordinates of the vertices p_0, p_1, p_2 are distinct. Suppose that a polygonal m-braid b intersects Δ such that $b \cap \Delta$ is an edge of Δ, say $|p_0, p_1|$, and let b' be the geometric m-braid obtained from b by replacing $|p_0, p_1|$ with $|p_0, p_2| \cup |p_2, p_1|$. We say that b' is obtained from b, or b is obtained from b', by an *elementary move* along Δ.

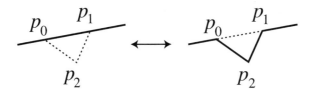

FIGURE 1.7. Elementary move

Two polygonal m-braids are said to be *combinatorially equivalent* if there is a finite sequence of polygonal m-braids between them such that each polygonal m-braid is obtained from the previous one by an elementary move.

THEOREM 1.13 (Polygonal Approximation).

(1) *Every geometric m-braid is equivalent to a polygonal m-braid.*
(2) *Two polygonal m-braids are equivalent as geometric m-braids if and only if they are combinatorially equivalent.*

Proof. The first assertion is easy, for any polygonal m-braid in a brick regular neighborhood $N(b)$ of b is equivalent to b.

The if part of the second assertion is obvious. We prove the only if part; that is, two polygonal m-braids b_0 and b_1 which are equivalent as geometric m-braids are combinatorially equivalent.

First, we consider the special case where b_0 is the trivial 1-braid $\{q_1\} \times I$ and b_1 is a polygonal 1-braid. Let $p_1 = (x_1, t_1), \ldots, p_s = (x_s, t_s)$ be the vertices of the polygonal 1-braid b_1, where x_i $(i = 1, \ldots, s)$ is an interior point of D^2 and t_i $(i = 1, \ldots, s)$ is a number in I such that $0 = t_1 < t_2 < \cdots < t_s = 1$. Note that $p_1 = (q_1, 0)$ and $p_s = (q_1, 1)$. If $s = 2$, then $b_1 = b_0$. Hence we assume that $s > 2$. For each $i = 1, \ldots, s - 1$, if $x_i \neq q_1$ and $x_{i+1} \neq q_1$, then let p'_i be the point $(q_1, (t_i + t_{i+1})/2)$ and apply an elementary move along the 2-simplex $\Delta = |p_i, p'_i, p_{i+1}|$. Then we may assume that $x_i = q_1$ for odd i and $x_j \neq q_1$ for even j. This 1-braid is transformed into the trivial 1-braid by elementary moves along $\Delta = |p_i, p_{i+1}, p_{i+2}|$ for odd i's. Therefore we have the result in this special case.

Next, we consider the case where b_1 is in a brick regular neighborhood $N(b_0) = N(a_1) \cup \cdots \cup N(a_m)$ of b_0. Let $D_1 \times I_1, \ldots, D_s \times I_s$ be the cylinders which cover $N(a_1)$ as in the definition of a brick regular neighborhood, where $(I_1, \ldots, I_s)$ is a division of I and a_1 is the first string of b_0. Let $I_1 = [0, t_1]$. Then $\mathrm{int} N(a_1) \cap (D^2 \times \{t_1\}) = (\mathrm{int} D_1 \cap \mathrm{int} D_2) \times \{t_1\}$. By elementary moves in $N(a_1)$, deform b_1 so that the intersection of the first string of b_1 with $(D^2 \times \{t_1\})$ is the same as that of b_0. By the previous case, we see that the restriction of b_1 to $D_1 \times I_1$ can be transformed into that of b_0 by elementary moves. Repeating this procedure, we see that b_1 is combinatorially equivalent to b_0.

Now, we consider the general case where b_0 is equivalent to b_1. We notice the following fact (whose proof is left to the reader):

- For two geometric m-braids b' and b'' and their brick regular neighborhoods $N(b')$ and $N(b'')$, if a geometric m-braid b is contained in $N(b') \cap N(b'')$, then b is equivalent to a polygonal m-braid contained in $N(b') \cap N(b'')$.

Using this and compactness of $[0, 1]$, we see that b_0 and b_1 are combinatorially equivalent. ∎

EXERCISE 1.14. Verify the fact in the proof of Theorem 1.13.

By a *smooth geometric m-braid* we mean a geometric m-braid b which is a smooth submanifold of $D^2 \times I$. (It is common to assume that $pr_1(b \cap pr_2^{-1}(y)) = Q_m$ for any point $y \in I$ which is sufficiently close to ∂I.) Two smooth geometric m-braids b and b' are *smoothly equivalent* if there is a (smooth) one-parameter family of smooth geometric m-braids from b to b'.

THEOREM 1.15 (Smooth Approximation).

(1) *Every geometric m-braid is equivalent to a smooth geometric m-braid.*
(2) *Two smooth geometric m-braids are equivalent as geometric m-braids if and only if they are smoothly equivalent.*

1.7. Braid Diagrams

Assume that the cylinder $D^2 \times I$ is in $\mathbb{R}^3$. For convenience, we assume that it is the product $D^3_{xyz} = D^2_{xy} \times I_z$ of $D^2_{xy} = \{(x, y) \mid x \in [0, 1], y \in [0, 1]\}$ and $I_z = \{z \mid z \in [0, 1]\}$. Moreover, $Q_m = \{q_1, \ldots, q_m\}$ are points of D^2_{xy} with $q_i = (i/(m +$

1), 1/2). (These particular coordinates of the points of Q_m are not important. All we need here is that their x-coordinates satisfy $x(q_1) < \cdots < x(q_m)$.)

Let b be a polygonal (or smooth geometric) m-braid and let $\pi : b \to D^2_{zx}$ be the restriction of the projection $D^3_{xyz} \to D^2_{zx}; (x, y, z) \mapsto (z, x)$, where $D^2_{zx} = I_z \times I_x$.

Modifying the braid b slightly up to equivalence, we assume that the multiple point set $\{q \in D^2_{zx} \mid \sharp(\pi^{-1}(q)) \geq 2\}$ consists of a finite number of transverse double points. Then we call π a *generic projection of b of the first order*. Furthermore if the z-coordinates of the double points are all distinct, then we call it a *generic projection of b of the second order*, or a *generic braid projection* (see Fig. 1.8).

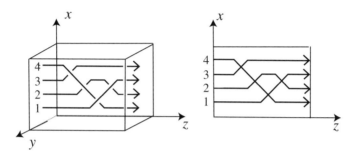

FIGURE 1.8. Generic braid projection

DEFINITION 1.16. A *braid diagram* or a *diagram of b* is the image $\pi(b)$ in the 2-disk D^2_{zx} by a generic braid projection of b equipped with over/under information over the double point singularities. (The over/under information is indicated by braking the under edge at each crossing point; see Fig. 1.9)

Note that two geometric m-braids with the same braid diagram are equivalent.

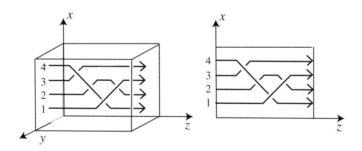

FIGURE 1.9. Braid diagram

1.8. Presentation of the Braid Group

For each $j \in \{1, \ldots, m - 1\}$, let σ_j and σ_j^{-1} be geometric m-braids whose diagrams are as in Fig. 1.10. We use the same symbols σ_j and σ_j^{-1} for the m-braids represented by them. The m-braid σ_j^{-1} is the inverse of σ_j in the braid group B_m.

Every m-braid b is represented by a polygonal (or smooth) braid b, and we may assume that the projection $\pi : b \to D^2_{zx}$ is a generic braid projection. Then

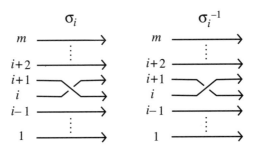

FIGURE 1.10. Standard generator

it can be described by a product of $\sigma_1, \ldots, \sigma_{m-1}$ and their inverses $\sigma_1^{-1}, \ldots, \sigma_{m-1}^{-1}$. Fig. 1.11 is an example of such a word expression. We call such a word a *braid word expression* or *braid word description of b*. The braids $\sigma_1, \ldots, \sigma_{m-1}$ generate the braid group B_m. We call them *standard generators* of the braid group B_m.

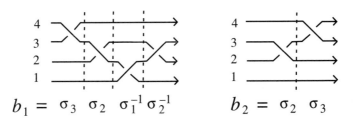

FIGURE 1.11. Braid word expression

THEOREM 1.17. *The braid group B_m has a presentation whose generators are*

$$\sigma_1, \ldots, \sigma_{m-1}$$

and the defining relations are

$$\left\{ \begin{array}{ll} \sigma_i\sigma_j\sigma_i = \sigma_j\sigma_i\sigma_j & (\text{if } |i-j| = 1) \\ \sigma_i\sigma_j = \sigma_j\sigma_i & (\text{if } |i-j| > 1) \end{array} \right.$$

for $i, j \in \{1, \ldots, m-1\}$.

This presentation of B_m is called *Artin's presentation* or the *standard presentation*.

Proof. It is easily verified that the relations hold in the m-braid group. We prove that if two polygonal m-braids b_0 and b_1 with generic braid projections are equivalent, then there is a finite sequence of polygonal m-braids with generic braid projections such that the braid word of each m-braid is obtained from the previous one by a consequence of the relations. By Theorem 1.13, there exists a finite sequence of elementary moves transforming b_0 to b_1. Modifying the sequence, we may assume that the polygonal m-braids in the sequence also have generic braid projections. Consider an elementary move in the sequence. If the 2-simplex is vertical to the zx-plane, then the move does not change the braid diagram and its word expression. If the 2-simplex is not vertical to the zx-plane, then divide the 2-simplex into some small 2-simplices such that the elementary moves along them, applied in an appropriate order, make a sequence of polygonal m-braids with generic braid projections such that the braid word of each polygonal m-braid is obtained from the previous one by a consequence of the relations. Some examples of an elementary move which contributes a braid word is given in Fig. 1.12. The details are left to the reader as an exercise. ∎

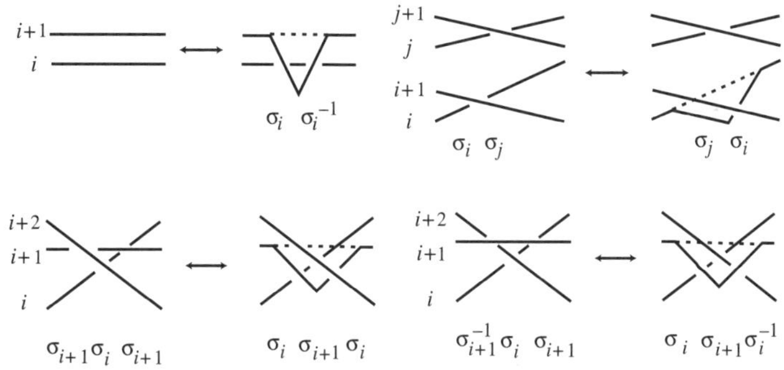

FIGURE 1.12. Some elementary moves contributing braid words

EXERCISE 1.18. Complete the proof of the theorem.

Theorem 1.17 is also proved by use of the configuration space. See [**17, 42, 57, 174, 557**].

CHAPTER 2

Braid Automorphisms

2.1. The Isotopies Associated with a Braid

A geometric m-braid b in $D^2 \times I$ induces an isotopy $a = \{a_t : Q_m \to D^2\}_{t \in I}$ of Q_m in D^2 such that

$$a_t(Q_m) = pr_1(b \cap pr_2^{-1}(t)) \quad \text{for } t \in I.$$

This isotopy has a property that

- $a_0(Q_m) = a_1(Q_m) = Q_m$ and $a_t(Q_m) \subset \text{int} D^2$.

Conversely, for any isotopy with this property, there is a geometric m-braid b as above.

DEFINITION 2.1. The *isotopy of Q_m associated with b* is the isotopy $a = \{a_t : Q_m \to D^2\}_{t \in I}$ with $a_t(Q_m) = pr_1(b \cap pr_2^{-1}(t))$ for $t \in I$.

DEFINITION 2.2. An *isotopy of D^2 associated with b* is an ambient isotopy $\{\phi_t\}$ of D^2 such that

(1) $\phi_t|_{\partial D^2} = \text{id}$,
(2) $\phi_t(Q_m) = a_t(Q_m) \ (= pr_1(b \cap pr_2^{-1}(t)))$ for $t \in I$.

DEFINITION 2.3. An *isotopy of $D^2 \times I$ associated with b* is an ambient isotopy $\{h_t\}$ of $D^2 \times I$ such that

(1) $h_1(Q_m \times I) = b$,
(2) h_t is an I-level-preserving homeomorphism of $D^2 \times I$ for each $t \in [0,1]$, and
(3) $h_t|_{D^2 \times \{0\} \cup \partial D^2 \times [0,1]} = \text{id}$ for each $t \in [0,1]$.

LEMMA 2.4. *Let b be a geometric m-braid. There is an isotopy of D^2 associated with b, and an isotopy of $D^2 \times I$ associated with b.*

Proof. Let b^{-1} be the inverse geometric m-braid in $D^2 \times [1,2]$. The concatenation product $b \cdot b'$ of b and b^{-1} in $D^2 \times ([0,1] \cup [1,2])$ is equivalent to the trivial geometric m-braid $Q_m \times [0,2]$. Let $\{b_t\}$ be a continuous sequence of geometric m-braids in $D^2 \times [0,2]$ with $b_0 = Q_m \times [0,2]$ and $b_1 = b \cdot b'$. Let $\{g_t\}$ be an isotopy of $D^2 \times [0,2]$ as in the braid isotopy extension theorem (Theorem 1.11). The restriction of this isotopy to $D^2 \times [0,1]$ gives an isotopy of $D^2 \times I$ associated with b. Now, let $\{h_t\}$ be an isotopy of $D^2 \times I$ associated with b. Since h_1 is I-level preserving, there is an isotopy $\phi = \{\phi_t : D^2 \to D^2\}$ of D^2 with

$$h_1(x, z) = (\phi_z(x), z) \quad \text{for} \quad x \in D^2, z \in I.$$

This is an ambient isotopy of D^2 associated with b. ∎

DEFINITION 2.5. A *homeomorphism of D^2 associated with b* is the terminal map $\varphi = \phi_1 : D^2 \to D^2$ of an isotopy $\{\phi_t\}$ of D^2 associated with b.

A homeomorphism of D^2 associated with a standard generator σ_i is a disk twist along a straight segment γ_i connecting q_i and q_{i+1}, that is a homeomorphism of D^2 which twists a disk neighborhood of γ_i by 180°-rotation counter-clockwise using its collar neighborhood; see Fig. 2.1.

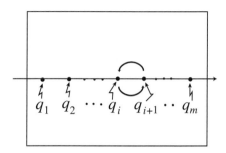

FIGURE 2.1. A homeomorphism associated with σ_i

An isotopy $\{\phi_t\}$ of D^2 associated with b is not unique and neither is the homeomorphism φ of D^2. However φ is uniquely determined as an element of the mapping class group, which will be observed later.

2.2. Mapping Class Group

Let $\mathcal{M}(D^2, Q_m)$ be the group of all homeomorphisms $f : (D^2, Q_m) \to (D^2, Q_m)$ such that $f|_{\partial D^2} = \mathrm{id}$. Here the product $fg \in \mathcal{M}(D^2, Q_m)$ is the composition $g \circ f$, namely, applying f first and then g. Two elements f and g of $\mathcal{M}(D^2, Q_m)$ are said to be *isotopic* if there is an isotopy $\{f_t\}$ with $f_0 = f$, $f_1 = g$ such that $f_t \in \mathcal{M}(D^2, Q_m)$ for all $t \in [0, 1]$.

DEFINITION 2.6. The *mapping class group of* (D^2, Q_m) (*relative to the boundary*) is the group of isotopy classes of $\mathcal{M}(D^2, Q_m)$. We denote it by $M(D^2, Q_m)$.

THEOREM 2.7. *The map*
$$\Phi : B_m \to M(D^2, Q_m), \quad [b] \mapsto [\varphi]$$
is well-defined, where φ is a homeomorphism of D_2 associated with a geometric m-braid b. Moreover this map Φ is an isomorphism.

Proof. First we check the well-definedness. Let b be a geometric m-braid and let φ and ψ be homeomorphisms of D^2 associated with b. Let $\{\varphi_t\}$ and $\{\psi_t\}$ be isotopies of $D^2 \times I$ associated with b such that $\varphi = \varphi_1$ and $\psi = \psi_1$. For each $s \in [0, 1]$, let $f_s : D^2 \to D^2$ be the map defined by
$$f_s(x) = \varphi_1 \circ \varphi_s^{-1} \circ \psi_s(x).$$
Then $\{f_s\}$ is an isotopy of D^2 with $f_0 = \varphi_1$ and $f_1 = \psi_1$ such that $f_s \in \mathcal{M}(D^2, Q_m)$. Thus $[\varphi] = [\psi]$ in $M(D^2, Q_m)$. If b and b' are equivalent geometric m-braids, then by the braid isotopy extension theorem (Theorem 1.11) we may assume that $b = b'$. Thus Φ is well-defined as a map.

Let $b_1 \subset D^2 \times I_1$ and $b_2 \subset D^2 \times I_2$ be geometric m-braids and $\{h_t\}$ and $\{g_t\}$ be isotopies of $D^2 \times I_1$ and of $D^2 \times I_2$ associated with b_1 and b_2, respectively.

Assume that $I_1 = [0, 1/2]$ and $I_2 = [1/2, 1]$ and put $I = [0, 1]$. Then the product $b_1 \cdot b_2$ of b_1 and b_2 is the union of b_1 and b_2 in $D^2 \times I$. Using the isotopies $\{h_t\}$ and $\{g_t\}$, we can construct an isotopy $\{f_t\}$ of $D^2 \times I$ associated with $b_1 \cdot b_2$ such that the homeomorphism $D^2 \to D^2$ determined from it is the composition of those determined from $\{h_t\}$ and $\{g_t\}$. Thus the map Φ is a homomorphism.

Let φ be an element of $M(D^2, Q_m)$. There is an isotopy $\{\varphi_t\}$ of D^2 such that $\varphi_0 = \text{id}$ and $\varphi_1 = \varphi$, and $\varphi_t|_{\partial D^2} = \text{id}$ for each $t \in [0, 1]$. Let b be a geometric m-braid determined from $\varphi_t(Q_m) = pr_1(b \cap pr_2^{-1}(t))$ in $D^2 \times I$ for each $t \in [0, 1]$. Then $\Phi(b) = [\varphi]$. Thus Φ is an epimorphism.

The injectivity of Φ is postponed until Sect. 2.5. ∎

2.3. Hurwitz Arc System

Let q_0 be a boundary point of D^2, and let m be a non-negative integer.

DEFINITION 2.8. A *Hurwitz path system* (with base point q_0) is an ordered m-tuple $\mathcal{A} = (\alpha_1, \ldots, \alpha_m)$ of simple paths $\alpha_i : [0, 1] \to D^2$ such that

(1) the intersection of the image of α and ∂D^2 is q_0 and this is the terminal point $\alpha_i(1)$ of α_i for each i,
(2) the intersection of the images of α_i and α_j is q_0 for $i \neq j$,
(3) the images of $\alpha_1, \ldots, \alpha_m$ appear in this order around the point q_0.

The *starting point set* of $\mathcal{A}$ is the set $\{\alpha_1(0), \ldots, \alpha_m(0)\}$.

Two Hurwitz path systems $\mathcal{A}$ and $\mathcal{A}'$ with the same starting point set are said to be *equivalent* if there is a one-parameter family of Hurwitz path systems $\mathcal{A}_t$ (whose base points are q_0) with $\mathcal{A}_0 = \mathcal{A}$ and $\mathcal{A}_1 = \mathcal{A}'$ such that the starting point sets of $\mathcal{A}_t$ for all t are identical.

DEFINITION 2.9. A *Hurwitz arc system* (with base point q_0) is an ordered (and oriented) m-tuple $\mathcal{A} = (a_1, \ldots, a_m)$ of simple arcs on D^2 such that

(1) the intersection of a_i and ∂D^2 is q_0 and this is the terminal point of a_i,
(2) the intersection of a_i and a_j is q_0 for $i \neq j$,
(3) $a_1, \ldots, a_m$ appear in this order around the base point q_0.

The *starting point set* of $\mathcal{A}$ is the set of the initial points of $a_1, \ldots, a_m$.

Two Hurwitz arc systems $\mathcal{A}$ and $\mathcal{A}'$ with the same starting point set are *equivalent* if there is a one-parameter family of Hurwitz arc systems $\mathcal{A}_t$ (whose base points are q_0) with $\mathcal{A}_0 = \mathcal{A}$ and $\mathcal{A}_1 = \mathcal{A}'$ such that the starting point sets of $\mathcal{A}_t$ for all t are identical.

Fix a Hurwitz arc system $\mathcal{A}_0 = (a_1^0, \ldots, a_m^0)$ whose starting point set is a given set of m interior points of D^2, say Q_m. Let $f : (D^2, Q_m) \to (D^2, Q_m)$ be a homeomorphism whose restriction to ∂D^2 is the identity map, i.e., $f \in \mathcal{M}(D^2, Q_m)$. The image $f(\mathcal{A}_0) = (f(a_1^0), \ldots, f(a_m^0))$ is another Hurwitz arc system whose starting point set is Q_m. Obviously, if $[f] = [g] \in M(D^2, Q_m)$, then $f(\mathcal{A}_0)$ is equivalent to $g(\mathcal{A}_0)$. Thus we have a map

$$M(D^2, Q_m) \to \{\text{Hurwitz arc systems with starting point set } Q_m\}/_\sim.$$

LEMMA 2.10. *The map*

$$M(D^2, Q_m) \to \{\text{Hurwitz arc systems with starting point set } Q_m\}/_\sim$$

is a bijection.

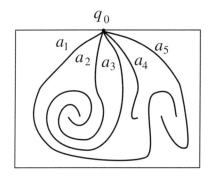

FIGURE 2.2. Hurwitz arc system

Proof. We construct an inverse map. For a Hurwitz arc system $\mathcal{A} = (a_1, \ldots, a_m)$ whose starting point set is Q_m, we construct a homeomorphism $f \in \mathcal{M}(D^2, Q_m)$ with $f(\mathcal{A}_0) = \mathcal{A}$ as follows: Consider the map from $\partial D^2 \cup a_1^0 \cup \cdots \cup a_m^0$ to $\partial D^2 \cup a_1 \cup \cdots \cup a_m$ such that it maps ∂D^2 to ∂D^2 identically, each a_i^0 to a_i homeomorphically. Let E^0 and E be the 2-disks obtained from D^2 by cutting along $a_1^0, \ldots, a_m^0$ and $a_1, \ldots, a_m$, respectively. We have already defined a homeomorphism from ∂E^0 to ∂E. This is extended to a homeomorphism $E^0 \to E$, and we have a desired homeomorphism f. Two homeomorphisms f and f' obtained this way are isotopic, since their restrictions to $\partial D^2 \cup a_1^0 \cup \cdots \cup a_m^0$ are isotopic relative to $\partial D^2 \cup Q_m$, and this isotopy is extended to an isotopy of E^0 to E. Note that two homeomorphisms $E^0 \to E$ whose restrictions to the boundary are the same are isotopic relative to the boundary. Therefore f and f' are isotopic. ∎

We notice that the bijection in this lemma depends on the choice of $\mathcal{A}_0$.

2.4. Hurwitz Arc System of a Braid

Assume that D^2 is a 2-disk $D_{xy}^2 = \{(x, y) \mid x \in [0, 1], y \in [0, 1]\}$ in the xy-plane $\mathbb{R}^2$ and $Q_m = \{q_1, \ldots, q_m\}$ are points with $q_i = (i/(m+1), 1/2)$ as before. Take a base point q_0 in the edge of the square D^2 defined by $y = 1$; see Fig. 2.3.

A *standard Hurwitz arc system* is a Hurwitz arc system $\mathcal{A}_0 = (a_1^0, \ldots, a_m^0)$ such that each a_i^0 is a straight segment from the point q_i to the base point q_0.

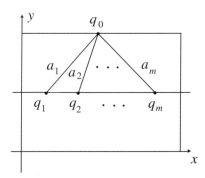

FIGURE 2.3. Standard Hurwitz arc system

DEFINITION 2.11. A *Hurwitz arc system associated with b* is the image $\varphi(\mathcal{A}_0)$ of the standard Hurwitz arc system $\mathcal{A}_0$ by a homeomorphism $\varphi : D^2 \to D^2$ associated with b.

Using the standard Hurwitz arc system $\mathcal{A}_0$, we have a bijection

$$M(D^2, Q_m) \to \{\text{Hurwitz arc systems with starting point set } Q_m\}/_\sim$$

as in Lemma 2.10. Composing the map $\Phi : B_m \to M(D^2, Q_m)$ (in Theorem 2.7) and this bijection, we have a map

$$B_m \to \{\text{Hurwitz arc systems with starting point set } Q_m\}/_\sim .$$

This sends an m-braid $[b]$ to the equivalence class of a Hurwitz arc system associated with b.

2.5. Injectivity of Φ

We prove the injectivity of the epimorphism $\Phi : B_m \to M(D^2, Q_m)$ in Theorem 2.7.

Proof. Let b a geometric m-braid representing an m-braid $b = [b]$ in B_m such that $\Phi(b) = 1$ in $M(D^2, Q_m)$. Let $\phi = \{\phi_z\}_{z \in [0,1]}$ be an isotopy of D^2 associated with b and put $\varphi = \phi_1$, which is a homeomorphism of D^2 associated with b. By assumption it is isotopic to the identity, and the Hurwitz arc system $\varphi(\mathcal{A}_0)$ is equivalent to $\mathcal{A}_0$, where $\mathcal{A}_0 = (a_1^0, \ldots, a_m^0)$ is the standard Hurwitz arc system. Changing the representative b up to equivalence and changing the isotopy ϕ if necessary, we may assume that for a sufficiently small positive number ϵ, $pr_1(b \cap pr_2^{-1}(z)) = Q_m$ and $\phi_z = \mathrm{id}$ for all $z \in [1 - \epsilon, 1]$, and that $\varphi(\mathcal{A}_0) = \mathcal{A}_0$. Since $\varphi(q_i) = q_i$ for $i = 1, \ldots, m$, b is a pure m-braid. Without loss of generality, we may assume that the first string $b_{(1)}$ of b is the straight segment $\{q_1\} \times I$ in $D^2 \times I$. Here we use notations $b_{(1)}, \ldots, b_{(m)}$ for the strings of b (instead of $a_1, \ldots, a_m$) in order to avoid confusing them with the symbols $a_1^0, \ldots, a_m^0$ for the Hurwitz arc system.

Consider the one-parameter family of arcs $\phi_z(a_1^0)$ ($z \in [0,1]$) in D^2. This satisfies the conditions that $\partial \phi_z(a_1^0) = \{q_0, q_1\}$ for each $z \in I$ and that $\phi_0(a_1^0) = \phi_1(a_1^0) = a_1^0$. There is an ambient isotopy $\{h_t\}$ of $D^2 \times I$ such that

(1) $h_1(a_1^0 \times \{z\}) = \phi_z(a_1^0) \times \{z\}$,
(2) h_t is an I-level-preserving homeomorphism of $D^2 \times I$, and
(3) $h_t|_{\partial(D^2 \times I)} = \mathrm{id}$.

Using this isotopy, we may assume that $\phi_z(a_1^0) = a_1^0$ for all $z \in [0, 1]$. This implies that we may assume ϕ_z is the identity map on $\partial D^2 \cup a_1^0$. Let $D^{2'}$ be the 2-disk obtained from D^2 by cutting along a_1^0 and b' the geometric $(m-1)$-braid $b \setminus b_{(1)}$ in $D^{2'} \times I$. Applying the same argument for this braid, we may assume that the second string $b_{(2)}$ of b (the first string of b') is the trivial segment $\{q_2\} \times I$. Repeating this procedure, we see that b is equivalent to the trivial geometric m-braid $Q_m \times I$. This completes the proof of the injectivity of Φ. $\blacksquare$

2.6. Artin's Braid Automorphism

Let D^2, $Q_m = \{q_1, \ldots, q_m\}$ and q_0 be as in Sect. 2.4.

Let $\mathcal{A} = (a_1, \ldots, a_m)$ be a Hurwitz arc system with starting point set Q_m. For each $i \in \{1, \ldots, m\}$, consider a loop c_i in $D^2 \setminus Q_m$ with base point q_0 such that it goes along a_i, turns around the initial point of a_i in the positive direction and comes back along a_i. Let η_i be the element of $\pi_1(D^2 \setminus Q_m, q_0)$ represented by this loop c_i. The fundamental group is a free group of rank m generated by $\eta_1, \ldots, \eta_m$. We call the ordered m-tuple $(\eta_1, \ldots, \eta_m)$ the *Hurwitz generator system* of $\pi_1(D^2 \setminus Q_m, q_0)$ associated with $\mathcal{A}$. In particular, it is called the *standard generator system* of $\pi_1(D^2 \setminus Q_m, q_0)$ if $\mathcal{A}$ is the standard Hurwitz arc system $\mathcal{A}_0$.

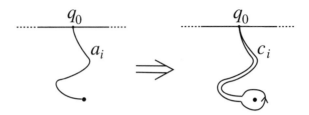

FIGURE 2.4. A loop c_i

Using the standard generating system, we identify the fundamental group $\pi_1(D^2 \setminus Q_m, q_0)$ with the free group F_m generated by $\{\eta_1, \ldots, \eta_m\}$.

DEFINITION 2.12. The *braid automorphism*, or *Artin's automorphism*, of F_m of an m-braid $b = [b]$ is the automorphism of $F_m = \pi_1(D^2 \setminus Q_m, q_0)$ induced by a homeomorphism of D^2 associated with b. We denote it by $\mathrm{Artin}(b)$.

By Theorem 2.7, we see that Artin's automorphism, $\mathrm{Artin}(b)$, is well-defined. A quick method to calculate Artin's automorphism is given in Sect. 27.1.

A homeomorphism of D^2 associated with a standard generator σ_i is a disk twist as in Fig. 2.1, which affects a_i's and η_i's as in Fig. 2.5. The induced automorphism of $\pi_1(D^2 \setminus Q_m, q_0) = F_m$ is given by

$$\mathrm{Artin}(\sigma_i)(\eta_k) = \begin{cases} \eta_k & \text{if} \quad k \neq i, i+1, \\ \eta_i\, \eta_{i+1}\, \eta_i^{-1} & \text{if} \quad k = i, \\ \eta_i & \text{if} \quad k = i+1. \end{cases}$$

Similarly, Artin's automorphism of σ_i^{-1} is given by

$$\mathrm{Artin}(\sigma_i^{-1})(\eta_k) = \begin{cases} \eta_k & \text{if} \quad k \neq i, i+1, \\ \eta_{i+1} & \text{if} \quad k = i, \\ \eta_{i+1}^{-1}\, \eta_i\, \eta_{i+1} & \text{if} \quad k = i+1. \end{cases}$$

For a group G, we denote by $\mathrm{Aut}^{\mathrm{R}}(G)$ the group of automorphisms of G such that the product $fg \in \mathrm{Aut}^{\mathrm{R}}(G)$ is $g \circ f$, i.e., applying f first and then g. We have a homomorphism

$$\mathcal{M}(D^2, Q_m) \to \mathrm{Aut}^{\mathrm{R}}(\pi_1(D^2 \setminus Q_m, q_0)) = \mathrm{Aut}^{\mathrm{R}}(F_m)$$

that maps $[\varphi]$ to the induced automorphism $\varphi_\sharp$ of $\pi_1(D^2 \setminus Q_m, q_0) = F_m$. Then the composition of the two homomorphisms $\Phi : B_m \to \mathcal{M}(D^2, Q_m)$ and $\mathcal{M}(D^2, Q_m) \to \mathrm{Aut}^{\mathrm{R}}(F_m)$ is the homomorphism

$$\mathrm{Artin} : B_m \to \mathrm{Aut}^{\mathrm{R}}(F_m).$$

Since we know braid automorphisms for standard generators of B_m, when a braid is expressed by a braid word, we can obtain Artin's automorphism algebraically.

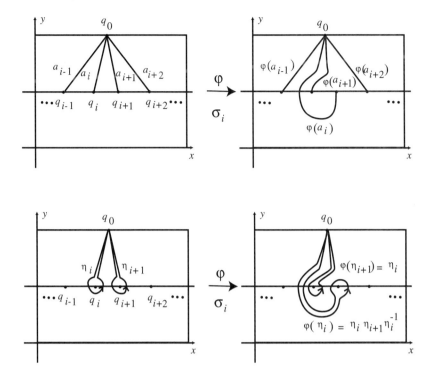

FIGURE 2.5. Artin's automorphism of σ_i

2.7. Slide Action of the Braid Group

For a group G, let $P^m(G)$ be the Cartesian product of m copies of G. A group homomorphism $f : G \to H$ induces a map $P^m(f) : P^m(G) \to P^m(H)$ by $P^m(f)(g_1, \ldots, g_m) = (f(g_1), \ldots, f(g_m))$. The *slide action* of the braid group B_m on $P^m(G)$ is determined by

$$\mathrm{slide}(\sigma_j)(g_1, \ldots, g_m) = (g_1, \ldots, g_{j-1}, g_j g_{j+1} g_j^{-1}, g_j, g_{j+2}, \ldots, g_m),$$

where we assume B_m acts from the left side; namely, $\mathrm{slide}(\beta_1 \beta_2)(g_1, \ldots, g_m) = \mathrm{slide}(\beta_1)(\mathrm{slide}(\beta_2)(g_1, \ldots, g_m))$.

DEFINITION 2.13. Two elements of $P^m(G)$ are *slide equivalent* if they are in the same orbit of the slide action of B_m.

Let $\mathcal{A}$ be a Hurwitz arc system in D^2 with starting point set Q_m and base point q_0, and let $\eta_1, \ldots, \eta_n$ be the generators of $\pi_1(D^2 \setminus Q_m, q_0)$ associated with $\mathcal{A}$.

PROPOSITION 2.14. *Let f be an automorphism of $F_m = \pi_1(D^2 \setminus Q_m, q_0)$. The following conditions are mutually equivalent:*

(1) *f is induced from a homeomorphism $g \in \mathcal{M}(D^2, Q_m)$.*
(2) *f is $\mathrm{Artin}(\gamma)$ for an m-braid γ.*
(3) *There is an m-braid γ such that*

$$\mathrm{slide}(\gamma)(\eta_1, \ldots, \eta_m) = P^m(f)(\eta_1, \ldots, \eta_m).$$

Furthermore m-braids γ appearing in (2) and (3) are the same.

Proof. The equivalence of (1) and (2) has been observed in Sect. 2.6. The equivalence of (2) and (3) is directly seen from the definition. ∎

EXERCISE 2.15. Complete the proof of Proposition 2.14.

DEFINITION 2.16. The *trace map*, trace : $P^m(G) \rightarrow G$, is a map sending $(g_1, \ldots, g_m)$ to the product $g_1 \ldots g_m \in G$.

The trace map is invariant under slide equivalence; namely, if $(g_1, \ldots, g_m)$ and $(g'_1, \ldots, g'_m)$ are slide equivalent, then $\text{trace}(g_1, \ldots, g_m) = \text{trace}(g'_1, \ldots, g'_m)$.

CHAPTER 3

Classical Links

3.1. Knots and Links

We will work in the PL category and omit the prefix PL.

DEFINITION 3.1. A *link* is a closed 1-manifold embedded in $\mathbb{R}^3$. A *knot* is a link with a single component (i.e., an embedded circle in $\mathbb{R}^3$). An *oriented link* is a link which is given an orientation as a 1-manifold.

A link is called a *topological link* or a *smooth link* when we work in the topological or smooth category.

Two links L and L' are *equivalent* if there exists an orientation-preserving homeomorphism $h : \mathbb{R}^3 \to \mathbb{R}^3$ such that $h(L) = L'$. When they are oriented links, we require that the induced homeomorphism $h_L : L \to L'$ preserves the orientations. A *link type* is the equivalence class of a link.

Two links are *ambient isotopic* if they are ambient isotopic as subsets of $\mathbb{R}^3$. When they are oriented links, it is assumed that the orientations are preserved.

PROPOSITION 3.2. *Two links are equivalent if and only if they are ambient isotopic.*

Proof. If two links L and L' are ambient isotopic, there is an ambient isotopy $\{h_t\}$ of $\mathbb{R}^3$ with $L' = h_1(L)$. Since h_1 is an orientation-preserving homeomorphism, L and L' are equivalent. Conversely, suppose that L and L' are equivalent links and let $h : \mathbb{R}^3 \to \mathbb{R}^3$ be an orientation-preserving homeomorphism with $L' = h(L)$. Define an ambient isotopy $\{h_t\}$ of $\mathbb{R}^3$ by $h_t(x) = th(x) + (1-t)x$ for $x \in \mathbb{R}^3$. This is an ambient isotopy connecting L and L'. (In the smooth category, modify $\{h_t\}$ so that it is smooth at $t = 0$ and $t = 1$.) ∎

A *polygonal link* is a link which is a disjoint union of some simple closed polygonal curves in $\mathbb{R}^3$. Since it is compact, it has a triangulation with a finite number of 1-simplices, called *edges*. Since we are working in the PL category, a link will be assumed to be a polygonal link.

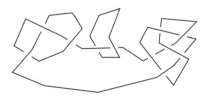

FIGURE 3.1. Polygonal link

27

A topological link is said to be *tame* if it is (topologically) equivalent to a polygonal link; otherwise it is *wild*. It is known that the following three families are in one-to-one correspondence to each other:

{smooth link types} $\longleftrightarrow$ {(PL) link types} $\longleftrightarrow$ {tame topological link types}.

This is not true in higher dimensional cases.

We call a closed 1-manifold embedded in a 3-sphere S^3 a *link in S^3*. Two links in S^3 are *equivalent* if there is an orientation-preserving homeomorphism $S^3 \to S^3$ which sends one to the other. When two links are oriented, we require that the homeomorphism restricted to the links is also orientation-preserving.

For a link L in S^3, by removing a point (or a 3-ball), we may regard it as a link in $\mathbb{R}^3$. Conversely, a link in $\mathbb{R}^3$ may be regarded as a link in S^3 by the Alexandroff one-point compactification of $\mathbb{R}^3$. It is common to assume that S^3 is $\{x \in \mathbb{R}^4 \mid |x| = 1\}$ and use a stereographic projection or its PL version.

EXERCISE 3.3. Show that the above correspondence induces a bijection between the family of link types in $\mathbb{R}^3$ and that in S^3.

3.2. Basic Symmetries

A *mirror image* of a link L is a link which is the image of L by an orientation-reversing homeomorphism of $\mathbb{R}^3$. It is denoted by L^*. When L is oriented, a mirror image is also oriented.

An *orientation-reversed link* of an oriented link L is a link which is obtained from L by reversing the orientation as a 1-manifold. It is denoted by $-L$.

DEFINITION 3.4. A link L is *invertible* if L is equivalent to $-L$. A link L is *positive amphicheiral* if L is equivalent to L^*. A link L is *negative amphicheiral* if L is equivalent to $-L^*$.

The same terminology will be applied to oriented surface links.

3.3. The Regular Neighborhood of a Link

Let L be a link and let $N(L)$ be a regular neighborhood of L in $\mathbb{R}^3$. There is a homeomorphism $f : N(L) \to D^2 \times L$ such that the core {center of D^2} $\times L$ is identified by L. Therefore $N(L)$ is often called a *tubular neighborhood*. And a homeomorphism $N(L) \to D^2 \times L$ is called a *trivialization* of $N(L)$.

A regular neighborhood (or a tubular neighborhood) $N(L)$ is unique in the following sense (cf. [**771**]). For two regular neighborhoods $N(L)$ and $N'(L)$, there is an ambient isotopy $\{h_t\}$ of $\mathbb{R}^3$ rel L with $h_1(N(L)) = N'(L)$ and their trivializations $f : N(L) \to D^2 \times L$ and $f' : N'(L) \to D^2 \times L$ such that $f = F' \circ h_1$.

We denote by $E(L)$ the space $\mathrm{cl}\,(\mathbb{R}^3 \setminus N(L))$ and call it an *exterior of L*. By uniqueness of a regular neighborhood, two exteriors of L are ambient isotopic in $\mathbb{R}^3$ rel L. On the other hand, $\mathbb{R}^3 \setminus L$ is called the *complement of L*.

Let K be a knot and let $N(K)$ be a tubular neighborhood with a trivialization $f : N(K) \to D^2 \times K$. A *meridian disk* of K is $f^{-1}(D^2 \times \{y\})$ for a point $y \in K$. The boundary of a meridian disk is called a *meridian*.

When K is oriented, we usually assume that a meridian disk is oriented as in Fig. 3.2. (This follows the "tangent followed by normal" rule.) Then the boundary

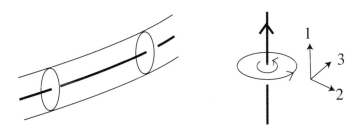

FIGURE 3.2. A tubular neighborhood and an oriented meridian

with induced orientation is a *positively oriented meridian*; otherwise it is a *negatively oriented meridian*.

A loop $f^{-1}(\{x\} \times K)$ for a point $x \in \partial D^2$ is called a *longitude* of K. When K is oriented, we assume that a longitude is oriented in the same direction.

A meridian is characterized by a condition that it is a simple loop in $\partial N(K)$ which is null-homotopic in $N(K)$. (For such a loop, there is a suitable trivialization $f : N(K) \to D^2 \times K$.) It is uniquely determined from K up to ambient isotopy of $\partial N(K)$ if K is oriented. A longitude is characterized by the condition that it is a simple loop in $\partial N(K)$ which is homotopic to K in $N(K)$. However it is not unique. By twisting along a meridian disk as in Fig. 3.3, the (free) homotopy class of a longitude in $\partial N(K)$ changes. The first homology group $H_1(E(K); \mathbb{Z})$ is an infinite cyclic group which is generated by the class of a meridian. Let $\pi_1(E(K)) \to H_1(E(K); \mathbb{Z}) \cong \mathbb{Z}$ be the Hurewicz homomorphism. A *preferred longitude of K* is a longitude whose image in $H_1(E(K); \mathbb{Z})$ is 0. Twisting as in Fig. 3.3 changes the homology class of a longitude by -1 or $+1$. Thus a preferred longitude of K exists. Moreover, since an ambient isotopy class of a simple loop in a torus is determined by the homology class (cf. Rolfsen[**761**]), a preferred longitude is uniquely determined.

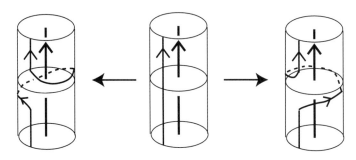

FIGURE 3.3. Twistings

EXERCISE 3.5. Prove that the first homology group $H_1(E(K); \mathbb{Z})$ of a knot exterior is an infinite cyclic group which is generated by the class of the meridian.

3.4. Trivial Links

A link L with d components is said to be *trivial* or *unknotted* if there is a 2-manifold M which is a union of d mutually disjoint embedded 2-disks such that $\partial M = L$. A trivial knot is also called an *unknot*.

PROPOSITION 3.6. *For a link L in $\mathbb{R}^3$, the following are equivalent.*

(1) *L is trivial.*

(2) *L is equivalent (or ambient isotopic) to a link which is contained in a 2-plane in $\mathbb{R}^3$.*

Proof. (1) $\Rightarrow$ (2) If L is trivial, then it bounds a 2-manifold M which is a union of mutually disjoint embedded 2-disks. Shrink each 2-disk into a 2-simplex. Then M can be moved into a 2-plane. The boundary is equivalent to L. (2) $\Rightarrow$ (1) Let L be in a 2-plane H. Each component of L bounds a 2-disk in H (this fact is well-known as the Schönflies theorem). Push off the 2-disks from the 2-plane so that they are disjoint and let M be the result. The boundary of M is equivalent to L. ∎

THEOREM 3.7 (Unknotting theorem). *A knot K is trivial if and only if $\pi_1(\mathbb{R}^3 \setminus K)$ is an infinite cyclic group. A link L is trivial if and only if $\pi_1(\mathbb{R}^3 \setminus L)$ is a free group.*

The first assertion is proved by use of Dehn's lemma [**724**]; see [**761**] for example. It is generalized to a link version in [**584**].

3.5. Split Union and Connected Sum

Let L_1 and L_2 be links in $\mathbb{R}^3$. Let S^2 be a 2-sphere embedded in $\mathbb{R}^3$. Put a copy of L_1 inside of S^2 and a copy of L_2 outside. Then the union $L_1 \cup L_2$ is regarded as a link in $\mathbb{R}^3$. A *split union* of L_1 and L_2 is such a link. S^2 is called a *splitting sphere of L*. A split union of L_1 and L_2 is uniquely determined up to equivalence.

Let L_0 be a split union of links L_1 and L_2 separated by a splitting sphere S^2. Let $f : [0,1] \times [0,1] \to \mathbb{R}^3$ be an embedding such that

(1) $f^{-1}(L_1) = \{0\} \times [0,1]$,

(2) $f^{-1}(L_2) = \{1\} \times [0,1]$,

(3) $f^{-1}(S^2) = [0,1] \times \{1/2\}$.

Then consider a new link $L = (L_0 \setminus f(\{0,1\} \times [0,1])) \cup f([0,1] \times \{0,1\})$. We call this link a *connected sum* of L_1 and L_2. The sphere S^2 is called a *decomposing sphere of L*. When L_1 and L_2 are oriented links and the image $f([0,1] \times [0,1])$ is an oriented 2-disk whose orientation is compatible with the orientations of L_1 and L_2, we call it an *oriented connected sum*.

If L_1 and L_2 are oriented knots, then an oriented connected sum is uniquely determined up to equivalence. However if they are links, a connected sum is not determined uniquely. We have to specify which components are connected by the operation.

3.6. Combinatorial Equivalence

Let L be a polygonal link. Let $\Delta = |p_0, p_1, p_2|$ be a linear 2-simplex in $\mathbb{R}^3$ such that the intersection of L and Δ is an edge of Δ, say $|p_0, p_1|$. Another polygonal link, say L', is obtained from L by replacing $|p_0, p_1|$ with $|p_0, p_2| \cup |p_2, p_1|$. We call such a replacement an *elementary move of type* I, and the inverse operation an *elementary move of type* II. See Fig. 1.7. When L is oriented, we give L' an orientation which is induced from $L \setminus |p_0, p_1|$.

Two links are said to be *combinatorially equivalent* if there is a finite sequence of links between them such that each link is obtained from the previous one by an elementary move.

Let D be a polyhedral 2-disk in $\mathbb{R}^3$ such that $L \cap D$ is an arc of ∂D. Replacing the arc $L \cap D$ by the complementary arc in ∂D, we have another link. Such a replacement is called a *cellular move*.

THEOREM 3.8. *For two links L and L', the following conditions are mutually equivalent.*

(1) *L and L' are equivalent.*
(2) *L and L' are ambient isotopic.*
(3) *There exists a finite sequence of cellular moves transforming L to L'.*
(4) *L and L' are combinatorially equivalent.*

It is obvious that $(4) \Rightarrow (3) \Rightarrow (2) \Rightarrow (1)$. We will prove the following lemma $((1) \Rightarrow (4))$ in Sect. 3.7. Then we have the theorem.

LEMMA 3.9. *Two equivalent links are combinatorially equivalent.*

3.7. Regular Projections

Let v be a vector of $\mathbb{R}^3$. Take a 2-plane H in $\mathbb{R}^3$ which is orthogonal to v and consider a projection $\pi : \mathbb{R}^3 \to H$ along v.

DEFINITION 3.10. A link L is *in general position* with respect to $\pi : \mathbb{R}^3 \to H$, or π is a *regular projection* of L, if the restriction map $\pi|_L : L \to H$ is a generic map; namely, it is an immersion and each multiple point is a transverse double point.

Any link L can be modified slightly up to equivalence, or up to combinatorial equivalence, so that π is a regular projection.

For a link L and a vector v, we denote by $L + v$ the link $\{x + v \in \mathbb{R}^3 \mid x \in L\}$.

LEMMA 3.11. *Let L be a link and v a unit vector such that a projection $\pi : \mathbb{R}^3 \to H$ along v is a regular projection of L. Then L and $L + \lambda v$ are combinatorially equivalent for any $\lambda \in \mathbb{R}$.*

Proof. For each double point p of $\pi(L)$, let d_p be one third of the distance between the two points of L that are the preimage of p. Let ϵ be the minimum among d_p for all double points. For any λ_1, λ_2 with $-\epsilon < \lambda_1 < \lambda_2 < \epsilon$, the union $L + \mu v$ for $\mu \in [\lambda_1, \lambda_2]$ is an embedded annuli. Using a triangulation of this set, we see that $L + \lambda_1 v$ and $L + \lambda_2 v$ are combinatorially equivalent. Since $\pi : \mathbb{R}^3 \to H$ is a regular projection of $L + \lambda v$ for every $\lambda \in \mathbb{R}$, we can take such a positive number ϵ for each $L + \lambda v$. Using compactness of an interval, we have that L and $L + \lambda v$ are combinatorially equivalent. ∎

Proof of Lemma 3.9. Let L and L' be links which are equivalent. We may assume that L is in general position with respect to a projection $\pi : \mathbb{R}^3 \to H$ along a vector v. Let $h : \mathbb{R}^3 \to \mathbb{R}^3$ be an orientation-preserving homeomorphism with $h(L) = L'$. Since L is compact, we may assume that h has a compact support; i.e, there exists a compact set K such that any point x with $h(x) \neq x$ is in K. Let λ be a real number such that $L + \lambda v$ is disjoint with K. By the previous lemma, there is a finite sequence of elementary moves transforming L to $L + \lambda v$. Apply the homeomorphism to L, $L + \lambda v$ and all the 2-simplices used for the elementary moves. They are mapped to L', $L + \lambda v$ and some combinatorial 2-disks along which cellular moves transform L' to $L + \lambda v$. Thus L' is combinatorially equivalent to $L + \lambda v$. Therefore L and L' are combinatorially equivalent. ∎

3.8. Link Diagrams

Let L be a link and let $\pi : \mathbb{R}^3 \to H$ be a regular projection of L along a vector v. For each double point p of the projection $\pi(L)$ of L, the preimage consists of two points of L, say x_1 and x_2. We call x_1 the *over crossing* and x_2 the *under crossing* of L if there is a positive real number λ with $x_1 - x_2 = \lambda v$. The edge of L containing x_1 is the *over edge* on p and the edge of L containing x_2 is the *under edge* on p. The image of the over edge (or under edge) is also called the *over edge* (or *under edge*) on p.

A *link diagram of L*, or a *diagram of L* for short, is the image $\pi(L)$ equipped with information about which edge is over or under for each crossing.

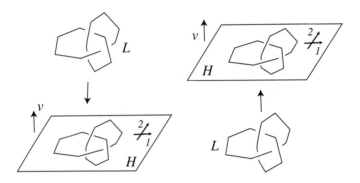

FIGURE 3.4. Link diagram

There are several methods to give over/under information. A standard method is as follows: Consider a small open interval containing x_2 in the under edge on p. Remove the portion corresponding to this open interval from the projection image $\pi(L)$; see Fig. 3.4.

Another way is to put symbols near the edges in order to indicate "over" and "under": for example, "o" and "u" (over and under), "a" and "b" (above and below), "+" and "−", etc.

EXERCISE 3.12. If two links L and L' have the same link diagram with respect to a projection $\pi : \mathbb{R}^3 \to H$, then they are equivalent. Prove this.

For an oriented plane H in $\mathbb{R}^3$, a normal vector v is *positive* if two vectors representing the orientation of H followed by the vector v coincide with the standard orientation of $\mathbb{R}^3$; otherwise v is *negative*. We always assume that the projection plane H of a projection $\pi : \mathbb{R}^3 \to H$ is oriented such that v is a positive normal vector.

PROPOSITION 3.13. *Let D and D' be link diagrams of two links L and L' with respect to a projection $\pi : \mathbb{R}^3 \to H$ along a vector v and another $\pi' : \mathbb{R}^3 \to H'$ along v', respectively. Suppose that there exists an orientation-preserving homeomorphism $g : H \to H'$ with $g(D) = D'$. Then L and L' are equivalent.*

Proof. Any point of $\mathbb{R}^3$ is written as $x + \lambda v$ uniquely for a point $x \in H$ and $\lambda \in \mathbb{R}$. The homeomorphism $g : H \to H'$ induces an orientation-preserving homeomorphism $h : \mathbb{R}^3 \to \mathbb{R}^3$ by $h(x + \lambda v) = g(x) + \lambda v'$. The image $h(L)$ of L by h, which is equivalent to L, is a link whose diagram with respect to π' is D'. By Exercise 3.12, $h(L)$ is equivalent to L'. ∎

By this proposition, once we have a link diagram of L, we may forget the vector v and the projection plane H. When they are not given, we may assume, for example, that $v = (0, 0, 1)$ and H is the xy-plane $\mathbb{R}^2$ in $\mathbb{R}^3$.

3.9. Reidemeister Moves

Let $\pi : \mathbb{R}^3 \to \mathbb{R}^2$ be a projection. Let D and D' be link diagrams in $\mathbb{R}^2$. Suppose that D and D' are the same outside of a 2-disk E in $\mathbb{R}^2$ and that $D \cap E$ and $D' \cap E$ are related by a move as in Fig. 3.5. Then we say that D' is obtained from D by a *Reidemeister move*.

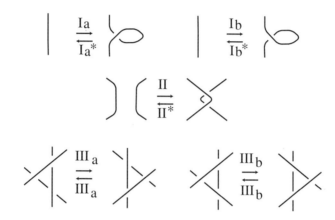

FIGURE 3.5. Reidemeister moves

DEFINITION 3.14. Two link diagrams D and D' are *equivalent* if there is a finite sequence $D = D_0, D_1, \ldots, D_m = D'$ of link diagrams such that for each $i \in \{1, \ldots, m\}$, D_i is obtained from D_{i-1} by an ambient isotopy of $\mathbb{R}^2$ or a Reidemeister move.

THEOREM 3.15. *Let D and D' be link diagrams of links L and L' respectively. The links L and L' are equivalent if and only if D and D' are equivalent.*

Proof. First we prove the if part. If D and D' are ambient isotopic, then L and L' are equivalent. This follows from Lemma 3.12. If D and D' are related by a single Reidemeister move, then by Lemma 3.12, we may assume that the restrictions of L and L' to $\pi^{-1}(\mathbb{R}^2 \setminus E)$ are the same. Then it is easily seen that L and L' are equivalent. Therefore we have the if part. We prove the only if part. Suppose that L and L' are equivalent. There is a finite sequence $L = L_0, \ldots, L_n = L'$ of links such that each L_i is obtained from L_{i-1} by an elementary move along a 2-simplex in $\mathbb{R}^3$, say Δ_i. By rotating $\mathbb{R}^3$ slightly, we may assume that the projection $\pi : \mathbb{R}^3 \to \mathbb{R}^2$ is a regular projection for every L_i, without changing the ambient isotopy types of D and D' in $\mathbb{R}^2$. Let $D_0, \ldots, D_n$ be the link diagrams of $L = L_0, \ldots, L_n = L'$. If the image $\pi(\Delta_i)$ is an arc (i.e. Δ_i is perpendicular to the plane $\mathbb{R}^2$), then D_i is the same as D_{i-1}. If $\pi(\Delta_i)$ is a triangle, divide Δ_i into small 2-simplices such that Δ-moves along them (applied in an appropriate order) induce a sequence of Reidemeister moves connecting D_{i-1} and D_i. ∎

REMARK 3.16. A Reidemeister move of type I_b is a consequence of moves of type I_a^* and type II; see Fig. 3.6. A Reidemeister move of type III_b is a consequence of moves of type III_a, type II and type II*; see Fig. 3.7. Therefore we have the following: Two link diagrams D and D' are *equivalent* if there is a finite sequence $D = D_0, D_1, \ldots, D_m = D'$ of link diagrams such that for each $i \in \{1, \ldots, m\}$, D_i is obtained from D_{i-1} by an ambient isotopy of $\mathbb{R}^2$ or a Reidemeister move of type I_a, I_a^*, II, II*, III_a or III_a^*.

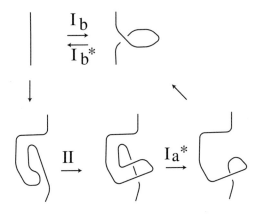

FIGURE 3.6

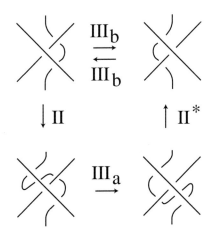

FIGURE 3.7

When we consider oriented links and oriented link diagrams, we need an oriented version of Reidemeister moves. For a move of type I, there exists two possible orientations. For a move of type II, there exist four orientations, and for a move of type III, there exist eight orientations. They are called *oriented Reidemeister moves*. Two oriented link diagrams D and D' are *equivalent* if there is a finite sequence $D = D_0, D_1, \ldots, D_m = D'$ of oriented link diagrams such that for each i ($i = 1, \ldots, m$), D_i is obtained from D_{i-1} by an ambient isotopy of $\mathbb{R}^2$ or an oriented Reidemeister move.

Theorem 3.15 is valid in the oriented case.

REMARK 3.17. By the previous remark, we may remove oriented versions of moves of type I_b, I_b^*, III_b, and III_b^* from our list of moves, if we want. Moreover any oriented version of a type III_a move is a consequence of a particular oriented version of a type III_a move illustrated as in Fig. 3.8 and the other oriented moves in our lists. Thus we have the following: Two oriented link diagrams D and D' are *equivalent* if there is a finite sequence $D = D_0, D_1, \ldots, D_m = D'$ of oriented link diagrams such that for each i ($i = 1, \ldots, m$), D_i is obtained from D_{i-1} by an ambient isotopy of $\mathbb{R}^2$ or an oriented Reidemeister move listed in Fig. 3.8.

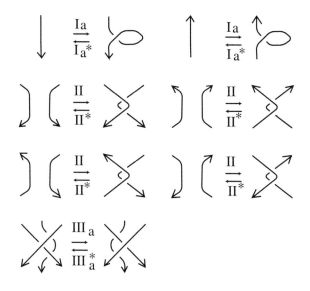

FIGURE 3.8. Oriented Reidemeister moves

EXERCISE 3.18. Verify that any oriented version of a type III_a move is a consequence of the oriented version as in Fig. 3.8 and oriented type II (type II*) moves.

3.10. The Group of a Link

Let L be a link in $\mathbb{R}^3$. The *link group of L*, or the *group of L*, is the fundamental group $\pi_1(\mathbb{R}^3 \setminus L)$ of the complement of L.

We explain how to get a presentation of the group of L from a link diagram.

Let $\pi : \mathbb{R}^3 = \mathbb{R}^2 \times \mathbb{R} \to \mathbb{R}^2 \times \{0\} = \mathbb{R}^2$ be a projection $(x, y, z) \mapsto (x, y, 0)$ or $(x, y, z) \mapsto (x, y)$. Let L be a link in general position with respect to π. A link diagram is the image $\pi(L)$ with over/under information. Let Σ be the set of double points. Let $a_1, \ldots, a_n$ be the connected components of $\pi(L) \setminus \Sigma$, each of which is an open arc or a loop embedded in $\mathbb{R}^2$. For each component a_i, fix a normal orientation, which will be denoted by a small arrow intersecting the curve a_i in a diagram.

For a curve γ in $\mathbb{R}^2$ which avoids Σ and intersects $\pi(L) \setminus \Sigma = a_1 \cup \cdots \cup a_n$ transversely, let $w(\gamma)$ denote a word in $\{x_1, \ldots, x_n\}$ such that the kth letter is x_i (or x_i^{-1}, resp.) if the kth intersection of γ with $\pi(L) \setminus \Sigma$ is in a_i and the orientation of γ matches the normal orientation of a_i (or does not, resp.).

Let $v_1, \dots, v_m$ be the double points. For each double point v_i ($i \in \{1, \dots, m\}$), consider a small oriented disk Δ_i in $\mathbb{R}^2$ whose center is v_i.

A *base relator of a double point* v is a word $w(\partial \Delta_i)$.

An *upper relator of* v is a word obtained from $w(\partial \Delta_i)$ by removing two letters that come from the lower arcs.

A *lower relator of* v is a word obtained from $w(\partial \Delta_i)$ by removing two letters that come from the upper arcs.

For a double point as in Fig. 3.9, a base relator is $x_5^{-1} x_1 x_8^{-1} x_3^{-1}$, an upper relator is $x_5^{-1} x_8^{-1}$, and a lower relator is $x_1 x_3^{-1}$.

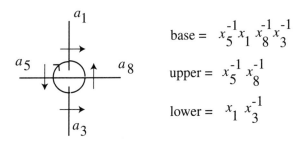

$$\text{base} = x_5^{-1} x_1 \; x_8^{-1} x_3^{-1}$$

$$\text{upper} = x_5^{-1} \; x_8^{-1}$$

$$\text{lower} = x_1 \; x_3^{-1}$$

FIGURE 3.9. Relators coming from a double point

An *upper presentation* (resp. *lower presentation*) associated with the diagram $\pi(L)$ is a presentation whose generators are $x_1, \dots, x_n$ and the relators are base relators and upper relators (resp. lower relators) of the double points.

REMARK 3.19. For an oriented 2-disk Δ_i, a base relator $w(\partial \Delta_i)$ is not determined uniquely. It depends on the starting point of $\partial \Delta_i$. If we change the orientation of Δ_i and hence the orientation of $\partial \Delta_i$, then the base relator changes. There are 8 possibilities of a base relator of a double point. However, the group presented by the upper (or lower) presentation does not change.

PROPOSITION 3.20. (1) *An upper presentation associated with a link diagram* $\pi(L)$ *presents the link group of* L. (2) *A lower presentation associated with a link diagram* $\pi(L)$ *presents the link group of* L.

Proof. We prove (1). For each double point $v_i \in \Sigma$, let B_i^3 be a 3-ball $\Delta_i \times [-1, 1] \subset \mathbb{R}^2 \times \mathbb{R} = \mathbb{R}^3$. We may assume that the restriction of L to each B_i^3 is as in Fig. 3.10(2) and the remainder is in $\mathbb{R}^2 \times \{0\}$. The fundamental group of $\mathbb{R}^2 \times [-1, \infty) \setminus (L \cup_{i=1}^n B_i)$ is a free group generated by $x_1, \dots, x_n$, where x_j is a meridian loop corresponding to a_j as in Fig. 3.10(2) with a base point in $\mathbb{R}^2 \times (1, \infty)$. Attach 2-disks $\Delta_i \times \{-1\}$ ($i \in \{1, \dots, n\}$) to this space; then the fundamental group has base relators for v_i. For each $i \in \{1, \dots, n\}$, consider a properly embedded 2-disk Δ_i' in B_i whose boundary is a loop as in Fig. 3.10(3). Attaching 2-disks Δ_i' ($i \in \{1, \dots, n\}$), we have upper relators. Now the space is homotopy equivalent to $\mathbb{R}^3 \setminus L$. Thus we have (1). (2) is proved similarly by taking a base point in $\mathbb{R}^2 \times (-\infty, -1)$. ∎

EXAMPLE 3.21. Let L be a knot whose projection is illustrated in Fig. 3.11(A). This knot is called a (*right-hand*) *trefoil*. The double point set Σ consists of three points, v_1, v_2, v_3. $\pi(L) \setminus \Sigma$ has 6 connected components $a_1, \dots, a_6$. We give normal orientations to them as in the figure. Base relators for v_1, v_2 and v_3 are $x_4^{-1} x_1^{-1} x_5 x_2$,

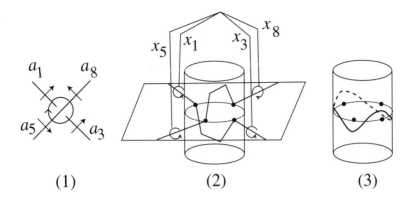

FIGURE 3.10

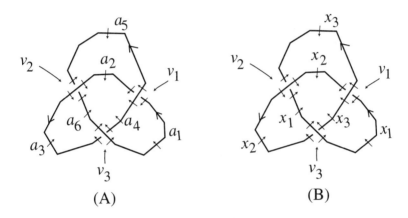

FIGURE 3.11

$x_2^{-1}x_5^{-1}x_3x_6$, and $x_6^{-1}x_3^{-1}x_1x_4$. Upper relators are $x_4^{-1}x_5$, $x_2^{-1}x_3$, and $x_6^{-1}x_1$. Thus an upper presentation is

$$\left\langle\ x_1, \ldots, x_6\ \middle|\ \begin{array}{c} x_4^{-1}x_1^{-1}x_5x_2,\ x_2^{-1}x_5^{-1}x_3x_6,\ x_6^{-1}x_3^{-1}x_1x_4, \\ x_4^{-1}x_5,\ x_2^{-1}x_3,\ x_6^{-1}x_1 \end{array}\ \right\rangle.$$

Using upper relators, we can eliminate x_2, x_4, x_6 and we have

$$\langle\ x_1, x_3, x_5\ |\ x_5^{-1}x_1^{-1}x_5x_3,\ x_3^{-1}x_5^{-1}x_3x_1,\ x_1^{-1}x_3^{-1}x_1x_5\ \rangle.$$

By the last relator, $x_5 = x_1^{-1}x_3x_1$. Thus we have

$$\langle\ x_1, x_3\ |\ x_1^{-1}x_3^{-1}x_1^{-1}x_3x_1x_3\ \rangle.$$

When we need only an upper presentation, we can reduce the number of the generators from an upper presentation. Assume that L is oriented. Give a normal orientation to each arc a_i induced from the orientation of $\pi(L)$ in $\mathbb{R}^2$. We use "tangent followed normal" rule. This means a vector of $\pi(L)$ giving the orientation of L followed by a normal vector is the standard orientation of $\mathbb{R}^2$. See Fig. 3.11 for example. Then for each double point, an upper relator implies $x_a = x_d$ if x_a and x_d

correspond to the arcs in the same upper arc of the diagram. Thus we can remove x_d in the first place by assuming $x_d = x_a$. Now the base relator implies $x_c = x_a^{-1}x_bx_a$, where we assume the arc corresponding to x_c is in the positive normal direction of the upper arc (i.e. on the left side of the upper arc corresponding to x_a). So we associate $x_1, \ldots, x_k$ with the upper arcs of the diagram, and associate a relator (a relation) as in Fig. 3.12 with each double point. For example, see Fig. 3.11(B). Then we have

$$\langle\, x_1, x_2, x_3 \,\big|\, x_2 = x_3^{-1}x_1x_3,\ x_1 = x_2^{-1}x_3x_2,\ x_3 = x_1^{-1}x_2x_1 \,\rangle.$$

From this presentation, we have

$$\langle\, x_1, x_2 \,\big|\, x_1x_2x_1 = x_2x_1x_2 \,\rangle.$$

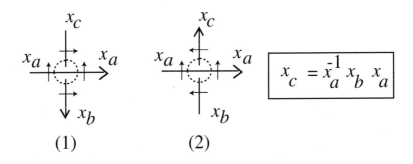

FIGURE 3.12. Relations in upper presentation

3.11. A Note on Knots as Embeddings

For any knot K, by definition, there exists an embedding $f : S^1 \to \mathbb{R}^3$ with $f(S^1) = K$. Such an embedding f is not unique.

DEFINITION 3.22. Two embeddings f and f' of S^1 into $\mathbb{R}^3$ are *knot equivalent* if there exists a homeomorphism $g : S^1 \to S^1$ and an orientation-preserving homeomorphism $h : \mathbb{R}^3 \to \mathbb{R}^3$ such that $h \circ f = f' \circ g$ (the following diagram is commutative).

$$
\begin{array}{ccc}
S^1 & \xrightarrow{\ f\ } & \mathbb{R}^3 \\
{\scriptstyle g}\downarrow & & \downarrow{\scriptstyle h} \\
S^1 & \xrightarrow[\ f'\ ]{} & \mathbb{R}^3
\end{array}
$$

Two embeddings f and f' yield equivalent knots if and only if they are knot equivalent. When oriented knots are considered, it is required that $g : S^1 \to S^1$ is orientation-preserving. In this case, we say that f and f' are *oriented knot equivalent*.

$$\left\{\begin{array}{c}\text{knot equivalence classes}\\\text{of embeddings from } S^1 \text{ into } \mathbb{R}^3\end{array}\right\} \longleftrightarrow \{\,\text{knot types}\,\}$$

$$\left\{\begin{array}{c}\text{oriented knot equivalence classes}\\\text{of embeddings from } S^1 \text{ into } \mathbb{R}^3\end{array}\right\} \longleftrightarrow \{\,\text{oriented knot types}\,\}$$

PROPOSITION 3.23. *Two embeddings $f : S^1 \to \mathbb{R}^3$ and $f' : S^1 \to \mathbb{R}^3$ are oriented knot equivalent if and only if there exists an orientation-preserving homeomorphism $h : \mathbb{R}^3 \to \mathbb{R}^3$ such that $h \circ f = f'$.*

This implies that, when we consider oriented knots, we do not need to take care of g in Definition 3.22.

Proof. The if part is trivial. We prove the only if part. Suppose $h \circ f = f' \circ g$ for some g and h. Since $g : S^1 \to S^1$ is orientation-preserving homeomorphism, there is an isotopy $\{g_t : S^1 \to S^1\}$ connecting the identity map and g. Using a tubular neighborhood $N(K)$ of $K = f(S^1)$ in $\mathbb{R}^3$, one can construct an ambient isotopy $\{\phi_t\}$ of $\mathbb{R}^3$ such that (i) for all t, $\phi_t(K) = K$ and $\phi_t(y) = y$ for $y \in \mathbb{R}^3 \setminus N(K)$ and (ii) $\phi_1 \circ f = f \circ g$. Replace h by $h \circ \phi_1^{-1}$. ∎

COROLLARY 3.24. *Two oriented knots are ambient isotopic (or equivalent as oriented knots) if and only if two embeddings yielding them are ambient isotopic.*

REMARK 3.25. Let us say that two knots K and K' are *isotopic* if there exist embeddings f and f' of S^1 into $\mathbb{R}^3$ yielding K and K' such that f and f' are isotopic as embeddings. This definition is not appropriate. In fact, *any knot is isotopic to a trivial knot in this sense.* [This is seen as follows. Deform K by an isotopy such that it is the union of a knotted arc in a cube and a trivial arc outside the cube as at the left in Fig. 3.13. Pulling the string in the cube tightly, we get the knotted part smaller and smaller toward a single point. Such a deformation is realized by an isotopy (which is not locally flat).] Thus we need to assume that the isotopy between f and f' is locally flat. In this case, using the isotopy extension theorem (cf. [**771**]), we see that K and K' are equivalent.

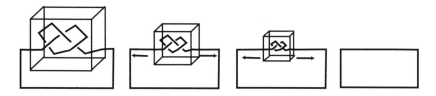

FIGURE 3.13. Isotopic deformation

CHAPTER 4

Braid Presentation of Links

4.1. Presenting Links by Braids

Let b be a geometric m-braid in $D^2 \times I$. Put $D^2 \times I$ in $\mathbb{R}^3$ and attach m arcs to b trivially as in Fig. 4.1. We have a link in $\mathbb{R}^3$. We say that such a link is in an *m-braid form*, or a *braid form*. When we consider an oriented link, we require that the geometric m-braid is oriented compatibly.

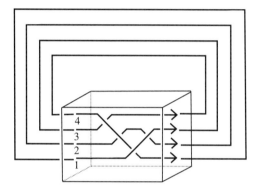

FIGURE 4.1. Braid form

Let b be a geometric $2m$-braid in $D^2 \times I$. Put $D^2 \times I$ in $\mathbb{R}^3$ and attach m arcs to the upper part of b and m arcs to the lower part trivially as in Fig. 4.2. Then we have a link in $\mathbb{R}^3$. We say that such a link is in a *2m-plat form*, or a *plat form*. When we consider an oriented link, a braid appearing in a plat form is oriented such that m strings are oriented in the same direction and the other m strings are oriented in the opposite direction.

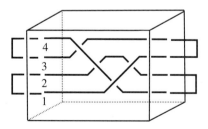

FIGURE 4.2. Plat form

EXERCISE 4.1. Every link type is represented by a link in a $2m$-plat form for some m. Prove this.

4.2. The Braiding Process

In this section we prove Alexander's Theorem [**5**].

THEOREM 4.2 (Alexander's Theorem). *Every (oriented) link type is represented by a link in a braid form.*

Let $\pi : \mathbb{R}^3 \to H$ be a projection onto a 2-plane H along a vector v, and let ℓ be a line in $\mathbb{R}^3$ which is parallel to v. For simplicity we assume that v is $e_z = (0, 0, 1)$ and H is the xy-plane $\mathbb{R}^2$. Then ℓ is the z-axis.

A 1-simplex $|p_0, p_1|$ in $\mathbb{R}^3$ is *in general position with respect to ℓ* if there exists no 2-plane in $\mathbb{R}^3$ containing ℓ and $|p_0, p_1|$. A 1-simplex $|q_0, q_1|$ in $\mathbb{R}^2$ is *in general position with respect to the origin O* of $\mathbb{R}^2$ if there exists no line in $\mathbb{R}^2$ containing O and $|q_0, q_1|$. Note that a 1-simple in $\mathbb{R}^3$ is in general position with respect to ℓ if and only if its image under π is a 1-simplex in $\mathbb{R}^2$ which is in general position with respect to O.

A 1-simplex $|q_0, q_1|$ in $\mathbb{R}^2$ which is in general position with respect to the origin O is *positive* if the 2-simplex $|O, q_0, q_1|$ defines the same orientation as the standard orientation of $\mathbb{R}^2$; otherwise it is *negative*. A 1-simplex in $\mathbb{R}^3$ which is in general position with respect to ℓ is a *positive 1-simplex* if the image under π is a positive 1-simplex; otherwise it is a *negative 1-simplex*.

DEFINITION 4.3. An oriented link L is a *closed braid about ℓ* if every edge is positive.

Proof of Theorem 4.2. If L is not oriented, give an orientation and suppose that L is an oriented link. We prove that L is equivalent to a closed braid about ℓ. Fix a triangulation of L. By a slight deformation of vertices of L, we can put L such that every edge of L is in general position with respect to ℓ and the projection π is a regular projection of L. If all edges are positive, it is a closed braid. So we suppose that there exists a negative edge. Let α be an arc in L which is a negative edge or a union of some negative edges of L. If (i) there is no crossing point on α or (ii) there is a single crossing point, then consider a polygonal arc α' in $\mathbb{R}^3$ such that every edge of α' is positive and a replacement of $\pi(\alpha)$ by $\pi(\alpha')$ yields a new oriented link diagram. Here we assume that all crossings on α' are over or under in case (i). In case (ii), we assume that all crossings on α' are over (resp. under) if the crossing point on α is under (resp. over). Note that L and L' are equivalent. This replacement reduces the number of negative edges of L. If there exists more than one crossing point on α, then divide the arc α into some arcs such that each subarc satisfies the above condition (we take a subdivision of the triangulation of L restricted to α). This increases the number of negative edges of L temporarily. Apply the above argument to each subarc inductively and let L' be the result. Then the number of negative edges of L' is smaller than that of L. By induction on the number of negative edges, we have the result. ∎

Fig. 4.3 shows how to deform a link (diagram) of a figure 8 knot into a link (diagram) which is a closed braid about ℓ, where the base point means the image of ℓ (which we usually assume to be the origin O). The dotted arcs in the figure are α (or its subarc) in the proof. The last step in the figure is a deformation of the link diagram by an ambient isotopy of $\mathbb{R}^2$.

Let L be an oriented link such that every edge of L is in general position with respect to ℓ. A point of L is a *critical point* if it is a common vertex of a positive

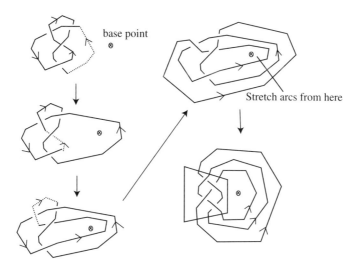

base point

Stretch arcs from here

FIGURE 4.3. Braiding of a figure 8 knot

edge and a negative edge. It is called a *minimal critical point* (or a *maximal critical point*) if a positive edge follows a negative edge (or a negative edge follows a positive edge) at the point along the orientation of L.

REMARK 4.4. In the proof of Theorem 4.2, we can choose an arc α arbitrarily such that it is a union of some negative edges of L. It is natural to choose α to be maximal so that it starts from a maximal critical point and terminates at a minimal critical point. In case there is a component of L consisting of negative edges only, we cannot take such a maximal arc α. Apply the argument in the proof for a non-maximal arc α on such a component of L and assume that there is a positive edge in each component. Then we can use induction on the number of such arcs α instead of the number of negative edges. This modification removes a nuisance to counting negative edges depending on a triangulation of L. Especially it can be applied to a smooth link.

For a smooth link L in $\mathbb{R}^3 \backslash \ell$, let $p : L \to S^1$ be the composition of the projection $\pi : \mathbb{R}^3 \backslash \ell \to \mathbb{R}^2 \backslash O$ and the projection $\mathbb{R}^2 \backslash O \to S^1; x \mapsto x/|x|$. Modifying L up to smooth equivalence, we assume that the map p is a Morse function into S^1. A regular point $x \in L$ of p is *positive* (or *negative*) if there is an open neighborhood of x in L which is mapped to an open arc in S^1 oriented in the same direction as S^1 (or opposite). When L consists of positive regular points only, we say that L is a *closed braid about* ℓ. Fig. 4.4 is a smooth version of Fig. 4.3.

EXERCISE 4.5. Prove that any oriented smooth link type is represented by a closed braid about ℓ.

4.3. Markov's Theorem

Let b be an m-braid. Consider a conjugate $\beta^{-1}b\beta$ of b by another m-braid β. Obviously the closure of $\beta^{-1}b\beta$ and the closure of b are equivalent as a link in $\mathbb{R}^3$.

DEFINITION 4.6. A *Markov move of type* I, or a *conjugation*, is a transformation $B_m \to B_m$ which takes a braid b to the conjugation by the m-braid β.

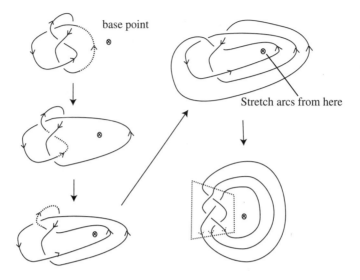

FIGURE 4.4. Smooth braiding of a figure 8 knot

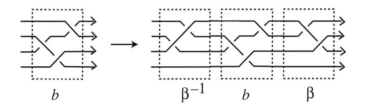

FIGURE 4.5. Conjugation (Markov I)

Let b be an m-braid. By introducing a trivial string as the last string, we regard b as an $(m+1)$-braid. The closure of $b\sigma_m$ and the closure of b are equivalent as a link in $\mathbb{R}^3$. Similarly, the closure of $b\sigma_m^{-1}$ and the closure of b are equivalent as a link in $\mathbb{R}^3$.

DEFINITION 4.7. A *Markov move of type* II, or a *stabilization*, is a transformation $B_m \to B_{m+1}$ which maps b to $b\sigma_m$ or $b\sigma_m^{-1}$. The former one is called a *positive stabilization* and the latter is a *negative stabilization*. The inverse operation of a stabilization is called a *destabilization*.

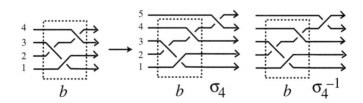

FIGURE 4.6. Stabilization (Markov II)

DEFINITION 4.8. Two braids b and b' are *Markov equivalent* if there is a finite sequence $b = b_0, b_1, \ldots, b_n = b'$ such that each b_i is obtained from b_{i-1} by Markov moves (conjugations, stabilizations and destabilizations).

THEOREM 4.9 (Markov's Theorem). *The closures of two braids b and b' are equivalent as a link in $\mathbb{R}^3$ if and only if they are Markov equivalent.*

A proof of this theorem is found in Birman's book [**42**]. In the next chapter, we introduce the idea of the proof.

CHAPTER 5

Deformation Chain and Markov's Theorem

We introduce the notion of a deformation chain. It was used in Birman's book [42] to prove Markov's theorem. An analogous notion will be used in order to prove Markov's theorem in dimension four. So it is worthwhile to understand this notion in the classical dimension. Skipping over this part is recommended to the reader who does not care about a proof of Markov's theorem in dimension three or in dimension four.

5.1. $\mathcal{E}$ Operation

Let $\pi : \mathbb{R}^3 \to \mathbb{R}^2$ and ℓ be as in Sect. 4.2. For a subset X of $\mathbb{R}^3 \setminus \ell$, put

$$J(X) = \{\lambda p \in \mathbb{R}^2 \mid p \in \pi(X), \lambda \geq 0\}.$$

Let V be a 2-simplex in $\mathbb{R}^3$ such that 1-faces A_1, A_2, A_3 of V are in general position with respect to ℓ. One of the following holds: (α) $J(A_a) = J(A_b) \cup J(A_c)$, or (β) $J(A_a) \cup J(A_b) \cup J(A_c) = \mathbb{R}^2$, where $\{a, b, c\}$ is the set $\{1, 2, 3\}$. In these cases, we call V a 2-*simplex of type* α or a 2-*simplex of type* β.

Recall the definition of an elementary move in Sect. 3.6.

DEFINITION 5.1. An elementary move along a 2-simplex V is an $\mathcal{E}$ *operation* if ∂V is in general position with respect to ℓ. It is also called an $\mathcal{E}_i$ *operation*, where $i \in \{1, 2\}$ is the number of 1-faces of V in $L \cap V$.

Let a link L' be obtained from a link L by an $\mathcal{E}$ operation along a 2-simple V. We denote the $\mathcal{E}$ operation by $\mathcal{F}$. We assume that each 1-face A of V contained in L (or in L', resp.) has an orientation induced from L (resp. L') and denote by $\tau(A)$ the sign that is $+$ or $-$ if A is a positive or negative 1-simplex (as defined in 4.2), respectively.

When the move $\mathcal{F}$ is an $\mathcal{E}_1$ operation, let A_1 be the 1-face of V in L and A_2, A_3 the 1-faces of V in L'. We call $\mathcal{F}$ an $\mathcal{E}_1^j$ *operation* $(j = 1, \cdots, 6)$ if signs of A_1, A_2 and A_3 are as in Table 5.1.

Fig. 5.1 shows the images of 2-simplices which induce $\mathcal{E}_1^j$ operations, where the dotted edges are the image of A_1, the solid edges are those of A_2, A_3, and $\otimes$ is the origin O. The case $j = 2$ or 4 may be divided into two cases as in the figure. Then there are 8 cases. In case $\mathcal{F}$ is $\mathcal{E}_2$, we call it an $\mathcal{E}_2^j$ *operation* $(j = 1, \cdots, 6)$ when its inverse is $\mathcal{E}_1^j$.

Let K, K' be triangulations of L, L' such that each 1-face of V is a 1-simplex of K or K' and K and K' are identical except for these three 1-simplices. Then we say that K' is obtained from K by an $\mathcal{E}$ *operation* along V.

TABLE 5.1

j	$\tau(A_1)$	$\tau(A_2), \tau(A_3)$	type of V	$h(K') - h(K)$
1	$+$	$+, +$	α	0
2	$+$	$+, -$	α	1
3	$-$	$-, -$	α	1
4	$-$	$+, -$	α	0
5	$+$	$-, -$	β	2
6	$-$	$+, +$	β	-1

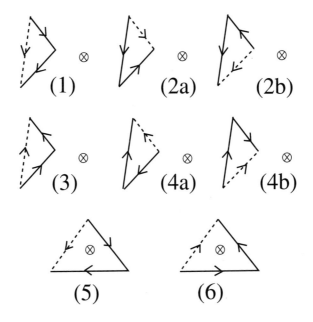

FIGURE 5.1

Let K be a triangulation of a link in $\mathbb{R}^3$. In case every 1-simplex of K is in general position with respect to ℓ, we define the *height* $h(K)$ of K by the number of negative 1-simplices of K. If a division K' is obtained from K by an $\mathcal{E}$ operation, then the difference $h(K') - h(K)$ is as in Table 5.1.

5.2. Deformation Chains

A sequence $L_0, L_1, \cdots, L_s$ of links in $\mathbb{R}^3$ is a *deformation chain* if each L_i $(i = 0, \cdots, s)$ is in general position with respect to ℓ and obtained from L_{i-1} $(i \neq 0)$ by an $\mathcal{E}, \mathcal{R}, \mathcal{W}^{\pm 1}$ or $\mathcal{S}^{\pm 1}$ operation (we exclude a $\mathcal{T}$ operation).

$\mathcal{R}$ **Operation.** Let L be a link in $\mathbb{R}^3$ and L' a link obtained from L by an $\mathcal{E}$ operation $\mathcal{F}$ along a 2-simplex V. If $\tau(A) = +$ for each 1-face of V, then $\mathcal{F}$ is called an $\mathcal{R}$ *operation*. In other words, an $\mathcal{R}$ operation is an $\mathcal{E}_1^1$ or $\mathcal{E}_2^1$ operation.

$\mathcal{W}$ **Operation.** Suppose that a link L_1 is obtained from a link L by an $\mathcal{E}_1^2$ operation $\mathcal{F}_1$ along a 2-simplex V_1. Let A_1, A_2 and A_3 be 1-faces of V_1 such that

$A_1 \subset L$, $A_2, A_3 \subset L_1$ and $\tau(A_3) = -1$. Suppose that a link L_2 is obtained from L_1 by an $\mathcal{E}_1^6$ operation $\mathcal{F}_2$ along a 2-simplex V_2 with $L_1 \cap V_2 = A_3$. If $L \cap V_1 * V_2 = A_1$, then the composition of $\mathcal{F}_1$ and $\mathcal{F}_2$ is called a $\mathcal{W}$ *operation*. See Fig. 5.2(W).

Here $X * Y$ means the join $\{\lambda x + (1 - \lambda)y \in \mathbb{R}^3 \mid x \in X, y \in Y, \lambda \in [0,1]\}$.

$\mathcal{T}$ **Operation.** Suppose that a link L_1 is obtained from a link L by an $\mathcal{E}_2^6$ operation $\mathcal{F}_1$ along a 2-simplex V_1. Let $A_1, \cdots, A_3$ be the 1-faces of V_1 such that $A_1, A_2 \subset L$ and $A_3 \subset L_1$. Let L_2 be a link obtained from L_1 by an $\mathcal{E}_1^6$ operation $\mathcal{F}_2$ along V_2 with $L_1 \cap V_2 = A_3$. The composition of $\mathcal{F}_1$ and $\mathcal{F}_2$ is called a $\mathcal{T}$ *operation*. (We do not require $L \cap V_1 * V_2 = A_1 \cup A_2$.)

$\mathcal{S}$ **Operation.** Let L be a link in $\mathbb{R}^3$ and A a 1-simplex in L which is in general position with respect to ℓ. Let $\mathbb{A} = \{A_1, \cdots, A_s\}$ be a sub-division of A. A family of 2-simplices $\mathbb{V} = \{V_1, \cdots, V_s\}$ in R^3 is called a *sawtooth on A avoiding L with basis* $\mathbb{A}$ if it satisfies the following conditions: (1) for each i ($i = 1, \cdots, s$), $L \cap V_i = A_i$ and V_i is of type β, and (2) $V_i \cap V_j = A_i \cap A_j$ for $i \neq j$. We say that $\mathbb{V}$ *avoids* a subset X of R^3 if $X \cap V_i = X \cap A_i$ ($i = 1, \cdots, s$). If A is a negative 1-simplex, then the composition of $\mathcal{E}_1^6$ operations along $V_1, \cdots, V_s$ is called an $\mathcal{S}$ *operation* along $\mathbb{V}$. It replaces A by $2s$ positive 1-simplices.

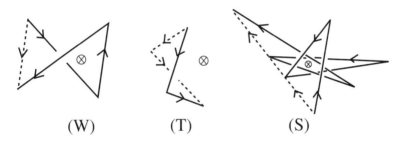

(W)   (T)   (S)

FIGURE 5.2

The inverse of a $\mathcal{Z}$ operation ($\mathcal{Z}$ is $\mathcal{W}$ or $\mathcal{S}$) is called a $\mathcal{Z}^{-1}$ *operation*. (In case $\mathcal{Z}$ is $\mathcal{E}, \mathcal{R}$ or $\mathcal{T}$, the inverse is also a $\mathcal{Z}$ operation.)

5.3. Deformation Chains at the Triangulation Level

We defined an $\mathcal{E}$ operation at the triangulation level. Other operations are also defined at the triangulation level.

Let L and L' be links as in the definition of an $\mathcal{R}$ operation. Let K and K' be their triangulations. We say that a K' is obtained from K by an $\mathcal{R}$ *operation along* V if K' is obtained from K by an $\mathcal{E}$ operation along V.

Let L, L_1 and L_2 be links as in the definition of a $\mathcal{W}$ or $\mathcal{T}$ operation. Let K, K_1 and K_2 be their triangulations. If K_1 is obtained from K by an $\mathcal{E}$ operation along V_1, and K_2 is obtained from K_1 by an $\mathcal{E}$ operation along V_2, then K_2 is said to be obtained from K by a $\mathcal{W}$ *or* $\mathcal{T}$ *operation*.

Let L and L' be links as in the definition of an $\mathcal{S}$ operation along $\mathbb{V}$. Let K and K' be their triangulations. We say that K' is obtained from K by an $\mathcal{S}$ *operation along* $\mathbb{V}$ if A is a negative 1-simplex of K, each 1-face of V_i ($i = 1, \cdots, s$) except A_i is a 1-simplex of K', and K and K' coincide on what remains.

At the triangulation level, we need the following two operations.

$\mathcal{A}$ **Operation and** $\mathcal{B}$ **Operation.** Let K be a triangulation of a link L and A a 1-simplex of K. Let K' be a triangulation of L obtained from K by replacing

$A = |p, q|$ with a pair of 1-simplices $|p, r|, |r, q|$ for some interior point r of A. We say that K' is obtained from K by an $\mathcal{A}$ operation (or $\mathcal{B}$ operation, resp.) on A if A is a positive edge (or negative edge, resp.).

A sequence $K_0, K_1, \cdots, K_s$ of triangulations of links in $\mathbb{R}^3$ is a *deformation chain* if $L_i = |K_i|$ $(i = 0, \cdots, s)$ is in general position with respect to ℓ and K_i $(i = 1, \cdots, s)$ is obtained from K_{i-1} by an $\mathcal{A}^{\pm 1}, \mathcal{B}^{\pm 1}, \mathcal{E}, \mathcal{R}, \mathcal{W}^{\pm 1}$ or $\mathcal{S}^{\pm 1}$ operation. (We exclude $\mathcal{T}$ operations.)

5.4. Height Reduction and Markov's Theorem

LEMMA 5.2 (Height Reduction Lemma I). *Let* K, K' *be a deformation chain of divisions of length one. If* $h(K) = h(K') \neq 0$, *then there is a deformation chain* $K, K_1, \cdots, K_s, K'$ *of length* $s + 1$ *for some* $s \geq 1$ *such that* $h(K_i) < h(K)$ $(i = 1, \cdots, s)$.

LEMMA 5.3 (Height Reduction Lemma II). *Let* K, K', K'' *be a deformation chain of divisions. If* $h(K') > h(K)$ *and* $h(K') > h(K'')$, *then there is a deformation chain* $K, K_1, \cdots, K_s, K''$ *of length* $s + 1$ *for some* $s \geq 1$ *such that* $h(K_i) < h(K')$ $(i = 1, \cdots, s)$.

Proofs of these two lemmas are given in Birman's book [**42**]. They are extremely long and not repeated here.

EXERCISE 5.4. Verify the height reduction lemmas for a few cases.

THEOREM 5.5. *Let* L *and* L' *be closed braids about* ℓ. *If* L *and* L' *are combinatorially equivalent as links in* $\mathbb{R}^3$, *then* L *and* L' *are Markov equivalent.*

Proof. Let K and K' be triangulations of L and L'. Since L and L' are combinatorially equivalent, there is a deformation chain from K to K'. Note that $h(K) = h(K') = 0$. Using the height reduction lemmas, we have a deformation chain from K to K' whose maximum height is 0. Thus each step in the sequence is an $\mathcal{A}^{\pm 1}, \mathcal{R}$ or $\mathcal{W}^{\pm 1}$ operation. Therefore we have a deformation chain from L to L', each step of which is an $\mathcal{R}$ or $\mathcal{W}^{\pm 1}$ operation. This implies that L and L' are Markov equivalent. ∎

By this theorem, we have Markov's theorem (Theorem 4.9).

5.5. Notes

There are various proofs of Alexander's theorem and Markov's theorem. For proofs of these theorems or a generalization for links in 3-manifolds, refer to [**5, 42, 451, 452, 667, 834, 902, 934, 963**]. Yamada's braiding method [**963**] can prove that the braid index of a link is equal to the minimum number of Seifert circles of diagrams representing the link. Vogel's method is a refinement of Yamada's and is very practical.

A remarkable application of braids in knot theory is the discovery of the Jones polynomial [**298, 299**]. This invariant is one of the necessities in knot theory and 3-dimensional manifold theory. We have not discussed the Jones polynomial because it is easily found in a lot of books on knot theory and our aim in Part 1 is to present some basics on braids and links for generalization into 4 dimensions.

In the bibliography of this book there are some articles on classical braids and research on links from the viewpoint of braid theory, including the Jones polynomial. However the bibliography is only a sample of the huge research in this field.

Part 2

Surface Knots and Links

CHAPTER 6

Surface Links

6.1. Surface Links

DEFINITION 6.1. A *surface link* is a closed surface embedded in $\mathbb{R}^4$ locally flatly.

We will work in the piecewise linear category. When we work in the topological or smooth category, it is called a *topological surface link* or a *smooth surface link*. In the smooth category, the local flatness is automatically satisfied.

REMARK 6.2. There are a lot of embedded surfaces in $\mathbb{R}^4$ which are not locally flat. An easy construction is as follows: Let K be a trefoil knot in a hyperplane H of $\mathbb{R}^4$. Take a point p_0 from a region of $\mathbb{R}^4 \setminus H$ and a point p_1 from the other region. Let $D_0 = K * p_0$ be the cone of K from p_0 and let $D_1 = K * p_1$ be that from p_1. They are 2-disks and the union forms a PL 2-sphere embedded in $\mathbb{R}^4$. It is not locally flat at p_0 and p_1. This construction is called *suspension* (cf. Artin[16] and Andrews-Curtis[8]). To investigate non-locally flat points is important for research on knot cobordism and vice versa. Refer to [173] for a relationship between them.

If F is a 2-sphere, it is also called a *2-knot* or a *2-dimensional knot*. If F is a connected surface, then it is called a *surface knot*. If each component of F is a 2-sphere, it is called a *2-link* or a *2-dimensional link*.

If F is (orientable and) oriented, then we call it an *oriented surface link*.

Two surface links F and F' are said to be *equivalent* if there exists an orientation-preserving homeomorphism $h : \mathbb{R}^4 \to \mathbb{R}^4$ such that $h(F) = F'$. When F and F' are oriented, it is assumed that $h|_F : F \to F'$ is an orientation-preserving homeomorphism. By $F \cong F'$, we mean that F and F' are equivalent. An equivalence class of a surface link is called a *surface link type*.

Two surface links F and F' are said to be *ambient isotopic* if there exists an ambient isotopy $\{h_t\}$ of $\mathbb{R}^4$ such that $h_1(F) = F'$. When F and F' are oriented, it is assumed that $h_1|_F : F \to F'$ is an orientation-preserving homeomorphism. Note that two surface links are equivalent if and only if they are ambient isotopic.

6.2. Trivial Surface Links

A *standard 2-sphere*, a *standard torus*, and *standard projective planes* in $\mathbb{R}^4$ are illustrated in Fig. 6.1 by motion pictures. (Motion pictures will be studied later.)

For surface knots and links, the notion of a split union and a connected sum is defined in a natural way analogous to the classical one.

DEFINITION 6.3. A surface knot is *trivial* (or *unknotted*) if it is obtained from some standard surfaces in $\mathbb{R}^4$ by taking a connected sum. A surface link is *trivial* (or *unknotted*) if it is obtained from some trivial surface knots by taking a split union.

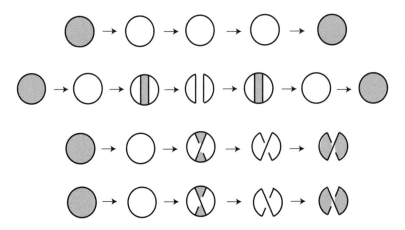

FIGURE 6.1. A standard 2-sphere, a standard torus, and standard
projective planes

PROPOSITION 6.4. *For an orientable surface link F in $\mathbb{R}^4$, the following are equivalent.*

(1) *F is trivial (unknotted).*
(2) *There is an embedded 3-manifold M in $\mathbb{R}^4$ with $\partial M = F$ such that each connected component of M is a handlebody.*
(3) *F is equivalent (or ambient isotopic) to a surface link which is contained in a 3-plane in $\mathbb{R}^4$.*

Here a *handlebody* means an orientable 3-manifold which is obtained from a 3-ball by attaching some 1-handles. The number of 1-handles is called the *genus* of the handlebody. In particular, a genus-0 handlebody is a 3-ball, and a genus-1 handlebody is a solid torus $D^2 \times S^1$.

Proof. (1) $\Rightarrow$ (2) and (1) $\Rightarrow$ (3) are obvious. (2) $\Rightarrow$ (1). Regard each handlebody of M as a 3-disk with some 1-handles. Deform it so that the 3-disk is a standard one and each 1-handle is a very thin 1-handle which is almost the same with its core attaching to the 3-disks. Arcs in $\mathbb{R}^4$ cannot link each other since their codimensions are three. Thus we can deform the arcs so that they are attached to the 3-disks trivially, and the 1-handles are trivial 1-handles. The boundary is a trivial surface link. (3) $\Rightarrow$ (1). Let F be a surface link in a hyperplane H. Push the components of F out from H into distinct parallel hyperplanes. If each component is a trivial knot, then F is a trivial link. Thus we assume that F is connected. We prove that F is unknotted by induction on the genus. If F is a 2-sphere, it bounds a 3-disk in H (the Schönflies theorem in dimension 3) and hence it is unknotted. If F is not a 2-sphere, there is a 2-handle $h = D^2 \times I$ in H attaching to F (i.e., $h \cap F = \partial D^2 \times I$) such that its attaching circle is not null-homotopic in F. Let F_1 be a surface in H obtained from F by surgery along h. If F_1 is connected, it has smaller genus than F. By the induction hypothesis, it is an unknotted surface knot. F is obtained from F_1 by surgery along a 1-handle. Thus it is unknotted (see Sect. 11.1). If F_1 is disconnected, let $F_1 = F_2 \cup F_3$. Push F_2 and F_3 into distinct parallel copies of H. By the induction hypothesis, they are unknotted. F is obtained from them by a connected sum. Thus F is unknotted. ∎

CONJECTURE 6.5. *An orientable surface knot F is trivial if and only if $\pi_1(\mathbb{R}^4 \setminus F)$ is an infinite cyclic group.*

This conjecture is called the *unknotting conjecture for surface knots*. This is true in the topological category ([**176**] for the 2-knot case, and [**270**] for the higher genus surface knot case). For non-orientable surface knots, see [**438, 454**] in the topological category. There is a counterexample in a case of a higher genus non-orientable surface [**160, 161**].

6.3. Combinatorial Equivalence

Let F be a surface link and B a 3-ball in $\mathbb{R}^4$. Suppose that $B \cap F$ is a 2-disk in ∂B. Replacing $B \cap F$ by the 2-disk $\mathrm{cl}\,(\partial B \setminus (B \cap F))$, we obtain a surface link from F. Such a replacement is called a *cellular move along B*. If F is oriented, then we assume that the new surface link is oriented by an orientation which is induced from $\mathrm{cl}\,(F \setminus (B \cap F))$.

LEMMA 6.6 (Cellular move lemma). *If a surface link F' is obtained from a surface link F by a cellular move along a 3-ball B in $\mathbb{R}^4$, then F and F' are ambient isotopic in $\mathbb{R}^4$ rel $\mathrm{cl}\,(\mathbb{R}^4 \setminus N(B))$, where $N(B)$ is a regular neighborhood of B.*

This lemma is a special case of a lemma so-called the *cellular move lemma*. For a proof of the lemma, refer to Rourke-Sanderson[**771**].

An *elementary move* is a cellular move such that the 3-ball B is a linear 3-simplex V in $\mathbb{R}^4$ and $F \cap V$ is the union of one, two or three 2-faces of V. This is a generalization of an elementary move for classical braids and links. It is said to be of *type* I, *type* II, or *type* III depending on whether $F \cap V$ is the union of one, two or three 2-faces of V, respectively. Suppose that $F \cap V$ is a 2-face of $V = |p_0, p_1, p_2, p_3|$, say $|p_0, p_1, p_2|$. Then an elementary move of type I along V changes F into

$$(F \setminus |p_0, p_1, p_2|) \cup |p_0, p_1, p_3| \cup |p_1, p_2, p_3| \cup |p_2, p_0, p_3|.$$

Two surface links are *combinatorially equivalent* if there exists a finite sequence of surface links $F = F_0, F_1, \ldots, F_n = F'$ such that each F_i is obtained from F_{i-1} by an elementary move.

THEOREM 6.7. *Let F and F' be surface links. The following conditions are mutually equivalent.*

(1) *F and F' are equivalent.*
(2) *F and F' are ambient isotopic.*
(3) *There is a finite sequence of surface links whose initial term is F and whose final term is F' such that each surface link is obtained from the previous one by a cellular move.*
(4) *F and F' are combinatorially equivalent.*

Proof. It is obvious that $(4) \Rightarrow (3) \Rightarrow (2) \Rightarrow (1)$. $((3) \Rightarrow (2)$ is the cellular move lemma.) By the following lemma, we have the theorem. ∎

LEMMA 6.8. *Two equivalent surface links are combinatorially equivalent.*

EXERCISE 6.9. (1) A regular projection of a surface link is defined in the next section. Prove Lemma 3.11 for a surface link F. [*Hint*: Since a regular projection may have branch points, we cannot take a positive number ϵ as in the proof of Lemma 3.11. First, translate the preimage in F of a regular neighborhood of each branch point into that in $F + \lambda v$. Then apply a similar argument.] (2) Prove Lemma 6.8.

CHAPTER 7

Surface Link Diagrams

7.1. Generic Maps

Let D^3 be the cube $\{(x, y, z) \in \mathbb{R}^3 \mid x, y, z \in [-1, 1]\}$, and let D_1, D_2 and D_3 be 2-disks which are the restrictions to D^3 of the xy-plane, yz-plane and zx-plane. Let O denote the origin.

Let $g : M \to \mathbb{R}^3$ be a map from a closed surface M to $\mathbb{R}^3$.

A point $y \in g(M)$ is called a *regular point* if there is a regular neighborhood N of y in $\mathbb{R}^3$ such that the restriction of g to the preimage $g^{-1}(g(M) \cap N)$ is an immersion and the triple $(N, g(M) \cap N, y)$ is homeomorphic to the triple (D^3, D_1, O).

A point $y \in g(M)$ is called a *double point in general position*, a *transverse double point* or simply a *double point* if there is a regular neighborhood N of y in $\mathbb{R}^3$ such that the restriction of g to the preimage $g^{-1}(g(M) \cap N)$ is an immersion and the triple $(N, g(M) \cap N, y)$ is homeomorphic to the triple $(D^3, D_1 \cup D_2, O)$.

A point $y \in g(M)$ is called a *triple point in general position*, or simply a *triple point*, if there is a regular neighborhood N of y in $\mathbb{R}^3$ such that the restriction of g to the preimage $g^{-1}(g(M) \cap N)$ is an immersion and the triple $(N, g(M) \cap N, y)$ is homeomorphic to the triple $(D^3, D_1 \cup D_2 \cup D_3, O)$.

A point $y \in g(M)$ is called an *elementary branch point*, or simply a *branch point*, if there is a regular neighborhood N of y in $\mathbb{R}^3$ such that $g^{-1}(y)$ consists of a single point $x \in M$, the restriction of g to $g^{-1}(g(M) \cap N)$ is an immersion except at x and the triple $(N, g(M) \cap N, y)$ is homeomorphic to a triple $(D^3, C * O, O)$, where C is an immersed circle in ∂D^3 with one transverse double point.

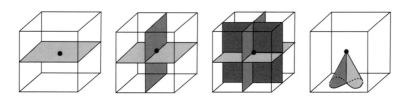

FIGURE 7.1. Regular point, double point, triple point, and branch point

DEFINITION 7.1. A map $g : M \to \mathbb{R}^3$ is a *generic map* or a *map in general position* if every point of $g(M)$ is a regular point, a double point (in general position), a triple point (in general position) or an elementary branch point.

REMARK 7.2. A point $y \in g(M)$ is called a *branch point of valency n* if there is a regular neighborhood N of y in $\mathbb{R}^3$ such that $g^{-1}(y)$ consists of a single point $x \in M$, the restriction of g to $g^{-1}(g(M) \cap N)$ is an immersion except at x and the triple $(N, g(M) \cap N, y)$ is homeomorphic to a triple $(D^3, C * O, O)$, where C

is an immersed circle in ∂D^3 with n transverse double points. If $n = 1$, then it is an elementary branch point. Suppose that a triangulation of M is fixed. For any positive number δ, we can deform g by moving the image of each vertex of M slightly in its δ-neighborhood in $\mathbb{R}^3$, without taking a subdivision of M, so that each point of $g(M)$ is a regular point, a double point, a triple point or a branch point of valency n for some n. Furthermore if we may take a subdivision, as we usually do in the PL category, then each branch point of valency n can be changed to n elementary branch points. Thus a map $g : M \to \mathbb{R}^3$ is changed into a generic map as a δ-approximation.

7.2. Surface Link Diagrams

Let $\pi : \mathbb{R}^4 \to H$ be a projection onto a hyperplane H of $\mathbb{R}^4$ along a vector $v \in \mathbb{R}^4$.

DEFINITION 7.3. A surface link F is *in general position with respect to* π, or π is a *generic projection of* L, if the map $\pi|_F : F \to H$ is a generic map. A *surface link diagram*, or simply a *diagram*, of a surface link F is the image $\pi(F)$ by a generic projection π equipped with "over/under" information on the multiple points with respect to the direction of v.

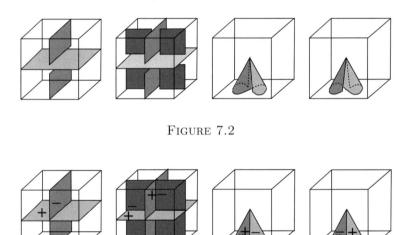

FIGURE 7.2

FIGURE 7.3

A method to describe over/under information is to remove a neighborhood of each under crossing point. Local pictures are as in Fig. 7.2. Such a diagram is called a *broken surface diagram*. Another method is to indicate by symbols "+" and "−" (Fig. 7.3), etc.

It is sometimes difficult to draw a whole picture of a surface link diagram especially when the configuration of the surface link is complicated. One has to draw some parts which are hiding behind other parts. A method to avoid this is to divide the diagram into some small blocks so that we can visualize each block. Another method is to use a series of slices by parallel hyperplanes of the 3-space $H = \mathbb{R}^3$, like a CAT scan. Then each object becomes a 1-dimensional object in a 2-plane so that we can describe it on paper completely. A *tomography* or a *movie*

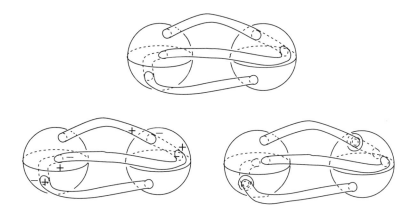

FIGURE 7.4. Projection and diagrams of a spun trefoil

of a surface link diagram is a series of classical link diagrams obtained by slicing a surface link diagram by parallel hyperplanes.

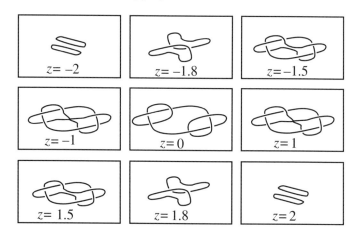

FIGURE 7.5. Tomograph of a spun trefoil

7.3. Elementary Moves to Diagrams

A surface link is presented by a surface link diagram. Of course such a presentation is not unique. We introduce here seven types of local moves on surface link diagrams. They do not change the surface link types.

Each move is illustrated without over/under information. (The over/under information in the tomography is an example.) The reader should consider all possible over/under information.

(1) A *type*-I *bubble move* is a local move which creates a bubble as in Fig. 7.6. In the tomography it is a Reidemeister move of type I, that is I_a or I_b, followed by its inverse.

(2) A *type*-II *bubble move* is a local move which creates a bubble as in Fig. 7.6. In the tomography it is a Reidemeister move of type II followed by its inverse II*.

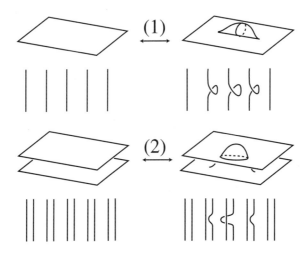

FIGURE 7.6. Type-I and type-II bubble moves

(3) A *type*-I *saddle move* is a local move which cancels a pair of branch points. In the tomography it is a Reidemeister move of type I, that is I_a^* or I_b^*, followed by its inverse.

(4) A *type*-II *saddle move* is a local move where a saddle point passes a sheet as in Fig. 7.7. In the tomography it is a Reidemeister move of type II* followed by its inverse II.

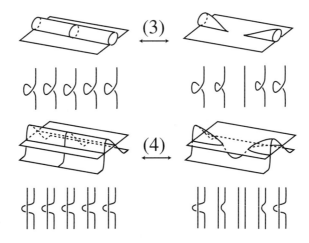

FIGURE 7.7. Type-I and type-II saddle moves

(5) A *type*-III *move* is a local move where a pair of triple points appear as in Fig. 7.8. In the tomography it is a Reidemeister move of type III followed by its inverse.

(6) A *tetrahedral move* is a local move where a sheet passes a triple point as in Fig. 7.9. In the tomography the order of four Reidemeister moves of type III changes.

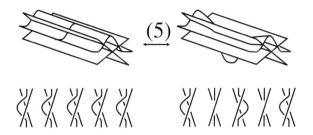

FIGURE 7.8. Type-III move

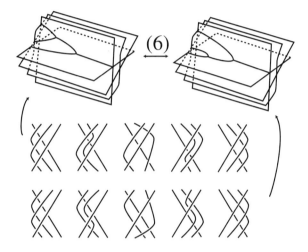

FIGURE 7.9. Tetrahedral move

(7) *Passing a branch point* is a local move where a branch point passes a third sheet as in Fig. 7.10.

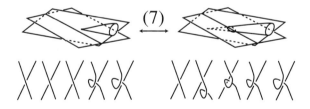

FIGURE 7.10. Passing a branch point

The above seven types of moves are called *Roseman moves*.

THEOREM 7.4. *Two surface link diagrams represent equivalent surface links if and only if they are related by a finite sequence of Roseman moves and ambient isotopies of $\mathbb{R}^3$.*

For a proof of this theorem, refer to Carter-Saito[**82**] and Roseman[**770**]. The reader who wishes to know more about surface link diagrams can consult Carter and Saito's book [**89**].

CHAPTER 8

Motion Pictures

8.1. Motion Pictures

Let v be a vector of $\mathbb{R}^4$ and let $\pi : \mathbb{R}^4 \to H$ be a projection onto a hyperplane H of $\mathbb{R}^4$ along v. For each $t \in \mathbb{R}$, let H_t be the unique hyperplane of $\mathbb{R}^4$ that is a parallel copy of H with $tv \in H_t$.

For a subset X of $\mathbb{R}^4$, the *motion picture* of X along v is a one-parameter family $\{X_t\}_{t \in \mathbb{R}}$ of subsets of H such that $X_t = \pi(X \cap H_t)$. Conversely, for a one-parameter family $\{X_t\}_{t \in \mathbb{R}}$ of subsets of H, the *trace*, or the *realization*, is a subset X of $\mathbb{R}^4$ such that $X_t = \pi(X \cap H_t)$.

We usually assume that $v = (0, 0, 0, 1)$ and H is the xyz-space determined by $t = 0$. Then $H_{t_0} = \{(x, y, z, t) \in \mathbb{R}^4 \mid t = t_0\}$. We denote this hyperplane by $\mathbb{R}^3[t_0]$. In general, for a subset A of $\mathbb{R}^3$ and a subset B of $\mathbb{R}^1$, we denote by AB (or by $A \times B$) the subset $\{(x, t) \in \mathbb{R}^3 \times \mathbb{R}^1 \mid x \in A, t \in B\}$ of $\mathbb{R}^3 \times \mathbb{R}^1 = \mathbb{R}^4$. When B consists of a single value t, we denote $A\{t\}$ by $A[t]$. In this situation, the *motion picture* of a subset X of $\mathbb{R}^4$ is a one-parameter family $\{\pi(X \cap \mathbb{R}^3[t])\}_{t \in R}$, where $\pi : \mathbb{R}^4 \to \mathbb{R}^3[0] = \mathbb{R}^3$.

A homeomorphism $f : \mathbb{R}^4 \to \mathbb{R}^4$ is said to be *height preserving* or *level preserving* if $f(\mathbb{R}^3[t]) = \mathbb{R}^3[t]$ for any $t \in \mathbb{R}$.

An ambient isotopy $\{h_s\}$ ($s \in [0, 1]$) of $\mathbb{R}^4$ is said to be *height preserving* or *level preserving* if each homeomorphism h_s ($s \in [0, 1]$) is height preserving.

8.2. Motion Pictures of Surface Links

Let F be a surface link and let $\pi : \mathbb{R}^4 \to \mathbb{R}^3[0] = \mathbb{R}^3, (x, y, z, t) \mapsto (x, y, z)$ be the projection.

A one-parameter family $\{\pi(F \cap \mathbb{R}^3[t])\}_{t \in \mathbb{R}}$ is the motion picture of F. A point $x \in F$ is called a *regular point* of the motion picture if there exists a regular neighborhood N in $\mathbb{R}^4$ and a height-preserving and orientation-preserving homeomorphism $f : \mathbb{R}^4 \to \mathbb{R}^4$ such that the pair $f(N, N \cap F)$ is the product $(D^3, D^1)[a, b] \subset \mathbb{R}^3 \times \mathbb{R}$ of the standard disk pair (D^3, D^1) and an interval $[a, b]$. Otherwise we call $x \in F$ a *critical point of F*.

Deform F slightly by an ambient isotopy of $\mathbb{R}^4$ such that all vertices of F are contained in distinct hyperplanes. Then a critical point must be a vertex of F. Hence the critical points are isolated and the number is finite. If $\mathbb{R}^3[t]$ has a critical point, then t is called a *critical value*; otherwise a *regular value*.

For a regular value $t \in \mathbb{R}$, the intersection $F \cap \mathbb{R}^3[t]$ is a link in the hyperplane $\mathbb{R}^3[t]$. When F is an oriented surface link, we give the link an orientation which is induced from the oriented surface $F \cap \mathbb{R}^3(-\infty, t]$. This enables us to recover the orientation of the surface link from the motion picture.

A critical point is called (1) a *maximal point* if a trivial loop shrinks toward the point and disappears, (2) a *minimal point* if a trivial loop appears from the point as t increases, or (3) a *saddle point* or a *hyperbolic point* if the motion picture about the point is equivalent to the one illustrated in Fig. 8.1(3). Maximal points, minimal points and saddle points are called *elementary critical points*.

The critical point illustrated in Fig. 8.2 is a non-elementary critical point.

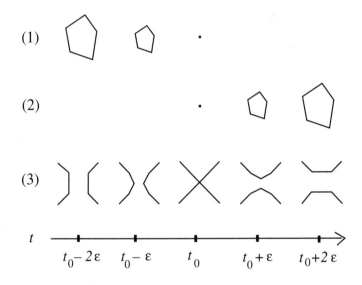

FIGURE 8.1. Maximal point, minimal point and saddle point

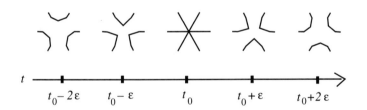

FIGURE 8.2. A non-elementary critical point

PROPOSITION 8.1. *Any surface link can be deformed up to equivalence such that it has only elementary critical points and such that it is a δ-approximation of the surface link for an arbitrarily given positive number δ.*

Proof. Let F be a polygonal surface link. Deform F so that all vertices have distinct fourth coordinates and the images of them under $\pi : \mathbb{R}^4 \to \mathbb{R}^3, (x, y, z, t) \mapsto (x, y, z)$ are also all distinct. Let v be a non-elementary critical point. Take a small cylindrical neighborhood $N(v) = V[a, b]$ of v where $V \subset \mathbb{R}^3$ is a convex neighborhood of $\pi(v)$ and $[a, b] \subset \mathbb{R}$ is a neighborhood of the fourth coordinate of v. We may assume that all vertices of F except v are outside of $\mathbb{R}^3[a, b]$ and that their images under π are outside of V. Then $F \cap \partial N(v)$ is a circle, say L, contained in $\partial V \times (a, b)$. The restriction of F to $N(v)$ is a cone $v * L$. Replacing

V if necessary, we may assume that the fourth coordinates of the vertices of L are all distinct. Let v' be a point such that the fourth coordinate is a little larger than b and $\pi(v') \in \mathrm{int}(V)$. Replacing the cone $v * L$ by $v' * L$, we have another surface link, say F'. F and F' are equivalent. F' has new vertices on L; however, they are regular points, saddle points or minimal points of F'. Since F is locally flat, F' is locally flat, and we see that v' is a maximal point of F'. This replacement is done in a small neighborhood of a vertex v. Applying the same argument for all non-elementary critical points of F, we have a desired surface link. ∎

EXAMPLE 8.2. By the argument above, the critical point illustrated in Fig. 8.2 is changed into a maximal point and three saddle points as in Fig. 8.3. Since the maximal point can be canceled with a saddle point, we have Fig. 8.4, where two saddle points occur.

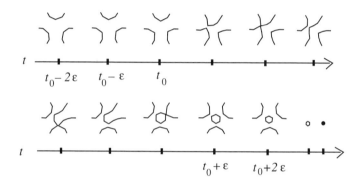

FIGURE 8.3

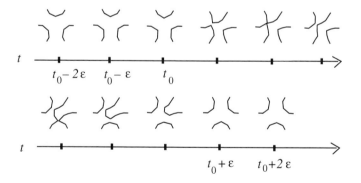

FIGURE 8.4

8.3. The Motion Picture of a Non-locally Flat Surface

Let F be a closed surface embedded in $\mathbb{R}^4$ which need not be locally flat. Let $\pi : \mathbb{R}^4 \to \mathbb{R}^3[0] = \mathbb{R}^3, (x, y, z, t) \mapsto (x, y, z)$ be the projection.

A vertex v of F is called a *non-locally flat maximal point* if a knotted circle shrinks toward the vertex and disappears. For example, see Fig. 8.5.

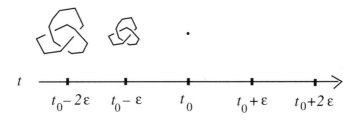

FIGURE 8.5. A non-locally flat maximal point

PROPOSITION 8.3. *Any closed surface embedded in $\mathbb{R}^4$ is deformed up to equivalence such that every critical point is an elementary critical point or a non-locally flat maximal point.*

Proof. Follow the proof of Proposition 8.1. ∎

8.4. Elementary Critical Bands

Let F be a surface link and let v be an elementary critical point which is (1) maximal, (2) minimal, or (3) saddle. In a small neighborhood of v, we can modify F, up to equivalence, as in Fig. 8.6, where (1) a trivial circle appears for a moment, bounds a 2-disk in an instant, say at $t = t_0$, and disappears; (2) there is nothing for a moment, a 2-disk appears suddenly, and the boundary of the 2-disk remains; (3) there exists a pair of trivial arcs for a moment, a band is attached to them in an instant, and the pair of arcs obtained by surgery along the band remains. The part of F corresponding to the 2-disk is called (1) a *maximal disk*, (2) *minimal disk*, or (3) *saddle band*, respectively. Such a disk (or band) is called an *elementary critical disk*.

Conversely, an elementary critical disk can be replaced by an elementary critical point.

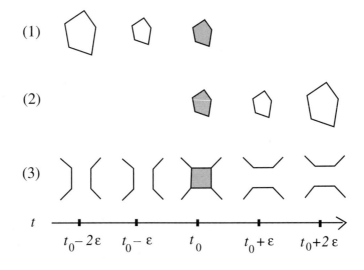

FIGURE 8.6. Maximal disk, minimal disk and saddle band

From Proposition 8.1, we have the following.

PROPOSITION 8.4. *Any surface link is deformed up to equivalence such that it has only elementary critical disks and it is a δ-approximation of the surface link for any arbitrarily given positive number δ.*

8.5. Trivial Disk Systems

Let L be a link in $\mathbb{R}^3$ with m components.

A *disk system in $\mathbb{R}^3$ with boundary L* is a union $D_1 \cup \cdots \cup D_m$ of mutually disjoint embedded 2-disks D_i in $\mathbb{R}^3$ whose boundary is L. By definition, L is a trivial link.

Assume that L is a link in $\mathbb{R}^3[0] \subset \mathbb{R}^4$. A *disk system in the upper half 4-space* $\mathbb{R}^3[0, \infty)$ *with boundary L* is a union $\widetilde{D}_1 \cup \cdots \cup \widetilde{D}_m$ of mutually disjoint, properly and locally flatly embedded 2-disks $\widetilde{D}_i$ in $\mathbb{R}^3[0, \infty)$ whose boundary is L.

A disk system F in $R^3[0, \infty)$ is *trivial* if there is no critical point except a single maximal point (or maximal disk) on each disk component $\widetilde{D}_i$ of F.

EXERCISE 8.5. Prove the following: Let t_0 be a positive number. A trivial disk system F in $\mathbb{R}^3[0, \infty)$ can be deformed, by an ambient isotopy rel $\mathbb{R}^3[0]$, so that

$$\pi(F \cap \mathbb{R}^3[t]) = \begin{cases} L & \text{for } t \in [0, t_0) \\ D_1 \cup \cdots \cup D_m & \text{for } t = t_0 \\ \emptyset & \text{for } t \in (t_0, \infty) \end{cases}$$

where $D_1 \cup \cdots \cup D_m$ is a disk system in $\mathbb{R}^3$ with boundary L. Moreover, when all maximal disks of the original F are in $\mathbb{R}^3[t_0]$, then we may assume the ambient isotopy of $\mathbb{R}^3[0, \infty)$ to be height preserving.

As a consequence, the boundary $L = \partial F$ of a trivial disk system in $\mathbb{R}^3[0, \infty)$ is a trivial link.

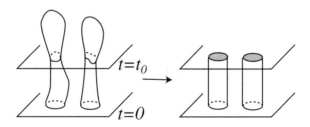

FIGURE 8.7. Trivial disk system

PROPOSITION 8.6. *Let F and F' be trivial disk systems in $\mathbb{R}^3[0, \infty)$ with the same boundary L. They are ambient isotopic in $\mathbb{R}^3[0, \infty)$ rel $\mathbb{R}^3[0]$.*

Proof. We may assume that F and F' satisfy the following (Exercise 8.5):

$$\pi(F \cap \mathbb{R}^3[t]) = \begin{cases} L & \text{for } t \in [0, 1) \\ D_1 \cup \cdots \cup D_m & \text{for } t = 1 \\ \emptyset & \text{for } t \in (1, \infty) \end{cases}$$

and

$$\pi(F' \cap \mathbb{R}^3[t]) = \begin{cases} L & \text{for } t \in [0,2) \\ D'_1 \cup \cdots \cup D'_m & \text{for } t = 2 \\ \emptyset & \text{for } t \in (2,\infty), \end{cases}$$

where $D_1 \cup \cdots \cup D_m$ and $D'_1 \cup \cdots \cup D'_m$ are disk systems in $\mathbb{R}^3$ with boundary L. By the following lemma (Lemma 8.7), there exists a family of mutually disjoint embedded 3-balls $B_1, \ldots, B_m$ in $\mathbb{R}^3[1,3]$ with

$$\partial B_i = D_i[1] \cup (\partial D_i)[1,2] \cup D'_i[2] \quad \text{for } i \in \{1, \ldots, m\}.$$

Applying the cellular move lemma (Lemma 6.6), we have a desired ambient isotopy.
∎

LEMMA 8.7 (Horibe and Yanagawa's Lemma). *Let $S_1, \ldots, S_m$ be mutually disjoint 2-spheres in $\mathbb{R}^3[0,1]$ such that for each $i \in \{1, \ldots, m\}$, $S_i = D_i[0] \cup (\partial D_i)[0,1] \cup D'_i[1]$, where D_i and D'_i are 2-disks in $\mathbb{R}^3$ with $\partial D_i = \partial D'_i$. Then there exist m mutually disjoint embedded 3-disks $B_1, \ldots, B_m$ in $\mathbb{R}^3[0,2]$ with $\partial B_i = S_i$.*

This is proved in the next section.

Proposition 8.6 is remarkable. Consider an analogy in the classical dimension as follows: Let $F^1 = D_1^1 \cup \ldots D_m^1$ and $F'^1 = D'^1_1 \cup \ldots D'^1_m$ be 'trivial 1-disk systems' in $\mathbb{R}^2[0,\infty)$ with the same boundary $L^0 \subset \mathbb{R}^2[0]$; i.e. each D_i or D'_i is a properly embedded arc in the upper half 3-space $\mathbb{R}^2[0,\infty)$ with a single maximal point. In Fig. 8.8, (1) and (2) are ambient isotopic in $\mathbb{R}^2[0,\infty)$ rel $\mathbb{R}^2[0]$, and so are (3) and (4). However (1) and (4) are never ambient isotopic rel $\mathbb{R}^2[0]$. Proposition 8.6 asserts that such a situation does not happen for 2-dimensional disk systems. Moreover, the following corollary is obviously false in the classical dimensional case.

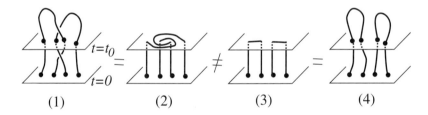

(1)   (2)   (3)   (4)

FIGURE 8.8

COROLLARY 8.8. *Let $F = S_1 \cup \cdots \cup S_m$ be a 2-link in $\mathbb{R}^4$ such that each component S_i has exactly two elementary critical points: one is maximal and the other is minimal. Then F is equivalent to a trivial 2-link.*

EXAMPLE 8.9. In the situation of Proposition 8.6, one cannot expect that the ambient isotopy is height preserving even if all maximal points (or maximal disks) of F and F' are in the same hyperplane. Consider the trivial disk systems illustrated in Fig. 8.9(1) and (2), where for (2) a component labeled 1 moves through the other component labeled 2. They are height preservingly ambient isotopic in $\mathbb{R}^3[0,\infty)$ rel $\mathbb{R}^3[0]$ to the surfaces illustrated in Fig. 8.10(3) and (4), respectively. The surfaces illustrated in (3) and (4) are not height preservingly ambient isotopic in $\mathbb{R}^3[0,\infty)$ rel $\mathbb{R}^3[0]$. In fact, there is no homeomorphism $\mathbb{R}^3[0,1] \to \mathbb{R}^3[0,1]$ rel $\mathbb{R}^3[0]$ which maps one to the other. [Let $\widetilde{D}_1 \cup \widetilde{D}_2$ and $\widetilde{D}_1 \cup \widetilde{D}_2'$ be the disk systems illustrated in (3) and (4), where $\widetilde{D}_1$ is the component labeled 1. Put $S = \widetilde{D}_2 \cup D[0]$ and $S' = \widetilde{D}_2' \cup D[0]$ where D is a 2-disk in $\mathbb{R}^3$ with $D = \pi(\widetilde{D}_2 \cap \mathbb{R}^3[1])$. If $\widetilde{D}_1 \cup \widetilde{D}_2$ is mapped to $\widetilde{D}_1 \cup \widetilde{D}_2'$ by a homeomorphism $\mathbb{R}^3[0,1] \to \mathbb{R}^3[0,1]$ rel $\mathbb{R}^3[0]$, there is a homeomorphism $h : \mathbb{R}^3[0,1] \setminus \widetilde{D}_1 \to \mathbb{R}^3[0,1] \setminus \widetilde{D}_1$ with $h(S) = S'$. However, $[S] = 0$ in $H_2(\mathbb{R}^3[0,1] \setminus \widetilde{D}_1; \mathbb{Z}) \cong \mathbb{Z}$ and $[S']$ is a generator. This is a contradiction.] This example is given in Kawauchi-Shibuya-Suzuki [**372**].

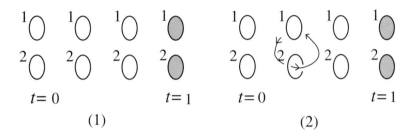

FIGURE 8.9

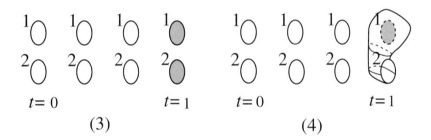

FIGURE 8.10. Kawauchi-Shibuya-Suzuki's example

8.6. Proof of Lemma 8.7

We give a proof of Lemma 8.7, which is a slightly modified version of a proof due to Kawauchi-Shibuya-Suzuki [**372**].

Proof. We construct a 3-disk B_m in $\mathbb{R}^3[0, \infty)$ with $\partial B_m = S_m$ such that B_m is disjoint from $S_1 \cup \cdots \cup S_{m-1}$. Deform D_m by an ambient isotopy $\{h_t\}$ of $\mathbb{R}^3$ rel ∂D_m such that $\widetilde{D}_m = h_1(D_m)$ intersects $D_1' \cup \cdots \cup D_m'$ in general position. Let n be the number of the components of $\widetilde{D}_m \cap (D_1' \cup \cdots \cup D_m')$. Each component is a simple loop or a proper arc. Let $t_i = i/(n+1)$ for $i \in \{0, \ldots, n+1\}$. Define $B_m \cap \mathbb{R}^3[0, t_1]$ by

$$\pi(B_m \cap \mathbb{R}^3[t]) = h_{t/t_1}(D_m) \quad \text{for } t \in [0, t_1].$$

Let $c \subset \widetilde{D}_m \cap (D_1' \cup \cdots \cup D_m')$ be an innermost loop or arc on some D_i', and let $\Delta \subset D_i'$ be the 2-disk cut off by c with $\text{int}\Delta \cap \widetilde{D}_m = \emptyset$. Do surgery along Δ, and $\widetilde{D}_m$ is divided into one 2-disk $\widetilde{D}_m^{(1)}$ and a 2-sphere Σ_1; see Fig. 8.11.

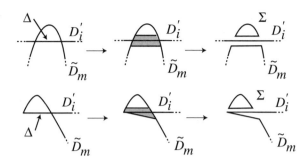

FIGURE 8.11

Define $B_m \cap \mathbb{R}^3[t_1, t_2]$ by

$$
\pi(B_m \cap \mathbb{R}^3[t]) = \begin{cases}
\widetilde{D}_m & \text{for } t \in [t_1, (t_1 + t_2)/2) \\
\widetilde{D}_m \cup N(\Delta; \mathbb{R}^3) & \text{for } t = (t_1 + t_2)/2 \\
\widetilde{D}_m^{(1)} \cup \Sigma_1 & \text{for } t \in ((t_1 + t_2)/2, t_2],
\end{cases}
$$

where $N(\Delta; \mathbb{R}^3)$ is a neighborhood of Δ along which the surgery was applied. Continuing this process, we obtain a 2-disk $\widetilde{D}_m^{(n)}$ and n spheres $\Sigma_1, \ldots, \Sigma_n$ such that they are disjoint from $D_1' \cup \cdots \cup D_m'$ except $\partial \widetilde{D}_m^{(n)} = \partial D_m'$. Define $B_m \cap \mathbb{R}^3[t_i, t_{i+1}]$ for $i \in \{2, \ldots, n\}$ similarly. We have defined $B_m \cap \mathbb{R}^3[0, 1]$, which is homeomorphic to a 3-disk with n open 3-disks removed. Rename $n+1$ mutually disjoint 2-spheres $\widetilde{D}_m^{(n)} \cup D_m', \Sigma_1, \ldots, \Sigma_n$ in $\mathbb{R}^3$ by $\Sigma_1, \ldots, \Sigma_{n+1}$ such that for each i, Σ_i is an innermost 2-sphere in $\mathbb{R}^3$ among $\{\Sigma_{i+1}, \ldots, \Sigma_{n+1}\}$. Let $\Delta_1^3, \ldots, \Delta_{n+1}^3$ be 3-disks in $\mathbb{R}^3$ bounded by $\Sigma_1, \ldots, \Sigma_{n+1}$, and let $\widetilde{\Delta}_1^3, \ldots, \widetilde{\Delta}_{n+1}^3$ be 3-disks in $\mathbb{R}^3[1, \infty)$ such that

$$
\pi(\widetilde{\Delta}_i^3 \cap \mathbb{R}^3[t]) = \begin{cases}
\Sigma_i & \text{for } t \in [1, i) \\
\Delta_i^3 & \text{for } t = i \\
\emptyset & \text{for } t \in (i, \infty).
\end{cases}
$$

Define $B_m \cap \mathbb{R}^3[1, \infty)$ by the union of $\widetilde{\Delta}_1^3, \ldots, \widetilde{\Delta}_{n+1}^3$. Now we have a 3-disk B_m with $\partial B_m = S_m$ and where B_m is disjoint from $S_1 \cup \cdots \cup S_{m-1}$.

Consider an ambient isotopy $\{g_t\}$ of $\mathbb{R}^3[0, \infty)$ rel $S_1 \cup \cdots \cup S_{m-1}$ and $D_m[0]$ such that it changes B_m into $D_m[0, \epsilon]$ for a sufficiently small positive number ϵ. Such an isotopy exists by the cellular move lemma along B_m. Construct a 3-disk B_{m-1}' with $\partial B_{m-1}' = S_{m-1}$; it is disjoint from $S_1 \cup \cdots \cup S_{m-2} \cup D_m[0, \epsilon]$ by the same argument as before. Apply the inverse ambient isotopy of $\{g_t\}$ to this 3-disk B_{m-1}' and let B_{m-1} be the result. Then $\partial B_{m-1} = S_{m-1}$, and B_{m-1} is disjoint from $S_1 \cup \cdots \cup S_{m-2} \cup B_m$. Repeat this procedure. ∎

CHAPTER 9

Normal Forms of Surface Links

9.1. Link Transformation Sequence

Let L be a link in $\mathbb{R}^3$. A 2-disk B in $\mathbb{R}^3$ is called a *band* attaching to L if $L \cap B$ is a pair of disjoint arcs in ∂B. The arcs are called the *attaching arcs* of B.

A *band set attaching to* L is a union $\mathcal{B} = B_1 \cup \cdots \cup B_m$ of mutually disjoint bands $B_1, \ldots, B_m$ attaching to L. Define a link $\mathrm{h}(L; \mathcal{B})$ by

$$\mathrm{h}(L; \mathcal{B}) = \mathrm{cl}\,(L \cup \partial \mathcal{B} \setminus (L \cap \mathcal{B})).$$

We say that the link $\mathrm{h}(L; \mathcal{B})$ is obtained from L by a *hyperbolic transformation*, or *surgery*, along $\mathcal{B}$.

Fig. 9.1 is an example of a hyperbolic transformation. In the figure, $(-)$ indicates that it is an attaching arc of a band, and $(+)$ indicates that it is an arc appearing after the transformation.

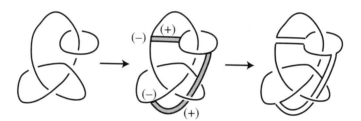

FIGURE 9.1. Hyperbolic transformation

If $\sharp(\mathcal{B}) = m$, $\sharp(L) = m + 1$ and $\sharp(\mathrm{h}(L; \mathcal{B})) = 1$ (where $\sharp$ means the number of connected components), then we say that $\mathrm{h}(L; \mathcal{B})$ is obtained from the link L by a *complete fusion*, and conversely L is obtained from the link $\mathrm{h}(L; \mathcal{B})$ by a *complete fission*.

Let $\{h_t\}$ be an ambient isotopy of $\mathbb{R}^3$. For a link L, we have an isotopy (a one-parameter family) $\{h_t(L)\}$ of links. We say that $h_1(L)$ is obtained from L by an *isotopic transformation* by an ambient isotopy $\{h_t\}$.

We often denote by $L \to L'$ a hyperbolic transformation or an isotopic transformation.

DEFINITION 9.1. A *link transformation sequence* is a finite sequence $L_0 \to \cdots \to L_n$ such that each $L_{i-1} \to L_i$ is a hyperbolic transformation or an isotopic transformation. A *hyperbolic transformation sequence* is a link transformation sequence such that each $L_{i-1} \to L_i$ is a hyperbolic transformation.

9.2. The Realizing Surface

Let $L \to L'$ be a hyperbolic transformation along a band set $\mathcal{B}$. A *realizing surface of* $L \to L'$ *in* $\mathbb{R}^3[a, b]$ is a surface F such that

$$\pi(F \cap \mathbb{R}^3[t]) = \begin{cases} L & \text{for } t \in [a, t_0), \\ L \cup \mathcal{B} & \text{for } t = t_0, \\ L' & \text{for } t \in (t_0, b], \end{cases}$$

where t_0 is a number between a and b.

Let $L \to L'$ be an isotopic transformation by an ambient isotopy $\{h_t\}$ of $\mathbb{R}^3$. A *realizing surface of* $L \to L'$ *in* $\mathbb{R}^3[a, b]$ is a surface F such that

$$\pi(F \cap \mathbb{R}^3[t]) = h_s(L) \quad \text{for } t \in [a, b],$$

where $s = (t - a)/(b - a)$.

We denote by $F(L \to L')_{[a,b]}$ a realizing surface of $L \to L'$ in $\mathbb{R}^3[a, b]$.

Let $L_0 \to L_1 \to \cdots \to L_n$ be a link transformation sequence. Let $[a, b]$ be an interval and let $(I_1, I_2, \ldots, I_n)$ be a division of $[a, b]$, namely, $I_i = [t_{i-1}, t_i]$ with $a = t_0 < t_1 < \cdots < t_n = b$.

DEFINITION 9.2. A *realizing surface of* $L_0 \to L_1 \to \cdots \to L_n$ *in* $\mathbb{R}^3[a, b]$ is a surface

$$F(L_0 \to L_1)_{I_1} \cup \cdots \cup F(L_{n-1} \to L_n)_{I_n}.$$

This surface is denoted by

$$F(L_0 \to L_1 \to \cdots \to L_n)_{[a,b]}.$$

We tend to omit the information on $a < t_1 < \cdots < t_n < b$ by the following:

EXERCISE 9.3. Two realizing surfaces in $\mathbb{R}^3[a, b]$ (with respect to distinct divisions of $[a, b]$) of the same link transformation sequence are ambient isotopic in $\mathbb{R}^3[a, b]$ rel $\mathbb{R}^3[a] \cup \mathbb{R}^3[b]$. Prove this.

9.3. Technical Lemmas on Realizing Surfaces

Here we introduce some lemmas on realizing surfaces. If the reader is not familiar with the motion picture method, the lemmas might seem weird. However the idea is simple, and they are useful to handle surfaces in 4-space. These lemmas will be used for a proof of Alexander's theorem in dimension 4.

LEMMA 9.4. *Let* $L_0 \to L_1$ *and* $L_1 \to L_2$ *be hyperbolic transformations along band sets* $\mathcal{B}_1$ *and* $\mathcal{B}_2$, *respectively. Suppose that* $\mathcal{B}_1$ *is disjoint from* $\mathcal{B}_2$. *Let* $L_0 \to L_2$ *be a hyperbolic transformation along* $\mathcal{B}_1 \cup \mathcal{B}_2$. *Then* $F = F(L_0 \to L_1 \to L_2)_{[a,b]}$ *and* $F' = F(L_0 \to L_2)_{[a,b]}$ *are ambient isotopic in* $\mathbb{R}^3[a, b]$ *rel* $\mathbb{R}^3[a] \cup \mathbb{R}^3[b]$.

Proof. Let t_1 and t_2 be numbers with $a < t_1 < t_2 < b$ and assume

$$\pi(F \cap \mathbb{R}^3[t]) = \begin{cases} L_0 & \text{for } t \in [a, t_1), \\ L_0 \cup \mathcal{B}_1 & \text{for } t = t_1, \\ L_1 & \text{for } t \in (t_1, t_2), \\ L_1 \cup \mathcal{B}_2 & \text{for } t = t_2, \\ L_2 & \text{for } t \in (t_2, b]. \end{cases}$$

A subset $\mathcal{B}_2[t_1, t_2]$ of $\mathbb{R}^3[a, b]$ is a union of mutually disjoint 3-balls. Applying the cellular move lemma along these 3-balls, we can deform F such that

$$\pi(F \cap \mathbb{R}^3[t]) = \begin{cases} L_0 & \text{for } t \in [a, t_1), \\ L_0 \cup \mathcal{B}_1 \cup \mathcal{B}_2 & \text{for } t = t_1, \\ L_2 & \text{for } t \in (t_1, b]. \end{cases}$$

This is a realizing surface of $L_0 \to L_2$. Thus F and F' are ambient isotopic in $\mathbb{R}^3[a, b]$ rel $\mathbb{R}^3[a] \cup \mathbb{R}^3[b]$ (recall Exercise 9.3). ∎

Let $L \to L'$ be a hyperbolic transformation along a band set $\mathcal{B}$. Let $\{h_t\}$ be an ambient isotopy of $\mathbb{R}^3$ which keeps L fixed setwise. The set $h_1(\mathcal{B})$ is a band set attaching to L. Let $L \to L''$ be a hyperbolic transformation along $h_1(\mathcal{B})$, and let $L'' \to L'$ be an isotopic transformation by an ambient isotopy $\{h_{1-t} \circ h_1^{-1}\}$. Let $F = F(L \to L')_{[a,b]}$ and $F' = F(L \to L'' \to L')_{[a,b]}$ be realizing surfaces of $L \to L'$ and $L \to L'' \to L'$ in $\mathbb{R}^3[a, b]$.

LEMMA 9.5 (Band deformation lemma I). *In the above situation, F and F' are ambient isotopic in $\mathbb{R}^3[a, b]$ rel $\mathbb{R}^3[a] \cup \mathbb{R}^3[b]$.*

Proof. Let $a < t_1 < t_0 < t_2 < b$ and assume that

$$\pi(F \cap \mathbb{R}^3[t]) = \begin{cases} L & \text{for } t \in [a, t_0), \\ L \cup \mathcal{B} & \text{for } t = t_0, \\ L' & \text{for } t \in (t_0, b]. \end{cases}$$

Consider a height preserving homeomorphism $G : \mathbb{R}^3[a, b] \to \mathbb{R}^3[a, b]$ such that for each $x \in \mathbb{R}^3$,

$$\pi(G(x[t])) = \begin{cases} h_{(t-a)/(t_1-a)}(x) & \text{for } t \in [a, t_1), \\ h_1(x) & \text{for } t \in [t_1, t_2], \\ h_{(b-t)/(b-t_2)}(x) & \text{for } t \in (t_2, b]. \end{cases}$$

The image $G(F)$ is ambient isotopic to F in $\mathbb{R}^3[a, b]$ rel $\mathbb{R}^3[a] \cup \mathbb{R}^3[b]$. On the other hand, $G(F)$ is a realizing surface of $L \to L'' \to L'$ in $\mathbb{R}^3[a, b]$, and hence ambient isotopic to F' in $\mathbb{R}^3[a, b]$ rel $\mathbb{R}^3[a] \cup \mathbb{R}^3[b]$. ∎

LEMMA 9.6. *Let $L_0 \to L_1$ and $L_1 \to L_2$ be hyperbolic transformations along band sets $\mathcal{B}_1$ and $\mathcal{B}_2$, respectively, where $\mathcal{B}_1$ may intersect $\mathcal{B}_2$. There is a hyperbolic transformation $L_0 \to L$ along a band set $\mathcal{B}$ and an isotopic transformation $L \to L_2$ for some link L such that $F = F(L_0 \to L_1 \to L_2)_{[a,b]}$ and $F' = F(L_0 \to L \to L_2)_{[a,b]}$ are ambient isotopic in $\mathbb{R}^3[a, b]$ rel $\mathbb{R}^3[a] \cup \mathbb{R}^3[b]$.*

Proof. Let c be a number between a and b such that $F = F(L_0 \to L_1)_{[a,c]} \cup F(L_1 \to L_2)_{[c,b]}$. Let $\{h_t\}$ be an ambient isotopy of $\mathbb{R}^3$ keeping L_1 fixed setwise such that $h_1(\mathcal{B}_2)$ is disjoint from $\mathcal{B}_1$. Let $L_1 \to L$ be a hyperbolic transformation along $h_1(\mathcal{B}_2)$. By the previous lemma, there is an isotopic transformation $L \to L_2$ such that $F(L_1 \to L_2)_{[c,b]}$ and $F(L_1 \to L \to L_2)_{[c,b]}$ are ambient isotopic in $\mathbb{R}^3[c, b]$ rel $\mathbb{R}^3[c] \cup \mathbb{R}^3[b]$. Since $h_1(\mathcal{B}_2)$ is disjoint from $\mathcal{B}_1$, using the cellular move lemma, we can change the level where $h_1(\mathcal{B}_2)$ appears into the same level as $\mathcal{B}_1$. The result is a realizing surface of $L_0 \to L \to L_2$, where $L_0 \to L$ is a hyperbolic transformation along $\mathcal{B} = \mathcal{B}_1 \cup h_1(\mathcal{B}_2)$. ∎

LEMMA 9.7. *Let $L_0 \to L_1$ and $L_1 \to L_2$ be isotopic transformations. There is an isotopic transformation $L_0 \to L_2$ such that $F = F(L_0 \to L_1 \to L_2)_{[a,b]}$ and $F' = F(L_0 \to L_2)_{[a,b]}$ are ambient isotopic in $\mathbb{R}^3[a,b]$ rel $\mathbb{R}^3[a] \cup \mathbb{R}^3[b]$.*

Proof. It is obvious. ∎

LEMMA 9.8. *Let $L_0 \to L_1$ be an isotopic transformation and let $L_1 \to L_2$ be a hyperbolic transformation along a band set $\mathcal{B}$. There is a hyperbolic transformation $L_0 \to L$ along a band set $\mathcal{B}'$ and an isotopic transformation $L \to L_2$ such that $F = F(L_0 \to L_1 \to L_2)_{[a,b]}$ and $F' = F(L_0 \to L \to L_2)_{[a,b]}$ are ambient isotopic in $\mathbb{R}^3[a,b]$ rel $\mathbb{R}^3[a] \cup \mathbb{R}^3[b]$.*

Proof. Let $a < t_1 < t_0 < t_2 < b$ and assume that

$$
\pi(F \cap \mathbb{R}^3[t]) = \begin{cases}
h_{(t-a)/(t_1-a)}(L_0) & \text{for } t \in [a, t_1), \\
L_1 & \text{for } t \in [t_1, t_0), \\
L_1 \cup \mathcal{B} & \text{for } t = t_0, \\
L_2 & \text{for } t \in (t_0, b].
\end{cases}
$$

Consider a height preserving homeomorphism $G : \mathbb{R}^3[a,b] \to \mathbb{R}^3[a,b]$ such that for each $x \in \mathbb{R}^3$,

$$
\pi(G(x[t])) = \begin{cases}
h^{-1}_{(t-a)/(t_1-a)}(x) & \text{for } t \in [a, t_1), \\
h^{-1}_1(x) & \text{for } t \in [t_1, t_2], \\
h^{-1}_{(b-t)/(b-t_2)}(x) & \text{for } t \in (t_2, b].
\end{cases}
$$

The image $G(F)$ is ambient isotopic to F in $\mathbb{R}^3[a,b]$ rel $\mathbb{R}^3[a] \cup \mathbb{R}^3[b]$. On the other hand, $G(F)$ is a realizing surface of $L_0 \to L \to L_2$ in $\mathbb{R}^3[a,b]$, where $L_0 \to L$ is a hyperbolic transformation along $h^{-1}_1\mathcal{B}$ and $L \to L_2$ is an isotopic transformation by an ambient isotopy $\{h^{-1}_{1-t} \circ h_1\}$. ∎

PROPOSITION 9.9. *Let $L = L_0 \to L_1 \to \cdots \to L_n = L'$ be a link transformation sequence. There is a sequence $L \to L'' \to L'$ such that $L \to L''$ is a hyperbolic transformation, $L'' \to L'$ is an isotopic transformation, and $F = F(L = L_0 \to L_1 \to \cdots \to L_n = L')_{[a,b]}$ and $F' = F(L \to L'' \to L')_{[a,b]}$ are ambient isotopic in $\mathbb{R}^3[a,b]$ rel $\mathbb{R}^3[a] \cup \mathbb{R}^3[b]$.*

Proof. This is a consequence of the lemmas above. ∎

Let $L \to L'$ be a hyperbolic transformation along a band set $\mathcal{B}$. Divide $\mathcal{B}$ into two band sets $\mathcal{B}_1$ and $\mathcal{B}_2$. Let $L_1 = \mathrm{h}(L; \mathcal{B}_1)$. Consider an ambient isotopy $\{h_t\}$ of $\mathbb{R}^3$ which keeps L_1 fixed setwise; the set $h_1(\mathcal{B}_2)$ is disjoint from $\mathcal{B}_1$. ($h_t(\mathcal{B}_2)$ may intersect $\mathcal{B}_1$ for $t \in (0,1)$.) Put $\mathcal{B}' = \mathcal{B}_1 \cup h_t(\mathcal{B}_2)$, let $L \to L''$ be a hyperbolic transformation along $\mathcal{B}'$, and let $L'' \to L'$ be an isotopic transformation by $\{h_{1-t} \circ h^{-1}_1\}$.

LEMMA 9.10 (Band deformation lemma II). *In the above situation, $F = F(L \to L')_{[a,b]}$ and $F' = F(L \to L'' \to L')_{[a,b]}$ are ambient isotopic in $\mathbb{R}^3[a,b]$ rel $\mathbb{R}^3[a] \cup \mathbb{R}^3[b]$.*

Proof. F is ambient isotopic to a realizing surface of the hyperbolic transformation sequence $L \to L_1 \to L'$ with band sets $\mathcal{B}_1$ and $\mathcal{B}_2$ (Lemma 9.4). By Lemma 9.5, it is ambient isotopic to a realizing surface of $L \to L_1 \to L'' \to L'$, where $L_1 \to L''$ is a hyperbolic transformation along $h_1(\mathcal{B}_2)$ and $L'' \to L'$ is an isotopic transformation by $\{h_{1-t} \circ h_1^{-1}\}$. By Lemma 9.4, it is ambient isotopic to F'. ∎

9.4. Closed Realizing Surface

Let F be a properly embedded surface in $\mathbb{R}^3[a, b]$ and let L_- and L_+ be links in $\mathbb{R}^3$ appearing as the boundary of F in $\mathbb{R}^3[a]$ and in $\mathbb{R}^3[b]$. When L_- and L_+ are trivial links, we can consider a closed surface $\widehat{F}$ such that

$$\widehat{F} = \widetilde{\mathcal{D}}_- \cup F \cup \widetilde{\mathcal{D}}_+,$$

where $\widetilde{\mathcal{D}}_-$ is a trivial disk system in $\mathbb{R}^3(-\infty, a]$ with boundary $L_-[a]$ and $\widetilde{\mathcal{D}}_+$ is a trivial disk system in $\mathbb{R}^3[b, \infty)$ with boundary $L_+[b]$. We call $\widehat{F}$ a *closure* of F. The process $F \mapsto \widehat{F}$ is called a *closing* or a *cap off*.

PROPOSITION 9.11. *Let $\widehat{F}_1$ and $\widehat{F}_2$ be closures of the same surface F in $\mathbb{R}^3[a, b]$. Then they are ambient isotopic in $\mathbb{R}^4$ rel $\mathbb{R}^3[a, b]$.*

Proof. This is a consequence of Proposition 8.6. ∎

By this proposition, when we consider a surface link up to equivalence or up to ambient isotopy, we do not need to specify the trivial disk systems $\widetilde{\mathcal{D}}_-$ and $\widetilde{\mathcal{D}}_+$ for closing.

DEFINITION 9.12. Let $L_0 \to L_1 \to \cdots \to L_n$ be a link transformation sequence such that L_0 and L_n are trivial links. A *closed realizing surface of $L_0 \to L_1 \to \cdots \to L_n$* is a closure of a realizing surface.

Let $F = \widetilde{\mathcal{D}}_- \cup F(L_0 \to L_1 \to \cdots \to L_n)_{[a,b]} \cup \widetilde{\mathcal{D}}_+$ be a closed realizing surface of a link transformation sequence $L_0 \to L_1 \to \cdots \to L_n$. Let t_1 and t_2 be real numbers such that $F = F(L_0 \to L_1)_{[a,t_1]} \cup F(L_1 \to \cdots \to L_{n-1})_{[t_1,t_2]} \cup F(L_{n-1} \to L_n)_{[t_2,b]}$.

LEMMA 9.13. *In the above situation, (1) if $L_0 \to L_1$ is an isotopic transformation, then F is ambient isotopic to a closed realizing surface $\widetilde{\mathcal{D}}'_- \cup F(L_1 \to \cdots \to L_n)_{[t_1,b]} \cup \widetilde{\mathcal{D}}_+$ of $L_1 \to \cdots \to L_n$ by an ambient isotopy of $\mathbb{R}^4$ rel $\mathbb{R}^3[t_1, \infty)$.*
(2) If $L_{n-1} \to L_n$ is an isotopic transformation, then F is ambient isotopic to a closed realizing surface $\widetilde{\mathcal{D}}_- \cup F(L_0 \to \cdots \to L_{n-1})_{[a,t_2]} \cup \widetilde{\mathcal{D}}'_+$ of $L_0 \to \cdots \to L_{n-1}$ by an ambient isotopy of $\mathbb{R}^4$ rel $\mathbb{R}^3(-\infty, t_2]$.

Proof. This is a consequence of Proposition 8.6. ∎

9.5. The Normal Form

THEOREM 9.14. *For any surface link F in $\mathbb{R}^4$, there is a hyperbolic transformation $L_- \to L_+$ along a band set $\mathcal{B}$ satisfying the following conditions:*
(1) L_- and L_+ are trivial links,
(2) F is equivalent to a closed realizing surface of $L_- \to L_+$.

Proof. By Proposition 8.4, we may assume that F has only elementary critical points. We may assume that F is in $\mathbb{R}^3(-1, 1)$. Consider an arc from each maximal point v to a point of $\mathbb{R}^3[1]$ such that the intersection with F is v and the intersection with $\mathbb{R}^3[t]$ is a single point or the empty set depending on whether $t \in [-1, 1]$ is above or below the fourth coordinate of v. Push v along the arc by an ambient isotopy into $\mathbb{R}^3[1]$. Applying this process to all maximal points, we may assume that all maximal points are in $\mathbb{R}^3[1]$. Similarly, all minimal points are assumed in $\mathbb{R}^3[-1]$. Change each saddle point into a saddle band, and we have a link transformation sequence $L_0 \to L_1 \cdots \to L_n$ such that L_0 and L_n are trivial links. By Propositions 9.9 and 9.13, we have the result. ∎

FIGURE 9.2

EXERCISE 9.15. In the situation of Theorem 9.14, the Euler characteristic $\chi(F)$ of the surface link F is given by

$$\chi(F) = \sharp(L_-) + \sharp(L_+) - \sharp(\mathcal{B}).$$

DEFINITION 9.16. A 2-knot is in a *normal form* if it is a closed realizing surface of a hyperbolic transformation sequence $O_- \to K \to O_+$ with band sets $\mathcal{B}_-, \mathcal{B}_+$ such that O_- and O_+ are trivial links and K is a knot, and such that $O_- \to K$ is a complete fusion and $K \to O_+$ is a complete fission (i.e., $\sharp(O_-) = \sharp(\mathcal{B}_-) + 1$ and $\sharp(O_+) = \sharp(\mathcal{B}_+) + 1$).

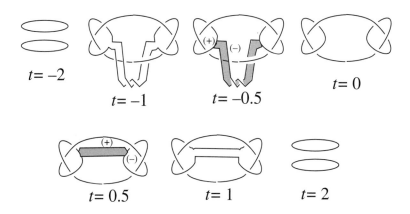

FIGURE 9.3. A 2-knot in a normal form

THEOREM 9.17. *Any 2-knot is equivalent to a 2-knot in a normal form.*

Proof. In the situation of Theorem 9.14, choose a subset $\mathcal{B}_-$ of $\mathcal{B}$ such that $L_- \to$ h$(L_-; \mathcal{B}_-)$ is complete fusion, and put $\mathcal{B}_+ = \mathcal{B} \setminus \mathcal{B}_-$. (For example, Fig. 9.3 is obtained from Fig. 9.2.) ∎

COROLLARY 9.18. *Any 2-knot F is deformed up to equivalence such that*

(1) *all minimal points (minimal disks) are in $\mathbb{R}^3[-3]$,*

(2) *all maximal points (maximal disks) are in* $\mathbb{R}^3[3]$,
(3) *all saddle points (saddle bands) are in* $\mathbb{R}^3[-2] \cup \mathbb{R}^3[2]$, *and*
(4) *the cross-section* $F \cap \mathbb{R}^3[0]$ *is a knot.*

An orientable surface knot F in $\mathbb{R}^4$ of genus g is in a *normal form* if it is a closed realizing surface of a hyperbolic transformation sequence

$$\mathcal{O}_- \to K_- \to L \to K_+ \to \mathcal{O}_+$$

with band sets $\mathcal{B}_{--}$, $\mathcal{B}_-$, $\mathcal{B}_+$ and $\mathcal{B}_{++}$ such that

(1) $\mathcal{O}_- \to K_-$ is a complete fusion and $K_+ \to \mathcal{O}_+$ is a complete fission,
(2) $K_- \to L$ is a complete fission and $L \to K_+$ is a complete fusion,
(3) $\sharp(L) = g + 1$.

Especially $\sharp(\mathcal{B}_-) = \sharp(\mathcal{B}_+) = g$.

THEOREM 9.19. *Any orientable surface knot is deformed in a normal form up to equivalence.*

EXERCISE 9.20. Prove Theorem 9.19. [*Hint*: This is not so obvious as Theorem 9.17. In the situation of Theorem 9.14, in general, one cannot divide $\mathcal{B}$ into four required band sets $\mathcal{B}_{--}$, $\mathcal{B}_-$, $\mathcal{B}_+$ and $\mathcal{B}_{++}$. For example, let L_- be a trivial knot, $\mathcal{B} = B_1 \cup \cdots \cup B_4$ a band set as in Fig. 9.4, and $L_+ = \mathrm{h}(L_-; \mathcal{B})$. A closed realizing surface of $L_- \to L_+$ is a surface knot of genus two. If band sets $\mathcal{B}_{--}$, $\mathcal{B}_-$, $\mathcal{B}_+$ and $\mathcal{B}_{++}$ are obtained by dividing $\mathcal{B}$, $\sharp(\mathcal{B}_-) = \sharp(\mathcal{B}_+) = 2$ and $\sharp(\mathcal{B}_{--}) = \sharp(\mathcal{B}_{++}) = 0$. However we cannot obtain a 3-component link L by surgery along any two bands of $\mathcal{B}$. Such a bad situation can be avoided by using the band deformation lemma II (see [**372**] for the details).]

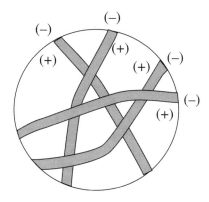

FIGURE 9.4. A band set

CHAPTER 10

Examples (Spinning)

10.1. Spinning Construction

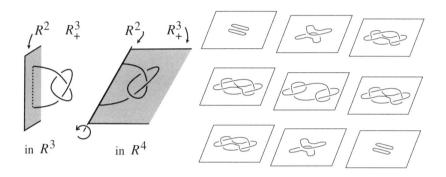

FIGURE 10.1. Artin's spinning construction

E. Artin [**16**] introduced a method to construct a 2-knot from a classical knot. For a properly embedded arc K_0 in $\mathbb{R}^3_+ = \{(x, y, z) \mid z \geq 0\}$, consider the subset

$$\{(x, y, z\cos\theta, z\sin\theta) \in \mathbb{R}^4 \mid (x, y, z) \in K_0, \theta \in [0, 2\pi)\}$$

of $\mathbb{R}^4$. This is a 2-sphere embedded in $\mathbb{R}^4$. We call this a *spun 2-knot*.

When we consider $\mathbb{R}^3_+$ and $\mathbb{R}^2$ as subsets of $\mathbb{R}^4$ by

$$\mathbb{R}^3_+ = \{(x, y, z, 0) \in \mathbb{R}^4 \mid z \geq 0\},$$

$$\mathbb{R}^2 = \{(x, y, 0, 0) \in \mathbb{R}^4\},$$

the spun knot is obtained by rotating K_0 around $\mathbb{R}^2$; see Fig. 10.1.

A spun trefoil in a normal form is illustrated in Fig. 10.2.

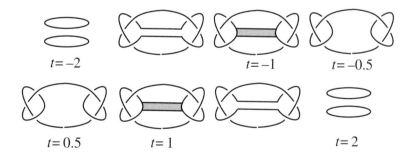

FIGURE 10.2. A spun trefoil in a normal form

A spun 2-knot can be defined as follows: Let (B, K_0) be a pair of a 3-ball B^3 and a properly embedded arc K_0. Define

$$\Sigma = B \times S^1/(x, \theta) \sim (x, \theta') \quad \text{for } x \in \partial B, \text{ and } \theta, \theta' \in S^1;$$

$$F = K_0 \times S^1/\sim .$$

Then Σ is a 4-sphere and F is a 2-sphere. Removing a point from Σ, we have a 2-knot F in $\mathbb{R}^4$. This is an alternative definition of a spun 2-knot.

EXERCISE 10.1. Verify that the two definitions give the same, up to equivalence, 2-knot.

10.2. Twist-Spinning Construction

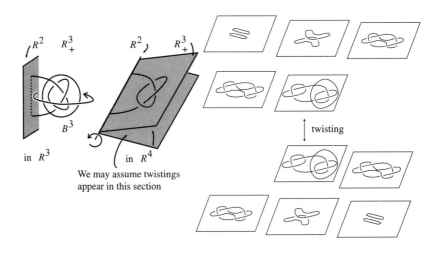

FIGURE 10.3. Zeeman's twist-spinning construction

E. C. Zeeman [**992**] introduced the notion of a twist-spun 2-knot. Consider a properly embedded arc K_0 in $\mathbb{R}^3_+$. Let B^3 be a 3-ball containing the knotted part of K_0 as in Fig. 10.3. A spun knot is obtained as a trace of K when we rotate the 3-space $\mathbb{R}^3_+$ around $\mathbb{R}^2$ as in Fig. 10.1. Now we modify this construction. While rotating the 3-space $\mathbb{R}^3_+$ around $\mathbb{R}^2$, twist the 3-ball B^3 d times. Then we have another 2-sphere in $\mathbb{R}^4$. We call this a *d-twist-spun 2-knot* of K_0 or of K, where K is a classical knot obtained from K_0 by connecting the boundary points by an arc in $\mathbb{R}^2$.

It is considered as a closed realizing surface of a link transformation sequence as in Fig. 10.4, where twisting is applied d times. It is ambient isotopic to a surface as in Fig. 10.5 (for simplicity, the $d = 1$ case is illustrated). Finally, a spun trefoil in a normal form is illustrated in Fig. 10.6.

A d-twist-spun 2-knot can be defined as follows: Let (B, K_0) be a pair of a 3-ball B^3 and a properly embedded arc K_0. Let A be an unknotted arc in B with $\partial A = \partial K_0$. Let $R_\theta : B \to B$ be rotation through the angle θ about A in the positive meridian direction. Define

$$\Sigma = B \times S^1/(x, \theta) \sim (x, \theta') \quad \text{for } x \in \partial B, \text{ and } \theta, \theta' \in S^1;$$

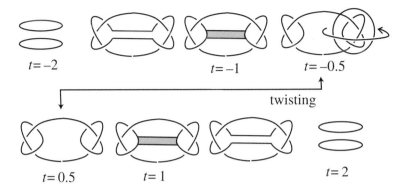

FIGURE 10.4. A twist-spun trefoil (1)

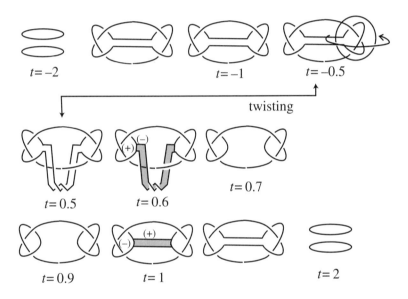

FIGURE 10.5. A twist-spun trefoil (2)

$$F = \cup_\theta (R_{d\theta}(K_0) \times \{e^{i\theta}\})/ \sim .$$

Then Σ is a 4-sphere and F is a 2-sphere. Removing a point from Σ, we have a 2-knot F in $\mathbb{R}^4$. This is an alternative definition of a d-twist-spun 2-knot. In fact, this is the original definition given by Zeeman.

For a knot K, we denote by $\tau_d(K)$ a d-twist-spun knot of K.

EXERCISE 10.2. Verify that the two definitions give the same, up to equivalence, 2-knot.

THEOREM 10.3 ([**992**]). *Let F be a d-twist-spun 2-knot of a knot K. Assume K is a knot in S^3. Let $M_d(K)$ denote a d-fold cyclic branched covering space of S^3 branching along K. Then F is a fibered 2-knot whose fiber is the punctured $M_d(K)$.*

Refer to Zeeman [**992**] for a proof and the terminology of this theorem. As a corollary, we have that a 1-twist-spun 2-knot is always trivial.

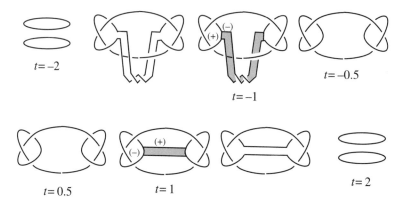

FIGURE 10.6. A 1-twist-spun trefoil in a normal form

For a surface link F we denote by F^* the mirror image, and by $-F$ the orientation-reversed one.

THEOREM 10.4 (Symmetry Theorem on Twist-Spun Knots [**529**]). *For any K and d,*

(1) $\tau_d(-K) \cong \tau_d(K)^*$.
(2) $\tau_d(K^*) \cong -\tau_d(K)$.
(3) $\tau_d(-K^*) \cong -\tau_d(K)^*$.

Refer to [**529**] for a proof.

10.3. Deform-Spinning Construction

R. H. Fox introduced another generalization of Artin's spinning, the so-called Fox's rolling-spinning. R. A. Litherland [**525**] established a systematic construction of 2-knots based on the deformation group of a knot. This includes all spinning constructions of Artin, Zeeman and Fox.

Let (B, K_0) be a pair of a 3-ball B and a properly embedded arc K_0. Let $f : B \to B$ be a homeomorphism with $f|_{\partial B} = \mathrm{id}$ and $f(K_0) = K_0$. Define

$$(\Sigma, F) = (B, K_0) \times [0, 1]/ \sim$$

where $\sim$ stands for

$$\begin{cases} (x, 0) \sim (f(x), 1) & \text{for } x \in B \\ (x, \theta) \sim (x, \theta') & \text{for } x \in \partial B \text{ and } \theta, \theta' \in [0, 1]. \end{cases}$$

Then Σ is a 4-sphere and F is a 2-sphere. Removing a point from Σ, we have a 2-knot F in $\mathbb{R}^4$. This is called a *deform-spun 2-knot* of (B, K_0) or a knot K, where K is a knot associated with (B, K_0).

We denote by $D_+(B, K_0)$ the group of all homeomorphisms $f : B \to B$ with $f|_{\partial B} = \mathrm{id}$ and $f(K_0) = K_0$, and by $\mathcal{D}_+(B, K_0)$ the group $\pi_0(D_+(B, K_0))$. An element of $\mathcal{D}_+(B, K_0)$ is called a *deformation*. In the next section, we study deformations in more detail.

REMARK 10.5. The original definition of the *deformation group* $D_+(B, K_0)$ by Litherland [**525**] is different from that given here. However it is also shown in [**525**] that they are the same for B a 3-ball and K_0 an arc.

EXERCISE 10.6. Prove that the knot type of a deform-spun 2-knot depends only on a deformation.

10.4. Deformations

Let (S^3, K) be a 1-knot and $N(K)$ be a tubular neighborhood of K in S^3. The boundary ∂E of the exterior $E = \mathrm{cl}\,(S^3 \setminus N(K))$ is identified with a torus $\mathbb{R}^2 / \langle \vec{\ell}, \vec{m} \rangle$, where vectors $\vec{\ell}$ and $\vec{m}$ correspond to a preferred longitude and a meridian on ∂E. We denote by $[\vec{x}]$ the point of $\mathbb{R}^2 / \langle \vec{\ell}, \vec{m} \rangle$ determined by $\vec{x}$. A collar of ∂E in E is identified with $\partial E \times [0,1]$ such that $\partial E = \partial E \times \{0\}$ and $N(K)$ is identified with $\partial E \times [-1, 0]/ \sim$, where $([\vec{x}], \lambda) \sim ([\vec{y}], \mu)$ iff $\lambda = \mu = -1$ and $\vec{x} - \vec{y} = s\vec{m}$ ($s \in \mathbb{R}$). For a vector $\vec{v} \in \mathbb{R}^2$, we define a homeomorphism $H(\vec{v}) : (S^3, K_0) \to (S^3, K_0)$ by

$$
\begin{cases}
H(\vec{v})(x) = x, & \text{for } x \in E \setminus \partial E \times [0,1] \\
H(\vec{v})([\vec{x}], \lambda) = ([\vec{x} + (1-\lambda)\vec{v}], \lambda), & \text{for } ([\vec{x}], \lambda) \in \partial E \times [0,1] \\
H(\vec{v})([\vec{x}], \lambda) = ([\vec{x} + \vec{v}], \lambda), & \text{for } ([\vec{x}], \lambda) \in \partial E \times [-1, 0]/ \sim.
\end{cases}
$$

Let (B, K_0) be a $(3,1)$-ball pair associated with (S^3, K); that is, (S^3, K) is decomposed into a pair of $(3,1)$-ball pairs, one of which is a standard ball pair and the other is homeomorphic to (B, K_0). We fix a homeomorphism ϕ from the ball pair in (S^3, K) to (B, K_0). Put

$$ r_\tau = \phi \circ H(-\vec{m}) \circ \phi^{-1} $$

and

$$ r_\rho = \phi \circ H(-\vec{\ell}) \circ \phi^{-1}. $$

They are elements of $D_+(B, K_0)$.

DEFINITION 10.7. The *twisting* and the *rolling deformations* are elements of $\mathcal{D}_+(B, K_0)$ represented by r_τ and r_ρ, respectively. They are denoted by τ and ρ.

Zeeman's d-twist spinning is a deform-spinning associated with a deformation $\tau^d \in \mathcal{D}_+(B, K_0)$, and Fox's rolling is that associated with ρ.

Let $\mathrm{Homeo}^*(S^3, K)$ be the group of all self-homeomorphisms of (S^3, K) which preserves the orientations of S^3 and K. A *symmetry* is a finite subgroup of $\mathrm{Homeo}^*(S^3, K)$. Let G be a symmetry. We may assume that $N(K) = \partial E \times [-1, 0]/ \sim$ and a collar $\partial E \times [0,1]$ are G-invariant and for each $g \in G$, $g([\vec{x}], \lambda) = ([\vec{x} + \vec{g}], \lambda)$, where $\vec{g}$ is a vector of $\mathbb{R}^2$. Put

$$ r_g = \phi \circ g \circ H(-\vec{g}) \circ \phi^{-1}. $$

A deformation represented by r_g is called a *g-symmetry deformation* and is denoted by σ_g. (A g-symmetry deformation σ_g depends on a choice of $\vec{g} \in \mathbb{R}^2$. There is a method to define $\vec{g} \in \mathbb{R}^2$ by use of the *slope* of G; cf. [**311**].)

REMARK 10.8. The deformation group $\mathcal{D}_+(S^3, K)$ is defined as the group of isotopy classes of orientation-preserving self-homeomorphisms of (S^3, K). $\mathcal{D}_+(S^3, K)$ is isomorphic to $\mathcal{D}_+(B, K_0)$. Let $\mathcal{P}(S^3, K)$ be the subgroup of $\mathcal{D}_+(S^3, K)$ generated by τ and ρ. Let $\mathrm{Sym}^*(S^3, K)$ be $\pi_0(\mathrm{Homeo}^*(S^3, K))$. Then there is an exact sequence

$$ 1 \to \mathcal{P}(S^3, K) \to \mathcal{D}_+(S^3, K) \to \mathrm{Sym}^*(S^3, K) \to 1. $$

For a proof, refer to [**311**].

10.5. Deform-Spun Projective Planes

Let B be the unit 3-ball in $\mathbb{R}^3$ and let K_0 be a proper arc in B whose endpoints are $(0, 1, 0)$ and $(0, -1, 0)$.

An *inverting homeomorphism* of (B, K_0) is a homeomorphism $h : B \to B$ such that $h(K_0) = K_0$ and $h|_{\partial B}$ is the π rotation about the z-axis. We denote by $D_-(B, K_0)$ the set of all inverting homeomorphisms and by $\mathcal{D}_-(B, K_0)$ the set of isotopy classes (rel ∂B) of inverting homeomorphisms. An element of $\mathcal{D}_-(B, K_0)$ is an *inverting deformation*.

For an inverting homeomorphism f, define

$$(\Sigma_0, F_0) = (B, K_0) \times [0, 1]/(x, 0) \sim (f(x), 1) \quad \text{for } x \in B.$$

Then Σ_0 is a 4-dimensional solid torus $B^3 \times S^1$ and F_0 is a Möbius strip. Embed Σ_0 into $\mathbb{R}^4$ in an 'obvious' way and close up F_0 with an unknotted disk in order to obtain a projective plane F in $\mathbb{R}^4$. (We assume that Σ_0 is embedded in $\mathbb{R}^4$ so that if K_0 is a trivial arc, then F is a standard projective plane of normal Euler number 2.) This is called a *deform-spun projective plane* of (B, K_0) or a knot K, where K is a knot associated with (B, K_0).

We say that inverting deformations g_1 and g_2 of (B, K_1) and (B, K_2) are *equivalent* if there exist representatives f_1 and f_2 and a homeomorphism $s : (B, K_1) \to (B, K_2)$ such that $s|_{\partial B} = \text{id}$ and $s \circ f_1 = f_2 \circ s$. It is easily seen that if g_1 and g_2 are equivalent, then deform-spun projective planes are equivalent.

LEMMA 10.9. *Let g and h be a deformation and an inverting deformation of (B, K_0), respectively. If there exist representatives r_g and r_h such that $r_h \circ r_g \circ r_h^{-1} = r_g^{-1}$, then for any deformation d, the deform-spun projective planes associated with $h \circ d$ and $h \circ g \circ d \circ g$ are equivalent.*

Proof. Let $s : (B, K_0) \to (B, K_0)$ be r_g^{-1}. Then $s \circ (r_h \circ r_d) = (r_h \circ r_g \circ r_d \circ r_g) \circ s$. Thus $h \circ d$ and $h \circ g \circ d \circ g$ are equivalent inverting deformations. ∎

COROLLARY 10.10. *Let d and h be a deformation and an inverting deformation of (B, K_0), respectively. Then deform-spun projective planes associated with $h \circ d$, $h \circ \tau^2 \circ d$ and $h \circ \rho^2 \circ d$ are equivalent.*

Proof. Let r_h be an inverting homeomorphism representing h and let u_h be a homeomorphism of (S^3, K) with $r_h = \phi \circ u_h \circ \phi^{-1}$. We may assume that u_h acts on $N(K) \cup \partial E \times [0, 1]$ by $u_h([\overrightarrow{x}], \lambda) = ([-\overrightarrow{x}], \lambda)$. Then $r_h \circ r_\tau \circ r_h^{-1} = r_\tau^{-1}$ and $r_h \circ r_\rho \circ r_h^{-1} = r_\rho^{-1}$. Thus $h \circ d$, $h \circ \tau \circ d \circ \tau$ and $h \circ \rho \circ d \circ \rho$ are equivalent. Since τ and ρ are central elements of the deformation group, we have the result. ∎

Let h_0 be the π rotation of B. Suppose that h_0 is an inverting homeomorphism of (B, K_0). Let C be a great circle on ∂B containing $(0, 0, 1)$, $(0, 0, -1)$ and $(0, 1, 0)$. The image $p(K_0 \cup \text{Fix}h_0 \cup C)$ is a θ-curve in 3-ball B/h_0, where $p : B \to B/h_0$ is the quotient map. Let $k(K_0)$ be a constituent knot of the θ-curve which contains $p(K_0 \cup \{(0, 0, -1)\})$.

THEOREM 10.11 (Price and Roseman [**741**]). *In the above situation, a deform-spun projective plane associated with h_0 is a connected sum of a standard projective plane (of normal Euler number 2) and a 2-twist-spun 2-knot of $k(K_0)$.*

THEOREM 10.12 ([**310**]). *In the above situation, a deform-spun projective plane associated with $\tau^m \circ \rho^n \circ h_0$ is a connected sum of a standard projective plane (of normal Euler number 2) and a 2-twist-spun 2-knot.*

THEOREM 10.13 ([**311**]). *For a ball pair (B, K_0) whose associated knot is a torus knot or hyperbolic knot, any deform-spun projective plane is a connected sum of a standard projective plane (of normal Euler number 2) and a 2-twist-spun 2-knot.*

A projective plane F in $\mathbb{R}^4$ is said to be *of Kinoshita type* if it is obtained as a connected sum of a 2-knot and a standard projective plane, [**404**]. If it is of Kinoshita type, then the order of a meridian element of $\pi_1(\mathbb{R}^4 \setminus F)$ is 2. It is an open question whether there exists a projective plane in $\mathbb{R}^4$ that is not of Kinoshita type. It is also unknown whether there exists a projective plane in $\mathbb{R}^4$ such that the order of a meridian element is not 2 (if this happens, the order is 4).

CHAPTER 11

Ribbon Surface Links

11.1. 1-Handle and 2-Handle Surgery

Let F be a surface link. A 1-*handle* attaching to F is a 3-ball h embedded in $\mathbb{R}^4$ such that $F \cap h$ is a pair of 2-disks in ∂h. We say that a surface link

$$\mathrm{h}(F; h) = \mathrm{cl}\,(F \cup \partial h \setminus F \cap h)$$

is obtained from F by a 1-*handle surgery*, or *surgery*, along a 1-handle h.

If F is oriented and the surgery result $\mathrm{h}(F; h)$ is orientable, then we give $\mathrm{h}(F; h)$ an orientation which is induced from $F \setminus F \cap h$. Such a 1-handle surgery is often called an *oriented* 1-*handle surgery*.

Let F be a surface link. A 2-*handle* attaching to F is a 3-ball h embedded in $\mathbb{R}^4$ such that $F \cap h$ is an annulus in ∂h. We say that a surface link

$$\mathrm{h}^2(F; h) = \mathrm{cl}\,(F \cup \partial h \setminus F \cap h)$$

is obtained from F by a 2-*handle surgery*, or *surgery*, along a 2-handle h.

The inverse operation of a 1-handle surgery is a 2-handle surgery, and vice versa.

Two 1-handles attaching to F are *equivalent* if they are ambient isotopic in $\mathbb{R}^4$ keeping F fixed setwise. Two 1-handles attaching to F are *equivalent in a subset N of* $\mathbb{R}^4$ if they are contained in N and one is ambient isotopic to the other by an ambient isotopy of N rel ∂N keeping F fixed setwise.

We say that a simple arc α in $\mathbb{R}^4$ *attaches* to F if $\alpha \cap F = \partial \alpha$. Two such arcs are said to be *equivalent* if they are ambient isotopic in $\mathbb{R}^4$ keeping F fixed setwise.

It is known that for a given simple arc α attaching to F, there are two equivalence classes in $N(\alpha)$ of 1-handles whose cores are α, where $N(\alpha)$ is a regular neighborhood of α such that $F \cap N(\alpha)$ is a pair of trivial 2-disks in $N(\alpha)$, [**59, 277**]. If F is oriented or $F \cap N(\alpha)$ is oriented, there is a unique equivalence class such that the 1-handle can be oriented coherently with the orientation of $F \cap N(\alpha)$.

EXERCISE 11.1. If two simple arcs α_0 and α_1 attaching to F are homotopic in $\mathbb{R}^4$ by a one-parameter family, say $\{\alpha_t\}$, such that for each $t \in [0, 1]$, $\alpha_t \cap F = \partial F$, then they are equivalent.

PROPOSITION 11.2. *Let F be an unknotted oriented surface knot. If a surface knot is obtained from F by an oriented 1-handle surgery, then it is also unknotted.*

Proof. Since $\pi(\mathbb{R}^4 \setminus F)$ is an infinite cyclic group generated by a meridian, any simple arc attaching to F is equivalent to a trivially attaching arc. Surgery along a 1-handle with such a trivial core arc changes F into the connected sum of F and a standard torus in $\mathbb{R}^4$, which is also unknotted. ∎

11.2. Ribbon Surface Links

DEFINITION 11.3. A surface link F is a *ribbon surface link* if it is obtained from a trivial 2-link F_0 by surgery along some (or no) mutually disjoint 1-handles attaching to F_0.

An orientable surface link is ribbon if and only if it bounds a 3-manifold immersed in $\mathbb{R}^4$ such that each component is a handlebody and the singularity consists of 'ribbon singularities'. (We will not use this fact and omit the definition of ribbon singularity.) This is the reason that we call it a ribbon surface link.

THEOREM 11.4. *An orientable surface link F is ribbon if and only if it is ambient isotopic to a surface link in a normal form which is symmetric with respect to the hyperplane $\mathbb{R}^3[0]$.*

Proof. Let F be in a normal form which is symmetric. It is a closed realizing surface of a hyperbolic transformation sequence

$$\mathcal{O}_+ \to K_+ \to L \to K_+ \to \mathcal{O}_+$$

with band sets $\mathcal{B}_{++}$, $\mathcal{B}_+$, $\mathcal{B}_+$ and $\mathcal{B}_{++}$. We may assume that $\mathcal{B}_{++}$ and $\mathcal{B}_+$ are disjoint. Thus, denoting $\mathcal{B}_{++} \cup \mathcal{B}_+$ by $\mathcal{B}$, F is ambient isotopic to a closed realizing surface of a hyperbolic transformation sequence

$$\mathcal{O}_+ \to L \to \mathcal{O}_+$$

with the same band set $\mathcal{B}$ such that

$$\pi(F \cap \mathbb{R}^3[t]) = \begin{cases} \mathcal{O}_+ & \text{for } t \text{ with } |t| \in [2,3] \\ \mathcal{O}_+ \cup \mathcal{B} & \text{for } t \text{ with } |t| = 2 \\ L & \text{for } t \text{ with } |t| \in [0,2]. \end{cases}$$

$\mathcal{B} \times [-2,2]$ is a union of mutually disjoint 3-disks in $\mathbb{R}^4$, each of which is attaching to F as a 2-handle. Let F_0 be the surgery result. Then F_0 is a trivial 2-link. This implies that F is obtained from F_0 by surgery along 1-handles. Thus F is ribbon.

Conversely, suppose that F is ribbon which is obtained from a trivial 2-link F_0 by surgery along 1-handles. We may assume that

$$\pi(F_0 \cap \mathbb{R}^3[t]) = \mathcal{O}_+ \quad \text{for } t \in [-3,3]$$

and that $F_0 \cap \mathbb{R}^3[3,\infty)$ and $F_0 \cap \mathbb{R}^3(-\infty,-3]$ are trivial disk systems. Since the inclusion induced homomorphism $i_\sharp : \pi_1(\mathbb{R}^3[0] \setminus \mathcal{O}_+) \to \pi_1(\mathbb{R}^4 \setminus F_0)$ is an isomorphism, we may assume that all 1-handles attaching to F_0 have cores in $\mathbb{R}^3[0]$. Since oriented 1-handles are uniquely determined from their cores up to equivalence, we may assume that each 1-handle is $B[-2,2]$ for a band B attaching to $\mathcal{O}_+$. The surgery result has a symmetric normal form. ∎

COROLLARY 11.5. *An oriented ribbon surface link is negative amphicheiral.*

A surface link is *simply knotted* if there is a surface link diagram whose singularity set consists of double points (without triple points and branch points).

PROPOSITION 11.6. *An orientable ribbon surface link is simply knotted.*

Proof. We may assume that a trivial 2-link F_0 is mapped by a projection $\pi : \mathbb{R}^4 \to \mathbb{R}^3$ to trivial 2-spheres in $\mathbb{R}^3$. Deform the 1-handles up to equivalence so that their images are mutually disjoint and intersect the 2-spheres transversely. ∎

In general, the converse is false. However, it is true when F is a 2-knot.

PROPOSITION 11.7. *A 2-knot is a ribbon 2-knot if and only if it is simply knotted.*

Proof. The only if part follows Proposition 11.6. The double point set of $\pi(F)$ consists of simple loops, called double loops. If there are no double loops, it is a trivial 2-knot. Suppose there are some double loops. Let C be an innermost loop on F among the preimage of the double loops, and let Δ be a 2-disk in F bounded by C. Do surgery on the diagram of F along the image $\Delta^* = \pi(\Delta)$. The result is a surface link diagram, which has fewer double loops. Repeat this procedure until there are no double loops. Then we see that F is obtained from a trivial 2-link by surgery along 1-handles. ∎

11.3. Slice versus Ribbon

A knot K in $\mathbb{R}^3$ is a *slice knot* if there is a properly embedded 2-disk D in $\mathbb{R}^3[0, \infty)$ with $\partial D = K[0]$. Such a 2-disk is called a *slice disk* for K.

A knot K in $\mathbb{R}^3$ is a *ribbon knot* if there is a properly embedded 2-disk D in $\mathbb{R}^3[0, \infty)$ with $\partial D = K[0]$ such that D has no minimal points. Such a 2-disk is called a *ribbon disk* for K.

A ribbon knot is a slice knot. The converse is unknown; cf. [**169**].

CONJECTURE 11.8 (Slice-ribbon conjecture). *A slice knot is a ribbon knot.*

LEMMA 11.9. *For a knot K, the following conditions are mutually equivalent.*
(1) *K is a ribbon knot.*
(2) *K bounds a singular disk M in $\mathbb{R}^3$ such that all singularities are* ribbon singularites *as in Fig. 11.1.*
(3) *K is obtained from a trivial link by surgery along some mutually disjoint band.*

The condition (2) is the reason that it is called a ribbon knot.

FIGURE 11.1. Ribbon Singularity

EXERCISE 11.10. Verify the above lemma and generalize it for a ribbon 2-knot.

CHAPTER 12

Presentations of Surface Link Groups

12.1. The Presentation from a Diagram

Let F be a surface link in $\mathbb{R}^4$. The *knot group of F* is the fundamental group of the complement $\mathbb{R}^4 \setminus F$.

When a surface link diagram is given, one can calculate the knot group as follows: Recall that a surface link diagram is the image $\pi(F)$ by a generic projection $\pi : \mathbb{R}^4 \to \mathbb{R}^3[0] = \mathbb{R}^3$ with over/under information. Let Σ_1 be the set of double points, and let Σ_2 be the set of triple points and branch points. The union $\Sigma = \Sigma_1 \cup \Sigma_2$ is the singular point set of the diagram. $\pi(F) \setminus \Sigma$ is a surface without boundary embedded in $\mathbb{R}^3$. Let $M_1, \ldots, M_n$ be the connected components of $\pi(F) \setminus \Sigma$. For each component M_i, fix a normal orientation, which will be denoted by a small arrow intersecting the sheet M_i in a diagram. (This is possible, because each M_i is orientable even if F is non-orientable.)

For a curve γ in $\mathbb{R}^3$ which avoids the singularity set Σ and intersects $\pi(F) \setminus \Sigma = M_1 \cup \cdots \cup M_n$ transversely, let $w(\gamma)$ denote the word in $\{x_1, \ldots, x_n\}$ such that the kth letter is x_i (or x_i^{-1}, resp.) if the kth intersection of γ with $\pi(F) \setminus \Sigma$ is in M_i and the orientation of γ matches the normal orientation of M_i (or does not, resp.).

Let $\alpha_1, \ldots, \alpha_m$ be the connected components of Σ_1, each of which is a closed curve or an open arc. For each component α_i ($i \in \{1, \ldots, m\}$), consider a small oriented disk Δ_i pierced by α_i.

A *base relator of a double curve α_i* is the word $w(\partial \Delta_i)$.

An *upper relator of α_i* is a word obtained from $w(\partial \Delta_i)$ by removing the two letters that come from the lower sheets.

A *lower relator of α_i* is a word obtained from $w(\partial \Delta_i)$ by removing the two letters that come from the upper sheets.

For a curve as in Fig. 12.1, a base relator is $x_5^{-1} x_1 x_8^{-1} x_3^{-1}$, an upper relator is $x_5^{-1} x_8^{-1}$, and a lower relator is $x_1 x_3^{-1}$.

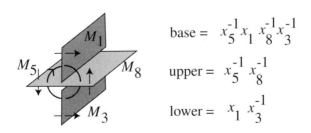

FIGURE 12.1. Relators coming from a double curve

An *upper presentation* (resp. *lower presentation*) associated with the diagram $\pi(F)$ is a presentation whose generators are $x_1, \ldots, x_n$ and the relators are base relators and upper relators (resp. lower relators) of the double curves.

PROPOSITION 12.1. (1) *An upper presentation associated with a surface link diagram $\pi(F)$ presents the surface link group of F. (2) A lower presentation associated with a surface link diagram $\pi(F)$ presents the surface link group of F.*

EXERCISE 12.2. Prove Proposition 12.1.

When F is an oriented surface link, this process is more simplified so that we can reduce the number of the generators from an upper presentation as follows: Give a normal orientation to each M_i by the "tangent followed by normal" rule. Then for each double arc, an upper relator implies $x_a = x_d$ if x_a and x_d correspond to the regions in the same upper sheet of the broken surface diagram. Thus we can assume that the generator set $\{x_i\}$ is in one-to-one correspondence to the set of the sheets of the broken surface diagram. Then the upper relations are given as in Fig. 12.2.

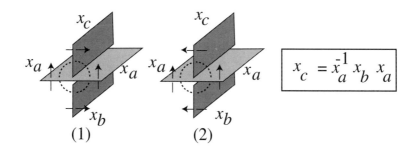

FIGURE 12.2. Relations in upper presentation for oriented surface links

12.2. The Presentation from a Motion Picture

Let $L \to L'$ be a hyperbolic transformation along a band set $\mathcal{B} = B_1 \cup \cdots \cup B_\ell$ attaching to L. Suppose that L and L' are in general position with respect to a projection $\pi' : \mathbb{R}^3 \to \mathbb{R}^2[0] = \mathbb{R}^2$ and the image of each band of $\mathcal{B}$ is as in Fig. 12.3 in the diagram on $\mathbb{R}^2$.

Let $E(L)$, $E(L \cup \mathcal{B})$ and $E(L')$ denote the exteriors $\mathrm{cl}(\mathbb{R}^3 \setminus N(L))$, $\mathrm{cl}(\mathbb{R}^3 \setminus N(L \cup \mathcal{B}))$ and $\mathrm{cl}(\mathbb{R}^3 \setminus N(L'))$. The middle column of Fig. 12.3 is a part of $E(L \cup \mathcal{B})$. Note that $E(L)$ is homotopy equivalent to $E(L \cup \mathcal{B}) \cup \mathcal{D}$, where $\mathcal{D}$ is a union of 2-disks each of which is as in the left column of the figure. $E(L')$ is homotopy equivalent to $E(L \cup \mathcal{B}) \cup \mathcal{D}'$, where $\mathcal{D}'$ is a union of 2-disks each of which is as in the right row of the figure. For a band B of $\mathcal{B}$, if y_1, y_2, y_3, y_4 are meridian loops of $E(L \cup \mathcal{B})$ as in the figure, there is a relator $y_1^{-1} y_2 y_3^{-1} y_4$. We denote this relator by $R(B)$. Attachment of a 2-disk of $\mathcal{D}$ induces a relator $y_1 y_4^{-1}$, which we denote by $R_-(B)$. Attachment of a 2-disk of $\mathcal{D}'$ induces a relator $y_1^{-1} y_2$, which we denote by $R_+(B)$. Relators $R(B)$, $R_-(B)$ and $R_+(B)$ are called a *base relator*, a *lower level relator* and an *upper level relator* of B, respectively. These relators imply $y_1 = y_2 = y_3 = y_4$.

Let $a_1, \ldots, a_n$ be the components of $\pi'(L) \setminus (\mathcal{B} \cup \Sigma(L)) \subset \mathbb{R}^2$, where $\Sigma(L)$ is the double point set of $\pi'(L)$. We assume that L is on the projection plane $\mathbb{R}^2$

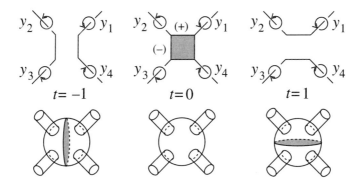

FIGURE 12.3. Relators coming from a saddle band

except neighborhoods of $\Sigma(L)$. Give a meridian loop to each a_i whose base point is far from the projection plane $\mathbb{R}^2$. Associating the meridian loop of a_i with x_i, we have a presentation of $\pi_1(E(L \cup \mathcal{B}))$ such that the generator set is $x_1, \ldots, x_n$ and the relators are $r_1, \ldots, r_m, r'_1, \ldots, r'_m, R_1, \ldots, R_\ell$, where $r_1, \ldots, r_m, r'_1, \ldots, r'_m$ are relators associated with crossing points, and $R_1, \ldots, R_\ell$ are base relators for saddle bands.

Let F be a realizing surface of $L \to L'$ in $\mathbb{R}^3[-1, 1]$. The exterior $E(F) = \mathrm{cl}(\mathbb{R}^3[-1, 1] \setminus N(F))$ is homotopy equivalent to $E(L \cup \mathcal{B})[-1, 1] \cup \mathcal{D}[-1] \cup \mathcal{D}'[1]$. Thus $\pi_1(E(F))$ has a presentation whose generator set is $x_1, \ldots, x_n$ and the relators are $r_1, \ldots, r_m, \ r'_1, \ldots, r'_m, \ R(B_1), \ldots, R(B_\ell), \ R_-(B_1), \ldots, R_-(B_\ell), R_+(B_1), \ldots, R_+(B_\ell)$. Since capping off by trivial disk systems do not change the group, we have the following.

PROPOSITION 12.3. *Let F be a closed realizing surface of a hyperbolic transformation $L \to L'$ along a band set $\mathcal{B} = B_1 \cup \cdots \cup B_\ell$. Then, in the above situation, we have a group presentation of the knot group of F whose generator set is $x_1, \ldots, x_n$ and the relators are $r_1, \ldots, r_m, r'_1, \ldots, r'_m, R(B_1), \ldots, R(B_\ell), R_-(B_1), \ldots, R_-(B_\ell), R_+(B_1), \ldots, R_+(B_\ell)$.*

By Theorem 9.14, this proposition is sufficient to calculate the group of a surface link. However the following is more practical.

COROLLARY 12.4. *Let F be a realizing surface of a hyperbolic transformation $L_- \to L \to L_+$ along band set $\mathcal{B}_-$ and $\mathcal{B}_+$. Then $\pi_1(\mathbb{R}^4 \setminus F)$ is obtained from $\pi_1(\mathbb{R}^3 \setminus L)$ by introducing the upper level relators (relations) for $\mathcal{B}_+$ and the lower level relators (relations) for $\mathcal{B}_-$.*

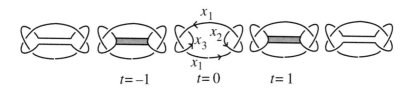

FIGURE 12.4. A spun trefoil

Fig. 12.4 is a motion picture of a spun trefoil. The knot group of a knot appearing at $t = 0$ has a presentation

$$\langle x_1, x_2, x_3 \mid x_1 x_2 x_1 = x_2 x_1 x_2,\ x_1 x_3 x_1 = x_3 x_1 x_3 \rangle.$$

The saddle band at $(t = 1)$-level introduces a relation $x_2 = x_3$ and the saddle band at $(t = -1)$-level introduces the same relation. Thus the group of a spun trefoil has

$$\langle x_1, x_2 \mid x_1 x_2 x_1 = x_2 x_1 x_2 \rangle.$$

12.3. The Elementary Ideals

Let $F = K_1 \cup \cdots K_\mu$ be an oriented μ-component surface link and let $G = \pi_1(\mathbb{R}^4 \setminus F)$ be the group of F. The first homology group $H_1(\mathbb{R}^4 \setminus F; \mathbb{Z}) = G/[G, G]$ is a free abelian group of rank μ, generated by $t_1, \ldots, t_\mu$, which we denote by $\mathbf{J}^\mu$, where t_i $(i \in \{1, \ldots, \mu\})$ is represented by an oriented meridian loop of K_i.

Let $\langle x_1, \ldots, x_m \mid r_1, \ldots, r_n \rangle$ be a presentation of the group G. Let $\phi : F_m \to G$ be the projection of the free group $F_m = \langle x_1, \ldots, x_m \rangle$ to G and let $\psi : G \to \mathbf{J}^\mu$ be the Hurewicz homomorphism. We denote by the same symbol the induced ring homomorphisms $\phi : \mathbb{Z}F_m \to \mathbb{Z}G$ and $\psi : \mathbb{Z}G \to \mathbb{Z}\mathbf{J}^\mu$. We usually denote $\mathbb{Z}\mathbf{J}^\mu$ by Λ, which is the Laurent polynomial ring $\mathbb{Z}[t_1, t_1^{-1}, \ldots, t_\mu, t_\mu^{-1}]$ of μ variables.

For each $j \in \{1, \ldots, m\}$, there is a unique map

$$\frac{\partial}{\partial x_j} : F_m \to \mathbb{Z}F_m$$

determined by the following conditions:

$$\frac{\partial x_i}{\partial x_j} = \delta_{ij}$$

and

$$\frac{\partial (uv)}{\partial x_j} = \frac{\partial u}{\partial x_j} + u \frac{\partial v}{\partial x_j}.$$

For calculation, the following exercise will be useful.

EXERCISE 12.5.   (1)

$$\frac{\partial 1}{\partial x_j} = 0 \quad \text{and} \quad \frac{\partial x_i^{-1}}{\partial x_j} = -\delta_{ij} x_i^{-1}.$$

(2) For a word $w = x_{j_1}^{\epsilon_1} x_{j_2}^{\epsilon_2} \cdots x_{j_m}^{\epsilon_m}$,

$$\frac{\partial w}{\partial x_j} = \epsilon_1 \delta_{jj_1} x_{j_1}^{(\epsilon_1 - 1)/2} + \epsilon_2 \delta_{jj_2} x_{j_1}^{\epsilon_1} x_{j_2}^{(\epsilon_2 - 1)/2} + \epsilon_3 \delta_{jj_3} x_{j_1}^{\epsilon_1} x_{j_2}^{\epsilon_2} x_{j_3}^{(\epsilon_3 - 1)/2} + \cdots.$$

The induced homomorphism $\mathbb{Z}F_m \to \mathbb{Z}F_m$ is also denoted by $\frac{\partial}{\partial x_j}$. This is called *free derivative* with respect to x_j.

DEFINITION 12.6. The *Alexander matrix*, A, of the group presentation is given by

$$A = (r_{ij}) = \left(\psi\phi\left(\frac{\partial r_i}{\partial x_j} \right) \right), \quad i \in \{1, \ldots, n\}, j \in \{1, \ldots, m\}.$$

Each entry r_{ij} is an element of Λ. For an integer d, the dth *elementary ideal* $E_d(A)$ of A is the Λ-ideal generated by the $(m - d)$-minors of A. We assume that $E_d(A) = \Lambda$ if $m - d < 1$ and $E_d(A) = 0$ if $m - d > n$. The dth *characteristic polynomial* $\Delta_d(A)$ of A is the greatest common divisor of the elements of $E_d(A)$.

The characteristic polynomial $\Delta_d(A)$ is a generator of the smallest principal Λ-ideal containing $E_d(A)$, which is determined uniquely up to multiplication of units of Λ. It is known that $E_d(A)$ and $\Delta_d(A)$ do not depend on a group presentation of G. Refer to [122] and [168] for details and examples.

DEFINITION 12.7. The *Alexander polynomial* of F is the first characteristic polynomial of an Alexander matrix A of the knot group of F. (Note that it is uniquely determined up to multiplication of units of Λ.)

REMARK 12.8. For a finitely generated Λ-module H, the dth *elementary ideal* $E_d(H)$ and the dth *characteristic polynomial* $\Delta_d(H)$ of H are defined to be those of a presentation matrix of H. (If H is presented by m generators and n relations over Λ, it is isomorphic to the cokernel of a homomorphism $f : \Lambda^n \to \Lambda^m$. A presentation matrix A of H means an (n, m)-matrix with $f(x) = xA$, where we consider the elements of Λ^n and Λ^m as row vectors.)

Let E be the exterior of F in $\mathbb{R}^4$ and let $p : \widetilde{E} \to E$ be the universal abelian covering, i.e., the covering corresponding to the kernel of the Hurewicz homomorphism $G = \pi_1(E, *) \to H_1(E; \mathbb{Z})$. $H_1(E)$ acts on $\widetilde{E}$ as the covering transformation group. The homology groups $H_1(\widetilde{E}; \mathbb{Z})$ and $H_1(\widetilde{E}, p^{-1}(*); \mathbb{Z})$ are regarded as Λ-modules. The former is called the *link module* of F and the latter is called the *Alexander module* of F. From the homology long exact sequence of the pair $(\widetilde{E}, p^{-1}(*))$, we have a short exact sequence

$$0 \to H_1(\widetilde{E}; \mathbb{Z}) \to H_1(\widetilde{E}, p^{-1}(*); \mathbb{Z}) \to \mathcal{E} \to 0,$$

where $\mathcal{E}$ is the kernel of the homomorphism $\Lambda \to \mathbb{Z}, t_i \mapsto 1$.

Let A be the Alexander matrix of a presentation $\langle x_1, \ldots, x_m \,|\, r_1, \ldots, r_n \rangle$ of G. Consider a chain complex

$$\Lambda^n \xrightarrow{f} \Lambda^m \xrightarrow{g} \Lambda \to 0,$$

where f is presented by A and g is presented by $(m, 1)$-matrix $(\psi\phi(x_i) - 1)_{1 \le i \le m}$. It is known that the link module is isomorphic to $\mathrm{Ker}\,g/\mathrm{Im}\,f$ and the Alexander module is isomorphic to $\mathrm{Coker}\,f$. So A is a presentation matrix of the Alexander module of F. The Alexander polynomial of F is the first characteristic polynomial of the Alexander module $H_1(\widetilde{E}, p^{-1}(*); \mathbb{Z})$, which is equal to the 0th characteristic polynomial of the link module $H_1(\widetilde{E}; \mathbb{Z})$ of F (cf. [369]).

In Sect. 30.3, we will calculate Alexander polynomials of 3-braid 2-knots by use of presentation matrices of the link modules.

Part 3

Surface Braids

CHAPTER 13

Branched Coverings

13.1. Branched Coverings

DEFINITION 13.1. A map $f : F \to M$ from a 2-manifold F to another M is a *branched covering map* of degree m if there exists a finite set Σ (which may be the empty set) consisting of interior points of M such that

$$f|_{F \setminus f^{-1}(\Sigma)} : F \setminus f^{-1}(\Sigma) \to M \setminus \Sigma$$

is a covering map of degree m and for each point $x \in f^{-1}(\Sigma)$, the map f about x is locally equivalent to the map $z \mapsto z^q$ ($z \in \mathbb{C}$) about $0 \in \mathbb{C}$ for some positive integer q. The integer q is called the *local degree*, or the *branching index*, and is denoted by $\deg(f; x)$.

When F and M are oriented, we assume that $f : F \to M$ is an *oriented branched covering map*; namely, for each point $x \in F \setminus f^{-1}(\Sigma)$, the map f carries the local orientation of F at x to the local orientation of M at $f(x)$.

For convenience, we define the local degree, $\deg(f; x)$, at a point $x \in F \setminus f^{-1}(\Sigma)$ to be 1.

A point $x \in F$ with $\deg(f; x) > 1$ is called a *singular point* of f; otherwise x is a *regular point*. A point $y \in M$ is called a *branch point* of f if there is a singular point $x \in F$ with $f(x) = y$; otherwise y is a *regular point*.

The reader might notice that the set Σ in the definition of a branched covering map is not determined from the branched covering map f. In fact, for any interior point y of M missing Σ, $\Sigma \cup \{y\}$ also satisfies the above condition with branching index $q = 1$ at y. Conversely, if Σ contains a regular point of f, we may remove it from Σ. Thus we usually assume that Σ has no regular points; namely, Σ is the set of branch points of f. In this case, we call Σ the *branch point set*.

EXAMPLE 13.2. (1) Identify $\mathbb{R}^2$ with $\mathbb{C}$ in a standard way. For a positive integer m, a map $\rho_m : \mathbb{R}^2 \to \mathbb{R}^2$, $z \mapsto z^m$ is a branched covering map of degree m. The singular point set and the branch point set consist of the origin.

(2) Let F and M be a 2-sphere $\{(x, y, z) \in \mathbb{R}^3 \,|\, x^2 + y^2 + z^2 = 1\}$. Let $\rho_m \times \mathrm{id} : \mathbb{R}^3 \to \mathbb{R}^3$ be a map which is the product of ρ_m and the identity map of $\mathbb{R}$. The restriction $f : F \to M$ of this map is a branched covering map of degree m. If $m = 1$, it is a homeomorphism and a covering map. If $m \geq 2$, the branch point set Σ consists of the north and south pole, where the local degree of f is m.

EXAMPLE 13.3. Let F be a closed oriented surface of genus k in $\mathbb{R}^3$ which is symmetric with respect to a 180°-rotation along a line as in Fig. 13.2. Let M be the quotient space of F by the involution; M is a 2-sphere. The projection map $f : F \to M$ is a branched covering map of degree 2. There are $2k + 2$ branch points.

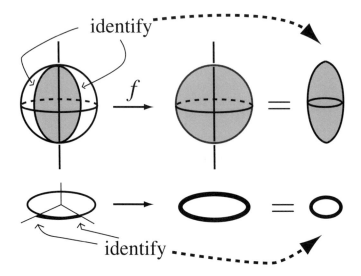

FIGURE 13.1. A branched covering of a sphere

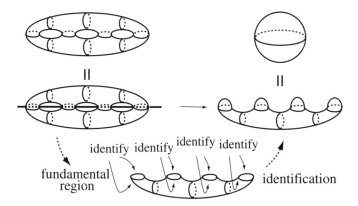

FIGURE 13.2. A branched covering of a sphere

13.2. Types of Branch Points

Let $f : F \to M$ be a branched covering map of degree m. For a branch point $y \in M$, let $\{x_1, \ldots, x_k\}$ be the preimage of y. Changing indices, we may assume that $\deg(f; x_1) \geq \cdots \geq \deg(f; x_k)$.

DEFINITION 13.4. The *branch type* of f at y is the k-tuple

$$(\deg(f; x_1), \ldots, \deg(f; x_k)).$$

It is obvious that

$$\sum_{i=1}^{k} \deg(f, x_i) = m.$$

When the degree m is specified, we often drop 1s from the branch type in order to reduce its length. The branch type is recovered from the reduced one by adding some 1s so that the sum is m. The reduced one is also called the *branch type*. See Fig. 13.3.

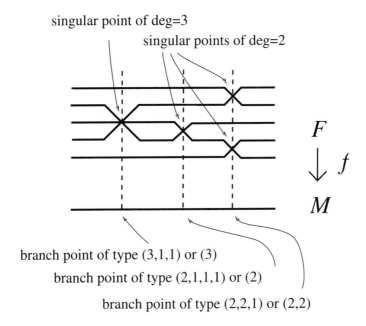

singular point of deg=3

singular points of deg=2

F

$\downarrow f$

M

branch point of type (3,1,1) or (3)

branch point of type (2,1,1,1) or (2)

branch point of type (2,2,1) or (2,2)

FIGURE 13.3. Branch types

13.3. Riemann-Hurwitz Formula

We study the relation between the Euler characteristics of the source and the target of a branched covering map.

THEOREM 13.5 (The Riemann-Hurwitz Formula). *Let $f : F \to M$ be a branched covering map of degree m such that F and M are compact. The Euler characteristics $\chi(F)$ and $\chi(M)$ satisfy the following:*

$$\chi(F) = m\chi(M) - \sum_{x \in F}(\deg(f;x) - 1).$$

In the summation above, we may assume that x runs over all singular points of f, since $\deg(f;x) = 1$ for a regular point x.

Proof. Let $\Sigma = \{y_1, \ldots, y_n\} \subset M$ be the branch point set and let $N(\Sigma) = N(y_1) \cup \cdots \cup N(y_n)$ be its regular neighborhood. Put $M_0 = M \setminus \text{int} N(\Sigma)$ and $F_0 = f^{-1}(M_0)$. Since M_0 is M with n open disks removed, $\chi(M_0) = \chi(M) - n$. The restriction $f|_{F_0} : F_0 \to M_0$ is a covering map of degree m and hence $\chi(F_0) = m\chi(M_0) = m\chi(M) - mn$. For each branch point $y \in \Sigma$, the preimage $f^{-1}(N(y))$ consists of k 2-disks in F around $x_1, \ldots, x_k$, where $\{x_1, \ldots, x_k\}$ is the preimage of y. Recall that $\sum_{i=1}^{k} \deg(f;x_i) = m$. Hence $\sum_{i=1}^{k}(\deg(f;x_i) - 1) = m - k$, and $k = m - \sum_{x \in f^{-1}(y)}(\deg(f;x) - 1)$. The preimage $f^{-1}(N(\Sigma))$ is the union of such 2-disks, and the total of the numbers of the 2-disks is $mn - \sum_{x \in f^{-1}(\Sigma)}(\deg(f;x) - 1)$. Therefore we have $\chi(F) = m\chi(M) - \sum_{x \in f^{-1}(\Sigma)}(\deg(f;x) - 1)$. ∎

It is convenient to regard the value $\deg(f; x) - 1$ as an index. For each point $x \in F$, we put $\tau(f; x) = \deg(f; x) - 1$ and call it the *singular index* of f at x. For a point $y \in M$, we define the *singular index*, $\tau(f; y)$, of f at y by

$$\sum_{x \in f^{-1}(y)} \tau(f; x).$$

Then a point $x \in F$ is a singular point if and only if $\tau(f; x) \neq 0$. A point $y \in M$ is a branch point if and only if $\tau(f; y) \neq 0$.

In terms of the singular index, the Riemann-Hurwitz Formula is

$$\chi(F) = m\chi(M) - \sum_{x \in F} \tau(f; x).$$

The formula is also stated as

$$\chi(F) = m\chi(M) - \sum_{y \in M} \tau(f; y).$$

13.4. Monodromy

Let $f : F \to M$ be a branched covering map of degree m such that M is connected. Let Σ be the branch point set and take a point y_0 of $M \setminus \Sigma$. The preimage $f^{-1}(y_0)$ consists of m points of F. We denote it by Q_m here.

Let $\alpha : [0, 1] \to M \setminus \Sigma$ be a path with $\alpha(0) = \alpha(1) = y_0$, namely, a loop in $M \setminus \Sigma$ with base point y_0. Since $f|_{F \setminus f^{-1}(\Sigma)} : F \setminus f^{-1}(\Sigma) \to M \setminus \Sigma$ is a covering map, for a point $q \in Q_m$, there is a unique lift $\alpha_q : [0, 1] \to F \setminus f^{-1}(\Sigma)$ with $\alpha_q(0) = q$. The terminal point $\alpha_q(1)$ is a point of Q_m, which we denote by $\alpha_\sharp(q)$. Then we have a bijection

$$\alpha_\sharp : Q_m \to Q_m$$

satisfying the following properties:

(1) If α and α' are homotopic as loops in $M \setminus \Sigma$ with base point y_0, then $\alpha_\sharp = \alpha'_\sharp$.
(2) Let $\alpha \cdot \beta$ be the product of two loops α and β; then $(\alpha \cdot \beta)_\sharp = \beta_\sharp \circ \alpha_\sharp$.

Thus we have a homomorphism

$$\rho : \pi_1(M \setminus \Sigma, y_0) \to \mathcal{S}(Q_m),$$

where $\mathcal{S}(Q_m)$ is the group of permutations of Q_m with product $fg = g \circ f$. This homomorphism is called the *monodromy representation* of $\pi_1(M \setminus \Sigma, y_0)$ of the branched covering map f, and $\rho([\alpha]) \in \mathcal{S}(Q_m)$ is called the *monodromy* induced from $[\alpha]$.

Identify $Q_m = f^{-1}(y_0)$ with $\{1, \ldots, m\}$ and fix the identification. Then ρ is regarded as a homomorphism $\rho : \pi_1(M \setminus \Sigma, y_0) \to \mathcal{S}(\{1, \ldots, m\}) = \mathcal{S}_m$, where $\mathcal{S}_m$ is the symmetry group. This is also called a *monodromy representation* of $\pi_1(M \setminus \Sigma, y_0)$ of the branched covering map f, which is nothing more than a monodromy map of the covering map $f|_{F \setminus f^{-1}(\Sigma)} : F \setminus f^{-1}(\Sigma) \to M \setminus \Sigma$. This homomorphism depends on the identification of Q_m with $\{1, \ldots, m\}$. When such an identification is not given, a monodromy representation is just determined up to inner-automorphisms of $\mathcal{S}_m$.

13.5. Simple Branched Coverings

DEFINITION 13.6. A *simple* branched covering map is a branched covering map $f : F \to M$ such that for every branch point y, there exists exactly one singular point x in $f^{-1}(y)$ and the local degree $\deg(f; x)$ is 2. In other words, the branch type of every branch point is $(2, 1, \ldots, 1)$ (or (2) in the reduced notation).

We say that a singular point x is *simple* if the singular index $\tau(f; x)$ is 1 (i.e. the local degree is 2), and a branch point y is *simple* if $\tau(f; y) = 1$. We may state the definition of a simple branched covering map as follows: A branched covering map is simple if every branch point is simple.

PROPOSITION 13.7. *Let $f : F \to M$ be a simple branched covering map of degree m such that F and M are compact. The Euler characteristics $\chi(F)$ and $\chi(M)$ satisfy the following:*

$$\chi(F) = m\chi(M) - n,$$

where n is the number of branch points of f.

Proof. It is a direct consequence of Theorem 13.5. ∎

Simple branched coverings play a quite essential part in the theory of branched coverings. And they are easy to describe and handle. It is well-known that the set of simple branched coverings is open and dense in the whole space of branched covering maps, provided the source F and the target M are given. Any branched covering map is approximated by a simple branched covering map, and once it is deformed into a simple one, then it is stable; i.e. any approximation is still simple. Refer to [**36**], [**37**] for example, for the precise statement, a modern proof and related topics.

A surface braid is a kind of generalization of a branched covering and a simple surface braid corresponds to a simple branched covering. However there exist infinitely many surface braids which are never approximated by simple surface braids (cf. [**320**] or Sect. 19.4). This makes the theory of surface braids more difficult than branched coverings.

CHAPTER 14

Surface Braids

14.1. Surface Braids

Let D_1^2, D_2^2 be 2-disks and $pr_i : D_1^2 \times D_2^2 \to D_i^2$ $(i = 1, 2)$ the ith factor projection. Let Q_m be a set of m interior points of D_1^2.

DEFINITION 14.1. A *surface braid* of degree m (or a *surface m-braid*) is an oriented 2-manifold S embedded properly and locally flatly in $D_1^2 \times D_2^2$ such that the restriction map $pr_2|_S : S \to D_2^2$ of the second factor projection $pr_2 : D_1^2 \times D_2^2 \to D_2^2$ is a branched covering map of degree m and $\partial S = Q_m \times \partial D_2^2$.

We call the branched covering map $pr_2|_S : S \to D_2^2$ the *associated branched covering map* of S.

For a point $x \in S$, the *local degree* of S at x means the local degree of $pr_2|_S$ at x which is denoted by $\deg(S; x)$. A *singular point* of S is a singular point of the branched covering $pr_2|_S$, i.e., $\deg(S; x) \neq 1$. A *branch point* of S is a branch point of $pr_2|_S$. We denote by $\Sigma(S)$ the set of branch points of S.

DEFINITION 14.2. Two surface braids S and S' are *equivalent* if there is an ambient isotopy $\{h_u\}_{u \in [0,1]}$ of $D_1^2 \times D_2^2$ satisfying the following conditions:

(1) $h_1(S) = S'$,
(2) for each $u \in [0,1]$, h_u is fiber-preserving; i.e., there is a homeomorphism $\underline{h}_u : D_2^2 \to D_2^2$ such that $pr_2 \circ h_u = \underline{h}_u \circ pr_2$, and
(3) $h_u|_{D_1^2 \times \partial D_2^2} = $ id for $u \in [0,1]$.

We usually regard equivalent surface braids as the same.

Comparing the definition of a surface braid (Definition 14.1) with that of a classical braid (Definition 1.1), the reader will find that a surface braid is a generalization of a classical braid. Note that the map $pr_2|_S : S \to D_2^2$ is a "branched" covering map. If we restrict this map to be an ordinary covering map, then the surface braid S is equivalent to the trivial surface braid $Q_m \times D_2^2$ and there is nothing interesting. However, this trivial case is still important in the study of classical braid theory and surface braid theory. This will be discussed later in detail.

Notice that the equivalence relation on a surface braid (Definition 14.2) is quite different from the equivalence relation on a classical braid. This will also be discussed later.

14.2. Motion Pictures

In order to visualize a surface braid, the motion picture method is useful.

Let S be a surface braid of degree m in $D_1^2 \times D_2^2$. We identify D_2^2 with the product $I_3 \times I_4$ of the unit intervals I_3 and I_4 and identify $D_1^2 \times D_2^2 = (D_1^2 \times I_3) \times I_4$. We fix these identifications.

Consider a motion picture of S by assuming I_4 is the time direction. For each $t \in I_4$, let b_t be the subset of $D_1^2 \times I_3$ such that $b_t \times \{t\} = S \cap (D_1^2 \times I_3) \times \{t\}$. Then $\{b_t\}$ $(t \in I_4)$ is a continuous sequence of geometric m-braids except for a finite number of values $t_1, t_2, \ldots, t_n \in I_4 = [0, 1]$ (possibly $n = 0$). For each exceptional value t $(t \in \{t_1, t_2, \ldots, t_n\})$, b_t is a geometric singular m-braid in $D_1^2 \times I_3$. The singular points of the singular braids correspond to the singular points of S. Since $\partial S = Q_m \times \partial D_2^2$, we see that the initial geometric m-braid b_0 and the terminal b_1 are trivial m-braids and that for any $t \in I_4$, $\partial b_t = Q_m \times \partial I_3$. This one-parameter family $\{b_t\}$ $(t \in [0, 1])$ is called a *motion picture*, a *braid motion picture*, or a *braid movie* of S.

Evidently a motion picture of S depends on the bi-parametrization of D_2^2 with $I_3 \times I_4$.

EXAMPLE 14.3. Fig. 14.1(A) is a motion picture of a surface braid of degree 2. There are two singular points, x_1 and x_2, with $\deg(S; x_1) = \deg(S; x_2) = 2$. By the Riemann-Hurwitz formula, we have $\chi(S) = 2 \times 1 - \{(2 - 1) + (2 - 1)\} = 0$. Since S is connected and has two boundary components, S is an annulus embedded in $D_1^2 \times D_2^2$.

Fig. 14.1(B) is also a surface braid of degree 2. Surprisingly, these two surface braids are equivalent. Do not worry if you have difficulty seeing this. After we learn chart description, it will become trivial (Example 18.16).

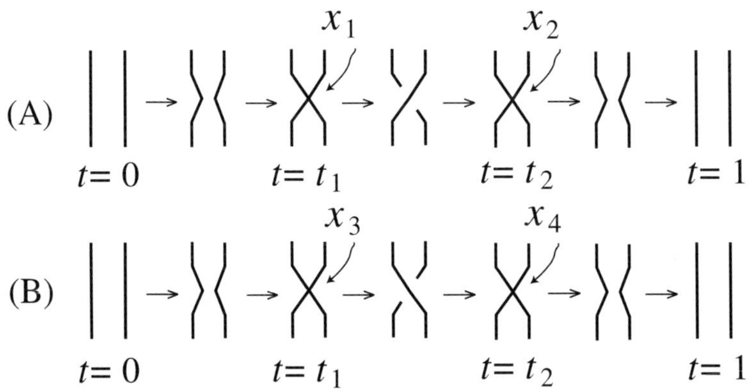

FIGURE 14.1

EXAMPLE 14.4. Fig. 14.2 is a motion picture of a surface braid of degree 3. There are three singular points, x_1, x_2 and x_3, with $\deg(S; x_1) = \deg(S; x_2) = 2$ and $\deg(S; x_3) = 3$. As an abstract surface, S is a 2-sphere with 3 holes.

EXAMPLE 14.5. Fig. 14.3 is a motion picture of a surface braid of degree 3 which has four singular points and four branch points. Two of the branch points are on the same interval, $I_3 \times \{t_1\} \subset I_3 \times I_4 = D_2^2$, and the other two are on $I_3 \times \{t_2\}$. Thus we have only two exceptional values, t_1 and t_2. Deforming the bi-parametrization $D_2^2 \cong I_3 \times I_4$, we can obtain another motion picture such that all singular points appear at distinct exceptional values in I_4. It is also possible that all singular points appear at the same exception value in I_4.

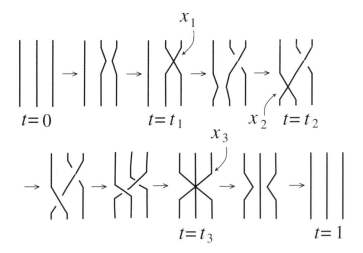

FIGURE 14.2

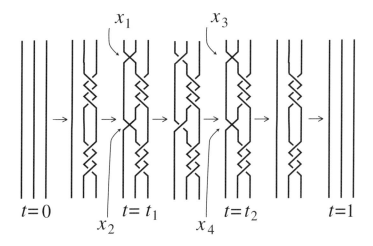

FIGURE 14.3

14.3. Trivial Surface Braids

DEFINITION 14.6. A surface braid of degree m is *trivial* if it is equivalent to the product $Q_m \times D_2^2$.

If a surface braid S is trivial, then the associated branched covering map $pr_2|_S$ is an ordinary covering map and there are no singular points and no branch points. There are no singular points in any motion picture $\{b_t\}_{t \in [0,1]}$ of S. Conversely, consider a motion picture $\{b_t\}_{t \in [0,1]}$ of a surface braid S and suppose that there are no singular points; i.e., each b_t is a non-singular braid. It is obvious that each b_t is a trivial braid. Can we say that S is trivial? The answer is "yes".

LEMMA 14.7. *For a surface braid S of degree m, the following conditions are mutually equivalent.*

(1) *S is trivial.*
(2) *The map $pr_2|_S : S \to D_2^2$ is an ordinary covering. So there are no singular/branch points.*
(3) *For a motion picture $\{b_t\}_{t \in [0,1]}$ of S, each b_t is a non-singular braid.*

Proof. (1) $\Rightarrow$ (2) and (2) $\Leftrightarrow$ (3) are obvious. We prove (2) $\Rightarrow$ (1). If $pr_2|_S$ has no branch point, then there exists a map $g : (D_2^2, \partial D_2^2) \to (C_m, Q_m)$ such that $g(y) = pr_1(S \cap pr_2^{-1}(y))$ for $y \in D_2^2$, where C_m is the configuration space $C_m(\text{int} D_1^2)$. It is known that the second homotopy group of the configuration space (C_m, Q_m) is trivial (cf. [**42, 140**]). Thus g is homotopic to the identity. The homotopy gives an isotopy of S to the trivial $Q_m \times D_2^2$. By a similar argument to the braid isotopy extension theorem (Theorem 1.11), this isotopy is extended to an ambient isotopy $\{h_t\}$ ($t \in [0,1]$) of $D_1^2 \times D_2^2$ such that h_t is fiber-preserving and $h_t|_{D_1^2 \times \partial D_2^2} = $ id for any $t \in [0,1]$. ∎

14.4. Simple Surface Braids

DEFINITION 14.8. A *simple surface braid* is a surface braid whose associated branched covering is simple.

Let $\{b_t\}_{t \in [0,1]}$ be a motion picture of a simple surface braid S of degree m. For each branch point $y \in D_2^2$, there is a unique singular point x with $pr_2(x) = y$ and the local degree at x is two. In other words, the branch type of y is $(2, 1, \ldots, 1)$. Modifying S slightly up to equivalence, one can deform the motion picture so that the restriction near each singular point looks like one in Fig. 14.4. (This fact will be seen later.) Thus, we may assume that a simple surface braid is a surface braid represented by a motion picture whose singular points are as in Fig. 14.4 and there is exactly one singular point in the preimage of each branch point.

Examples 14.3 and 14.5 are simple surface braids.

A singular point illustrated as (A) or (D) in Fig. 14.4 is called a *positive singular point*. A singular point illustrated as (B) or (C) in Fig. 14.4 is called a *negative singular point*.

PROPOSITION 14.9. *Let $\{b_t\}_{t \in [0,1]}$ be a motion picture of a simple surface braid S. The number of positive singular points is equal to that of negative ones. In particular the number of branch points of S is even.*

Proof. For a braid b, let $e(b) \in \mathbb{Z}$ denote the exponent sum of a braid word expression of b. A positive (or negative) singular point changes $e(b)$ by $+1$ (or -1, resp.). Since the initial braid b_0 and the terminal braid b_1 are trivial braids, $e(b_0) = e(b_1) = 0$. Hence the number of positive singular points is equal to that of the negative ones. Since the number of branch points is the same as the number of singular points, we have the latter assertion. ∎

The latter assertion of Proposition 14.9 is also seen from the Riemann-Hurwitz formula as follows: If a simple surface m-braid S has n branch points, then

$$\chi(S) = m - n.$$

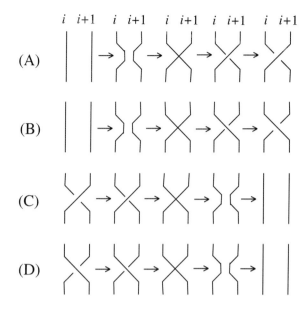

FIGURE 14.4. Singular points of a simple surface braid

The boundary of S is the union of m circles. Consider a closed oriented surface obtained from S by attaching m disks along the boundary, say $\widehat{S}$. Then

$$\chi(\widehat{S}) = m - n + m = 2m - n.$$

Since the Euler characteristic of a closed orientable surface is even, n is even.

14.5. Equivalence Relations

A homeomorphism $h : D_1^2 \times D_2^2 \to D_1^2 \times D_2^2$ is called a *fiber-preserving home-omorphism* if there is a homeomorphism $\underline{h} : D_2^2 \to D_2^2$ such that $pr_2 \circ h = \underline{h} \circ pr_2$. Furthermore if $\underline{h}$ is the identity, then h is called an *isomorphism*.

We here introduce four equivalence relations on surface braids (the second one is already defined).

DEFINITION 14.10. Let S and S' be surface braids of degree m.

(1) S and S' are *isomorphic* if they are ambient isotopic by an isotopy $\{h_u\}_{u \in [0,1]}$ of $D_1^2 \times D_2^2$ such that for each $u \in [0,1]$, h_u is an isomorphism and $h_u|_{D_1^2 \times \partial D_2^2} = \mathrm{id}$.

(2) S and S' are *equivalent* if they are ambient isotopic by an isotopy $\{h_u\}_{u \in [0,1]}$ of $D_1^2 \times D_2^2$ such that for each $u \in [0,1]$, h_u is fiber-preserving and $h_u|_{D_1^2 \times \partial D_2^2} = \mathrm{id}$.

(3) S and S' are *braid ambient isotopic* if they are ambient isotopic by an isotopy $\{h_u\}_{u \in [0,1]}$ of $D_1^2 \times D_2^2$ (called a *braid ambient isotopy*) such that for each $u \in [0,1]$, $h_u(S)$ is a surface braid and $h_u|_{D_1^2 \times \partial D_2^2} = \mathrm{id}$.

(4) S and S' are *braid isotopic* if there is an isotopy $\{S_u\}_{u \in [0,1]}$ (called a *braid isotopy*) with $S_0 = S$ and $S_1 = S'$ such that for each $u \in [0,1]$, S_u is a surface braid.

By definition, we have $(1) \Rightarrow (2) \Rightarrow (3) \Rightarrow (4)$. In classical braid theory, four equivalence relations corresponding to the above, replacing D_2^2 with I, are mutually equivalent (Theorem 1.12). However, this is not true for surface braids. In general neither $(2) \Rightarrow (1)$ nor $(3) \Rightarrow (2)$ holds. It is uncertain whether $(4) \Rightarrow (3)$ holds or not.

Let S be a surface braid in $D_1^2 \times D_2^2$. For a homeomorphism $g : D_2^2 \to D_2^2$ with $g|_{\partial D_2^2} = \mathrm{id}$, construct a fiber-preserving homeomorphism $h : D_1^2 \times D_2^2 \to D_1^2 \times D_2^2$ with $pr_2 \circ h = g \circ pr_2$ by defining h by $\mathrm{id}_{D_1^2} \times g$, i.e.,

$$h(x, y) = (x, g(y)) \quad \text{for } (x, y) \in D_1^2 \times D_2^2.$$

LEMMA 14.11. *In the above situation, the image $S' = h(S)$ of S is a surface braid which is equivalent to S.*

Proof. There is an ambient isotopy $\{g_u\}_{u \in [0,1]}$ of D_2^2 with $g_1 = g$ and $g_u|_{\partial D_2^2} = \mathrm{id}$ for $u \in [0,1]$. Construct an ambient isotopy $\{h_u\}_{u \in [0,1]}$ of $D_1^2 \times D_2^2$ by $h_u = \mathrm{id}_{D_1^2} \times g_u$ for $u \in [0,1]$. Then $h_1 = h$, and this ambient isotopy carries S onto S'. ∎

EXAMPLE 14.12. Let S be a surface braid in $D_1^2 \times D_2^2$ such that the branch point set $\Sigma(S)$ is not empty. In the situation of the above lemma, assume that $g(\Sigma(S)) \neq \Sigma(S)$. Then S' is not isomorphic to S, because $\Sigma(S') = g(\Sigma(S))$ is different from $\Sigma(S)$. By the lemma, S and S' are equivalent. Thus $(2) \Rightarrow (1)$ does not hold.

Two surface braids which differ locally as (A) and (B) in Fig. 14.5 are equivalent. This is easily seen by Lemma 14.11. This example shows that the equivalence relation "isomorphism" (the first one in Definition 14.10) is too strong. If one changes an identification of D_2^2 with a 2-disk slightly, then the surface braid is no longer regarded as the same one. Thus we usually use the equivalence relation "equivalence" (the second one in Definition 14.10).

On the other hand, when we use the motion picture method to describe a surface braid, "equivalence" is quite flexible, but unfortunately it is sometimes difficult to compare two motion pictures. For instance, the two surface braids which differ locally as in Fig. 14.1 are equivalent.

EXAMPLE 14.13. Let S be a surface braid of degree 4 whose motion picture looks like Fig. 14.6(A) locally. There are two singular points, x_1 and x_2, in the same fiber $D_1^2 \times \{y\}$. Let S' be a surface braid obtained from S by replacing the part of Fig. 14.6(A) with Fig. 14.6(B). Then the two singular points, x_1' and x_2', are in distinct fibers, say $D_1^2 \times \{y_1\}$ and $D_1^2 \times \{y_2\}$. It is obvious that S is braid ambient isotopic to S'. However it is not equivalent to S', because the number of branch points of S is smaller than that of S' by one. Thus $(3) \Rightarrow (2)$ does not hold.

A braid ambient isotopy or a braid isotopy between two surface braids yields a one-parameter family of branched covering maps. If this one-parameter family changes a branch point into some branch points, then we say that a *fission of a branch point*, or a *branch point fission* occurs. (The inverse is called a *fusion of a branch point*, or a *branch point fusion*.) In the previous example, the branch point y changes into two branch points, y_1 and y_2. Thus this is a fission of y. If the one-parameter family changes a singular point into some singular points, then we say that a *fission of a singular point* occurs. When a fission of a singular point occurs, a fission of a branch point automatically occurs.

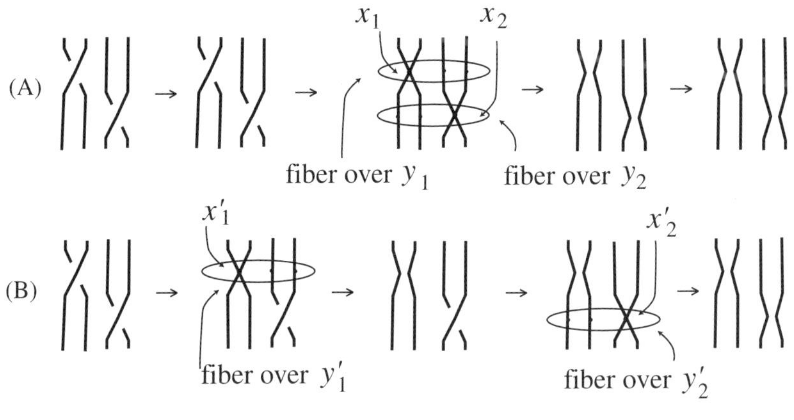

FIGURE 14.5. Equivalent surface braids

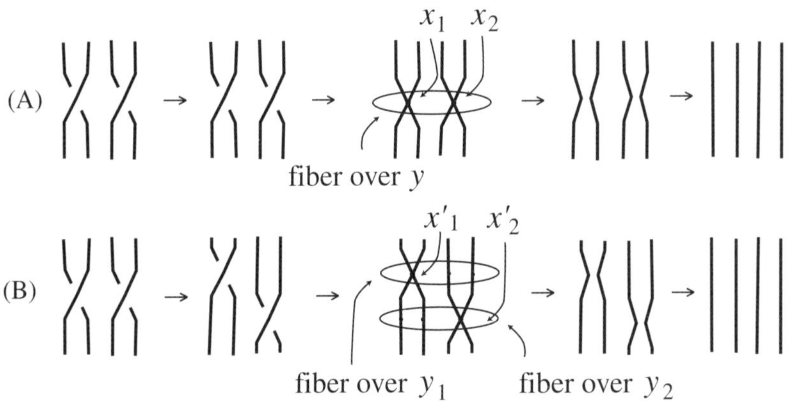

FIGURE 14.6

EXAMPLE 14.14. Let S be a surface braid of degree 3 whose motion picture contains a local part as in Fig. 14.7(A). Let S' be a surface braid obtained from S by replacing the part with Fig. 14.7(B). It is known that S is braid ambient isotopic to S' (Lemma 19.3). There is a singular point, x, of local degree 3 in a fiber $D_1^2 \times \{y\}$. This point is transformed into a pair of singular points, x_1' and x_2'; namely, a fission of a singular point x occurs. Consequently, the branch point y is also transformed into a pair of branch points y_1' and y_2'. S and S' are not equivalent, since the numbers of singular points (or branch points) of them are different. This is another example showing that $(3) \Rightarrow (2)$ does not hold.

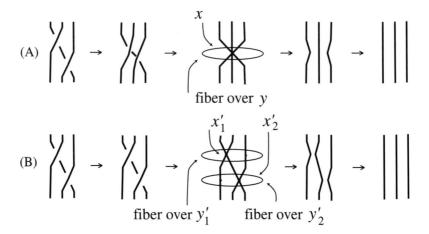

FIGURE 14.7

CHAPTER 15

Products of Surface Braids

15.1. Products of Motion Pictures

Let $\{b_t^{(1)}\}$ and $\{b_t^{(2)}\}$ be motion pictures of two surface braids S_1 and S_2 of the same degree.

DEFINITION 15.1. (1) The *vertical product* is a motion picture $\{\beta_t\}_{t\in[0,1]}$ such that
$$\beta_t = b_t^{(1)} \cdot b_t^{(2)}.$$
(2) The *horizontal product* is a motion picture $\{\beta_t\}_{t\in[0,1]}$ such that
$$\beta_t = \left\{ \begin{array}{ll} b_{2t}^{(1)} & \text{for } t \in [0, 0.5], \\ b_{2t-1}^{(2)} & \text{for } t \in (0.5, 1]. \end{array} \right.$$

For (1), each β_t is just a product of $b_t^{(1)}$ and $b_t^{(2)}$ in classical braid theory. For (2), $\{\beta_t\}_{t\in[0,1]}$ is the composition of motion pictures that is a motion picture of S_1 followed by a motion picture of S_2.

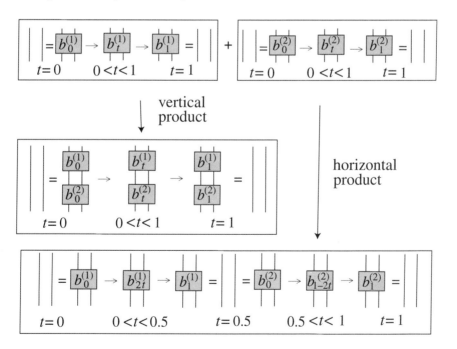

FIGURE 15.1. Vertical and horizontal products

15.2. Products of Surface Braids

Let S_i $(i = 1, 2)$ be a surface braid of degree m in $D_1^2 \times D_{2(i)}^2$, where $D_{2(i)}^2$ $(i = 1, 2)$ is a copy of D_2^2. Divide the 2-disk D_2^2 into two 2-disks E_1 and E_2 by a proper arc in D_2^2, and fix an identification map

$$D_{2(i)}^2 \cong E_i$$

for each i $(i = 1, 2)$. In other words, we assume that D_2^2 is obtained from $D_{2(1)}^2$ and $D_{2(2)}^2$ by identifying an arc in the boundary of $D_{2(1)}^2$ and an arc in the boundary of $D_{2(2)}^2$. (In this situation, we say that D_2^2 is a *boundary connected sum* of $D_{2(1)}^2$ and $D_{2(2)}^2$.) Using this identification map, we assume that S_i is a surface in $D_1^2 \times E_i$ which is contained in $D_1^2 \times D_2^2$. Then the union $S_1 \cup S_2$ in $D_1^2 \times (E_1 \cup E_2) = D_1^2 \times D_2^2$ is a surface braid of degree m.

DEFINITION 15.2. A *product* of S_1 and S_2, denoted by $S_1 \cdot S_2$, is a surface braid obtained as above.

Of course, the product $S_1 \cdot S_2$ depends on the division $D_2^2 = E_1 \cup E_2$ and the identification maps $D_{2(i)}^2 \cong E_i$. However such a surface braid $S_1 \cdot S_2$ is uniquely determined up to equivalence. Moreover, the equivalence class of $S_1 \cdot S_2$ depends only on the equivalence classes of S_1 and S_2. Thus we can define the product $[S_1] \cdot [S_2]$ of two equivalence classes $[S_1]$ and $[S_2]$ by the class $[S_1 \cdot S_2]$.

EXERCISE 15.3. Show that the product $[S_1] \cdot [S_2]$ of two equivalence classes $[S_1]$ and $[S_2]$ is well-defined. And $[S_1] \cdot [S_2] = [S_2] \cdot [S_1]$. [*Hint*: Use Lemma 14.11.]

It is easy to visualize a product $S_1 \cdot S_2$ when we use the motion picture method. By the following lemma, surface braids described by the vertical product and by the horizontal product of two motion pictures are equivalent.

LEMMA 15.4. *Let two surface braids S_1 and S_2 be described by motion pictures* $\{b_t^{(1)}\}$ *and* $\{b_t^{(2)}\}$.

(1) *The vertical product of* $\{b_t^{(1)}\}$ *and* $\{b_t^{(2)}\}$ *is a motion picture of a product* $S_1 \cdot S_2$.

(2) *The horizontal product of* $\{b_t^{(1)}\}$ *and* $\{b_t^{(2)}\}$ *is a motion picture of a product* $S_1 \cdot S_2$.

Proof. This follows from Exercise 15.3. ∎

15.3. The Surface Braid Monoid

We denote by $\mathcal{B}_m$ the set of equivalence classes of surface braids of degree m. In the previous section, we defined a product $S_1 \cdot S_2$ of two surface braids S_1 and S_2 of the same degree. The equivalence class of the product $S_1 \cdot S_2$ is uniquely determined from the equivalence classes of S_1 and S_2. Thus we have a binary operation

$$\cdot : \mathcal{B}_m \times \mathcal{B}_m \to \mathcal{B}_m, \quad ([S_1], [S_2]) \mapsto [S_1 \cdot S_2]$$

on the set $\mathcal{B}_m$.

This binary operation satisfies the following:

(1) *Commutativity*: $[S_1] \cdot [S_2] = [S_2] \cdot [S_1]$ for any S_1 and S_2.

(2) *Associativity*: $([S_1] \cdot [S_2]) \cdot [S_3] = [S_1] \cdot ([S_2] \cdot [S_3])$ for any S_1, S_2 and S_3.

(3) *Existence of the identity*: There exists an element E such that $E \cdot [S] = [S] \cdot E = [S]$ for any S.

Thus $\mathcal{B}_m$ forms a commutative monoid. The identity element E of $\mathcal{B}_m$ is the equivalence class of a trivial surface braid.

DEFINITION 15.5. The *surface braid monoid of degree m* is the commutative monoid $\mathcal{B}_m$ consisting of equivalence classes of surface braids of degree m.

For a surface braid S, we denote by $n(S)$ the number of branch points of S. Then we have a monoid homomorphism

$$n : \mathcal{B}_m \to \mathbb{N}_0, \quad [S] \mapsto n(S),$$

where $\mathbb{N}_0$ is the monoid of non-negative integers under addition.

PROPOSITION 15.6. *Every non-trivial element of the surface braid monoid $\mathcal{B}_m$ has no inverse element.*

Proof. By Lemma 14.7, if $[S]$ is a non-trivial element of $\mathcal{B}_m$, then $n(S) > 0$. If $[S]$ has an inverse element $[S']$, then $n(S) + n(S') = n([S \cdot S']) = n(E) = 0$. This is impossible, for $n(S')$ is a non-negative integer. ∎

DEFINITION 15.7. The *simple surface braid monoid of degree m* is the submonoid of $\mathcal{B}_m$ consisting of equivalence classes of simple surface braids of degree m. It is denoted by $S\mathcal{B}_m$.

By Lemma 14.7, we see that $\mathcal{B}_1$ and $S\mathcal{B}_1$ are the trivial monoid consisting of the identity element.

$\mathcal{B}_2$ and $S\mathcal{B}_2$ are the same, because any surface braid of degree 2 is simple. Recall that every simple surface braid has an even number of branch points (Proposition 14.9). Thus we have a monoid homomorphism

$$(n/2) : S\mathcal{B}_m \to \mathbb{N}_0, \quad [S] \mapsto n(S)/2.$$

Two surface braids of degree 2 are equivalent if and only if they have the same number of branch points (Proposition 17.18). Thus the monoid homomorphism

$$(n/2) : \mathcal{B}_2 = S\mathcal{B}_2 \to \mathbb{N}_0$$

is an isomorphism.

CHAPTER 16

Braided Surfaces

16.1. Braided Surfaces

We introduce the notion of a braided surface. This was introduced by L. Rudolph [**778**]. A surface braid is regarded as a special case of a braided surface.

Let D_1^2, D_2^2 be 2-disks and let $pr_i : D_1^2 \times D_2^2 \to D_i^2$ $(i = 1, 2)$ be the ith factor projection. Let Q_m be a set of m interior points of D_1^2.

DEFINITION 16.1. A *braided surface* of degree m is an oriented 2-manifold S embedded properly and locally flatly in $D_1^2 \times D_2^2$ such that the restriction map $pr_2|_S : S \to D_2^2$ of the second factor projection $pr_2 : D_1^2 \times D_2^2 \to D_2^2$ is a branched covering map of degree m and ∂B is a closed m-braid in $D_1^2 \times \partial D_2^2$.

A *pointed braided surface* is a braided surface with a base point $y_0 \in \partial D_2^2$ such that $pr_1(S \cap pr_2^{-1}(y_0)) = Q_m$.

Do not confuse a braided surface with a surface braid. A surface braid is a special case of a braided surface such that the boundary is the trivial closed braids $Q_m \times \partial D_2^2$.

The notions of the *associated branched covering map*, the *local degree*, a *singular point*, *branch point* and a *regular point* of S are defined in the same way as surface braids.

DEFINITION 16.2. A *simple braided surface* is a braided surface whose associated branched covering is simple.

16.2. The Motion Picture of a Braided Surface

Fix a bi-parametrization $D_2^2 \cong I_3 \times I_4$ of a 2-disk D_2^2.

Let S be a braided surface of degree m in a bidisk $D_1^2 \times D_2^2$ and consider a motion picture $\{b_t\}_{t\in[0,1]}$ of S defined by $b_t \times \{t\} = S \cap (D_1^2 \times I_3) \times \{t\}$ for $t \in I_4 = [0,1]$ as before. Here we assume an additional condition that

$$pr_1(S \cap pr_2^{-1}(y)) = Q_m \quad \text{for } y \in (\partial I_3) \times I_4 \subset D_2^2.$$

By this condition, we may say that a motion picture of a braided surface is a motion picture of a surface braid without assuming that the initial braid b_0 and the terminal braid b_1 are trivial braids.

Fig. 16.1 is an example of a motion picture of a braided surface of degree 3. It has a singular point of degree two.

16.3. Equivalence Relations on Braided Surfaces

The equivalence relations given in 14.5 are stated for braided surfaces as follows.

DEFINITION 16.3. Let S and S' be braided surfaces of degree m.

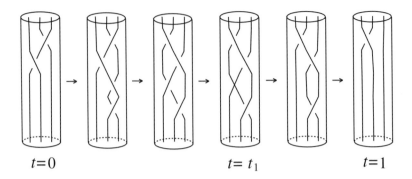

$t=0$ $t=t_1$ $t=1$

FIGURE 16.1. Example of a braided surface

(1) S and S' are *isomorphic in the strong sense* if they are ambient isotopic by an isotopy $\{h_u\}_{u\in[0,1]}$ of $D_1^2 \times D_2^2$ such that for each $u \in [0,1]$, h_u is an isomorphism and $h_u|_{D_1^2 \times \partial D_2^2} = \mathrm{id}$.
(2) S and S' are *equivalent in the strong sense* if they are ambient isotopic by an isotopy $\{h_u\}_{u\in[0,1]}$ of $D_1^2 \times D_2^2$ such that for each $u \in [0,1]$, h_u is fiber-preserving and $h_u|_{D_1^2 \times \partial D_2^2} = \mathrm{id}$.
(3) S and S' are *braid ambient isotopic in the strong sense* if they are ambient isotopic by an isotopy $\{h_u\}_{u\in[0,1]}$ of $D_1^2 \times D_2^2$ (called a *braid ambient isotopy*) such that for each $u \in [0,1]$, $h_u(S)$ is a braided surface and $h_u|_{D_1^2 \times \partial D_2^2} = \mathrm{id}$.
(4) S and S' are *braid isotopic in the strong sense* if there is an isotopy $\{S_u\}_{u\in[0,1]}$ (called a *braid isotopy*) with $S_0 = S$ and $S_1 = S'$ such that for each $u \in [0,1]$, S_u is a braided surface with $\partial S_u = \partial S$.

The equivalence relations are somehow inflexible. If S and S' are related by one of the above relations, then ∂S must coincide with $\partial S'$. So we introduce more elastic equivalence relations.

DEFINITION 16.4. Let S and S' be pointed braided surfaces of degree m with base point $y_0 \in \partial D_2^2$.

(1) S and S' are *isomorphic in the weak sense* if they are ambient isotopic by an isotopy $\{h_u\}_{u\in[0,1]}$ of $D_1^2 \times D_2^2$ such that for each $u \in [0,1]$, h_u is an isomorphism and $h_u|_{pr_2^{-1}(y_0)} = \mathrm{id}$.
(2) S and S' are *equivalent in the weak sense* if they are ambient isotopic by an isotopy $\{h_u\}_{u\in[0,1]}$ of $D_1^2 \times D_2^2$ such that for each $u \in [0,1]$, h_u is fiber-preserving and $h_u|_{pr_2^{-1}(y_0)} = \mathrm{id}$.
(3) S and S' are *braid ambient isotopic in the weak sense* if they are ambient isotopic by an isotopy $\{h_u\}_{u\in[0,1]}$ of $D_1^2 \times D_2^2$ (called a *braid ambient isotopy*) such that for each $u \in [0,1]$, $h_u(S)$ is a braided surface and $h_u|_{pr_2^{-1}(y_0)} = \mathrm{id}$.
(4) S and S' are *braid isotopic in the weak sense* if there is an isotopy $\{S_u\}_{u\in[0,1]}$ (called a *braid isotopy*) with $S_0 = S$ and $S_1 = S'$ such that for each $u \in [0,1]$, S_u is a pointed braided surface with base point y_0.

16.4. Braided Surfaces without Branch Points

LEMMA 16.5. *Let S be a braided surface in $D_1^2 \times D_2^2$. If the associated branched covering map has no branch points, then the boundary ∂S is a trivial closed m-braid in the solid torus $D_1^2 \times \partial D_2^2$.*

Proof. Take a point y_0 of ∂D_2^2, and put $Q_m' = pr_1(S \cap pr_2^{-1}(y_0))$. As in 14.3, we can define a map $g : (D_2^2, \{y_0\}) \to (C_m, Q_m')$ by $g(y) = pr_1(S \cap pr_2^{-1}(y))$ for $y \in D_2^2$. The boundary ∂S is a closed m-braid corresponding to a loop $(I, \partial I) \to (\partial D_2^2, \{y_0\}) \to (C_m, Q_m')$. The map g gives a homotopy between this loop and the trivial loop $(I, \partial I) \to (\{y_0\}, \{y_0\}) \to (C_m, Q_m')$. Thus ∂S is a trivial closed m-braid. ∎

LEMMA 16.6. *Let ℓ be a trivial closed m-braid in $D_1^2 \times \partial D_2^2$. There is a braided surface S with $\partial S = \ell$ whose associated branched covering map has no branch points.*

Proof. Take a point y_0 of ∂D_2^2, and put $Q_m' = pr_1(S \cap pr_2^{-1}(y_0))$. The closed m-braid ℓ determines a map $(\partial D_2^2, \{y_0\}) \to (C_m, Q_m')$. Since ℓ is trivial, we can extend it to a map $(D_2^2, \{y_0\}) \to (C_m, Q_m')$. This determines the desired braided surface S. ∎

LEMMA 16.7. *Let S and S' be braided surfaces in $D_1^2 \times D_2^2$ such that $\partial S = \partial S'$. If their associated branched covering maps have no branch points, then they are isomorphic in the strong sense.*

Proof. Consider an ambient isotopy $\{h_u\}_{u \in [0,1]}$ of $D_1^2 \times D_2^2$ such that each h_u is an isomorphism of $D_1^2 \times D_2^2$ and $h_1(\partial S) = Q_m \times \partial D_2^2$ in $D_1^2 \times \partial D_2^2$. (Since ∂S is a trivial closed braid, using a collar neighborhood of ∂D_2^2 in D_2^2, we can construct such an ambient isotopy.) The images $h_1(S)$ and $h_1(S')$ are surface braids without branch points. By the proof of Lemma 14.7, we see that $h_1(S)$ and $h_1(S')$ are isomorphic as surface braids and isomorphic in the strong sense as braided surfaces. Thus S and S' are isomorphic in the strong sense. ∎

DEFINITION 16.8. A pointed braided surface of degree m is *trivial* if it is equivalent in the weak sense to the trivial surface braid $Q_m \times D_2^2$.

LEMMA 16.9. *A braided surface S is trivial if and only if the associated branched covering map has no branch points.*

Proof. The only if part is obvious. We prove the if part. Since the associated branched covering has no branch points, by Lemma 16.5, the boundary ∂S is a trivial closed m-braid. Using a collar neighborhood of ∂D_2^2 in D_2^2, we can deform S up to isomorphism in the weak sense so that ∂S is the trivial closed m-braid $Q_m \times \partial D_2^2$. By Lemma 16.7 this braided surface is isomorphic to the trivial surface braid $Q_m \times D_2^2$. ∎

16.5. The Set A_m and Multiple Cones

Here we define a subset A_m of the braid group B_m and a multiple cone which is a special braided surface. They are used for the braid monodromy of a surface braid or a braided surface in the next chapter.

A closed braid ℓ in a solid torus $D^2 \times S^1$ is *completely split* if there exist c mutually disjoint convex disks in D^2, say $N_1, \ldots, N_c$, such that every open solid torus int$N_i \times S^1$ contains a component of ℓ, where c is the number of components of ℓ. A closed braid is *completely splittable* if it is equivalent to a completely split closed braid. Let A_m be the subset of the braid group B_m consisting of non-trivial elements b satisfying the condition that there exists a completely split closed m-braid ℓ in the solid torus $D_1^2 \times \partial D_2^2$ representing the closure $\hat{b}$ of b such that ℓ is a trivial link in the 3-sphere $\partial(D_1^2 \times D_2^2)$.

Let $\ell = \ell_1 \cup \cdots \cup \ell_c$ be a completely split closed m-braid in $D_1^2 \times \partial D_2^2$ representing $\hat{b}$ for some $b \in A_m$, and let $N_1, \ldots, N_c$ be mutually disjoint convex disks in D_1^2 with $\ell_i \subset \mathrm{int} N_i \times \partial D_2^2$. Let y be an interior point of D_2^2. Take points $x_1, \ldots, x_c$ in $\mathrm{int} N_1, \ldots, \mathrm{int} N_c$, respectively. For each $i \in \{1, \ldots, c\}$, consider a cone of ℓ from $p_i = (x_i, y) \in D_1^2 \times D_2^2$ and denote it by C_i. Since each component ℓ_i is a trivial knot in $\partial(D_1^2 \times D_2^2)$, the cone C_i is a locally flat embedded disk in $D_1^2 \times D_2^2$. Let $C(\ell)$ be the union of $C_1, \ldots, C_c$, which we call a *multiple cone* of ℓ. This is a braided surface of degree m whose branch point set is $\{y\}$. The branch type of the branch point y is $(\deg(\ell_1), \ldots, \deg(\ell_c))$, where $\deg(\ell_i)$ is the degree of the closed braid ℓ_i.

Let SA_m be the subset of B_m consisting of conjugates of the standard generators and their inverses. Each element of SA_m is called a *band*. It is easily seen that $SA_m \subset A_m$.

LEMMA 16.10. *In the above situation, the multiple cone $C(\ell)$ is a simple braided surface if and only if $b \in SA_m$.*

Proof. The if part is obvious. To prove the only if part, suppose that $C(\ell)$ is a simple braided surface. So the branch type of y is $(2, 1, \ldots, 1)$. Since ℓ_1 is a closed 2-braid which is a trivial knot in the 3-sphere, it is equivalent to the closure of the standard generator σ_1 or its inverse. The other components $\ell_2, \ldots, \ell_c$ are trivial closed 1-braids. Since ℓ is completely split, it is equivalent to the closure of a standard generator or its inverse. Thus $b \in SA_m$. ∎

LEMMA 16.11. *For given ℓ and y (as above), the multiple cone $C(\ell)$ is unique up to isomorphism in the strong sense as a braided surface.*

Proof. In the special case that $c = 1$ and $N_1 = D_1^2$, it is easily seen that the isomorphism class of $C(\ell)$ does not depend on a choice of the apex x_1. Using this special case, we see that when $c \geq 1$ and convex 2-disks $N_1, \ldots, N_c$ are provided, the isomorphism class of $C(\ell)$ does not depend on $x_i \in \mathrm{int} N_i$ ($i = 1, \ldots, c$). Suppose that $C(\ell)$ and $C'(\ell)$ are constructed by use of N_i ($i = 1, \ldots, c$) and N_i' ($i = 1, \ldots, c$) and points $x_i \in N_i$ ($i = 1, \ldots, c$) and $x_i' \in N_i'$ ($i = 1, \ldots, c$), respectively. By the previous case, we may assume that for each $i \in \{1, \ldots, c\}$, $x_i = x_i' \in \mathrm{int}(N_i \cap N_i')$. Then $C(\ell) = C'(\ell)$. ∎

16.6. Braided Surfaces with One Branch Point

Let S be a pointed braided surface of degree m whose boundary ∂S represents a braid $\beta \in B_m$.

LEMMA 16.12. *If S has only one branch point, then $\beta \in A_m$. Moreover, if S is a simple braided surface, then $\beta \in SA_m$.*

Proof. Let y be the branch point and let $N(y)$ be a sufficiently small regular neighborhood of y in D_2^2 so that the restriction of S over $N(y)$ is a multiple cone $C(\ell)$ for some closed braid ℓ in a solid torus $D_1^2 \times \partial N(y)$. Since S is locally flat and ℓ is non-trivial, the open braid of ℓ (which is unique up to conjugation in B_m)

is an element of A_m. Using the restriction of S over the annulus $\mathrm{cl}(D_2^2 \setminus N(y))$, we see that ℓ is equivalent to ∂S as a closed braid. Thus β is also an element of A_m. If S is simple, then by Lemma 16.10, $\beta \in SA_m$. ∎

CHAPTER 17

Braid Monodromy

17.1. Braid Monodromy

Let S be a surface braid of degree m in $D_1^2 \times D_2^2$ and let $\Sigma(S)$ be the branch point set of S. Take a point y_0 of ∂D_2^2 and fix it. For any loop $c : (I, \partial I) \to (D_2^2 \setminus \Sigma(S), y_0)$, we define a loop

$$\rho_S(c) : (I, \partial I) \to (C_m, Q_m)$$

in the configuration space $C_m = C_m(\mathrm{int} D^2)$ by

$$\rho_S(c)(t) = pr_1(S \cap pr_2^{-1}(c(t))),$$

where $pr_i : D_1^2 \times D_2^2 \to D_i^2$ ($i = 1, 2$) is the ith factor projection. Then we have a homomorphism

$$\rho_S : \pi_1(D_2^2 \setminus \Sigma(S), y_0) \to \pi_1(C_m, Q_m) = B_m$$

with $\rho_S([c]) = [\rho_S(c)]$.

DEFINITION 17.1. The *braid monodromy*, or the *braid monodromy representation*, of S is a homomorphism

$$\rho_S : \pi_1(D_2^2 \setminus \Sigma(S), y_0) \to \pi_1(C_m, Q_m) = B_m.$$

LEMMA 17.2. *Let S and S' be surface braids of degree m. If S and S' are isomorphic, then their braid monodromies ρ_S and $\rho_{S'}$ are the same.*

Proof. Let $\{h_u\}_{u \in [0,1]}$ be a fiber-preserving ambient isotopy of $D_1^2 \times D_2^2$ as in the definition of isomorphism (Definition 14.10) which carries S to S'. For a loop $c : (I, \partial I) \to (D_2^2 \setminus \Sigma(S), y_0)$, there is a one-parameter family of loops $\rho_{h_u(S)}(c) : (I, \partial I) \to (C_m, Q_m)$ ($u \in [0,1]$). Thus $\rho_S(c)$ and $\rho_{S'}(c)$ represent the same element of the braid group B_m. ∎

Let S and S' be surface braids of degree m and let ρ_S and $\rho_{S'}$ be their braid monodromies. We say that ρ_S and $\rho_{S'}$ are *equivalent* if there exists a homeomorphism $g : D_2^2 \to D_2^2$ with $g(\Sigma(S)) = \Sigma(S')$ and $g|_{\partial D_2^2} = \mathrm{id}$ such that $\rho_{S'} \circ g_* = \rho_S$, where $g_* : \pi_1(D_2^2 \setminus \Sigma(S), y_0) \to \pi_1(D_2^2 \setminus \Sigma(S'), y_0)$ is the isomorphism induced by g.

LEMMA 17.3. *Let S and S' be surface braids of degree m. If S and S' are equivalent, then their braid monodromies ρ_S and $\rho_{S'}$ are equivalent.*

Proof. Let $\{h_u\}_{u \in [0,1]}$ be a fiber-preserving ambient isotopy of $D_1^2 \times D_2^2$ as in the definition of equivalence (Definition 14.10) which carries S to S'. There is an ambient isotopy $\{\underline{h}_u\}$ of D_2^2 such that $pr_2 \circ h_u = \underline{h}_u \circ pr_2$. Let $g = \underline{h}_1$. Then $g(\Sigma(S)) = \Sigma(S')$ and $g|_{\partial D_2^2} = \mathrm{id}$. For a loop $c : (I, \partial I) \to (D_2^2 \setminus \Sigma(S), y_0)$, there is a one-parameter family of loops $\rho_{h_u(S)}(\underline{h}_u \circ c) : (I, \partial I) \to (C_m, Q_m)$ $(u \in [0, 1])$. Thus $\rho_S(c)$ and $\rho_{S'}(g \circ c)$ represent the same element of the braid group B_m. ∎

We can also define the braid monodromy for a pointed braided surface.

EXERCISE 17.4. Let S and S' be pointed braided surfaces. If they are isomorphic in the weak sense, then their braid monodromies ρ_S and $\rho_{S'}$ are the same. Prove this. [*Hint*: Follow the proof of Lemma 17.2.]

EXERCISE 17.5. Let S and S' be pointed braided surfaces. If they are equivalent in the weak sense, then their braid monodromies ρ_S and $\rho_{S'}$ are equivalent. Prove this. [*Hint*: Follow the proof of Lemma 17.3. Now $g|_{\partial D_2^2} = $ id is not satisfied. Modify g.]

17.2. Braid System

Let S be a surface braid, or a pointed braided surface, of degree m and $\Sigma(S)$ the branch point set of S. Take a point y_0 of ∂D_2^2 and fix it. If S is a pointed braided surface, then we assume that y_0 is the base point.

Let $\mathcal{A} = (a_1, \ldots, a_n)$ be a Hurwitz arc system whose starting point set is $\Sigma(S)$ and the terminal point is y_0, where n is the number of the branch points of S. Let $\eta_1, \ldots, \eta_n$ be the Hurwitz generator system of $\pi_1(D_2^2 \setminus \Sigma(S), y_0)$ associated with $\mathcal{A}$ (Sect. 2.6).

DEFINITION 17.6. The *braid system* of S associated with the Hurwitz arc system $\mathcal{A}$ is an ordered n-tuple

$$(\rho_S(\eta_1), \ldots, \rho_S(\eta_n)),$$

where ρ_S is the braid monodromy of S.

Notice that the braid system determines the braid monodromy, and vice versa, provided the Hurwitz arc system is specified.

For a set H and a positive integer n, let $P^n(H)$ denote the Cartesian product of n copies of H:

$$P^n(H) = \{(g_1, \ldots, g_n) \mid g_1, \ldots, g_n \in H\}.$$

An element of $P^n(H)$ is often written as $\overrightarrow{g}$. Suppose that H is a subset of a group, say G. For an element h of the group G, we denote by $P_h^n(H)$ a subset of $P^n(G)$ such that

$$P_h^n(H) = \{(g_1, \ldots, g_n) \in P^n(H) \mid g_1 \cdots g_n = h \in G\}.$$

Let A_m and SA_m be the subsets of the braid group B_m defined in Sect. 16.5.

LEMMA 17.7. *Let $\overrightarrow{b} = (b_1, \ldots, b_n)$ be a braid system of a pointed braided surface S of degree m whose boundary ∂S represents a braid $\beta \in B_m$. Then $\overrightarrow{b} \in P_\beta^n(A_m)$, namely,*

(1) *$b_i \in A_m$ for $i = 1, \ldots, n$, and*
(2) *$b_1 \cdots b_n = \beta$ in B_m.*

In particular, if S is a surface braid, then $\overrightarrow{b} \in P_1^n(A_m)$.

Proof. Let $\mathcal{A} = (a_1, \ldots, a_n)$ be a Hurwitz arc system with which the braid system $\overrightarrow{b}$ of S is associated. The restriction of S over a regular neighborhood of a_i in D_2^2 is a pointed braided surface with a single branch point. Thus $b_i \in A_m$ (Lemma 16.12). The boundary ∂D_2^2 is a loop in $D_2^2 \setminus \Sigma(S)$ with base point y_0, whose homotopy class is the product of the Hurwitz generator system $\eta_1, \ldots, \eta_n$. Thus $\beta = \rho_S([\partial S]) = \rho_S(\eta_1) \cdots \rho_S(\eta_n) = b_1 \cdots b_n$. ∎

LEMMA 17.8. *Let* $\overrightarrow{b} = (b_1, \ldots, b_n)$ *be a braid system of a pointed simple braided surface* S *of degree* m *whose boundary* ∂S *represents a braid* $\beta \in B_m$. *Then* $\overrightarrow{b} \in P_\beta^n(SA_m)$. *In particular, if* S *is a simple surface braid, then* $\overrightarrow{b} \in P_1^n(SA_m)$.

Proof. In the proof of Lemma 17.7, if the ith branch point is simple, then $b_i \in SA_m$ (Lemma 16.12). ∎

17.3. A Characterization of Braid Systems

Let ℓ be a closed m-braid in the solid torus $D_1^2 \times \partial D_2^2$ with $pr_1(\ell \cap pr_2^{-1}(y_0)) = Q_m$, where y_0 is a point of ∂D_2^2. Let $\beta \in B_m$ be an open braid of ℓ cut along the fiber $pr_2^{-1}(y_0)$ over y_0.

LEMMA 17.9. *For an element* $\overrightarrow{b} = (b_1, \ldots, b_n)$ *of* $P_\beta^n(A_m)$ *and a Hurwitz arc system* $\mathcal{A} = (a_1, \ldots, a_n)$ *in* D_2^2, *there is a pointed braided surface* S *such that the braid system associated with* $\mathcal{A}$ *is* $\overrightarrow{b}$ *and the boundary is* ℓ.

Proof. Let $\{y_1, \ldots, y_n\}$ be the starting point set of $\mathcal{A}$ with $\partial a_i = \{y_i, y_0\}$. For each i $(i = 1, \ldots, n)$, let $N(y_i)$ be a regular neighborhood of y_i in D_2^2. Put $X = \mathrm{cl}(D_2^2 \setminus \cup_{i=1}^n N(y_i))$. The restriction of a_i to X is a proper embedded arc, which we denote by a_i^*. One of the boundary points of a_i^* is y_0 and the other, which we denote by y_i^*, is a point on $\partial N(y_i)$. Let E be a disk obtained from X by cutting along $a_1^*, \ldots, a_n^*$. We define S over ∂D_2^2 by $S \cap (D_1^2 \times \partial D_2^2) = \ell$. We define S over a_i^* $(i = 1, \ldots, n)$ by $pr_1(S \cap pr_2^{-1}(y)) = Q_m$ for any point y of a_i^*. Since b_i is an element of A_m, there is a multiple cone in $D_1^2 \times N(y_i)$ with base point y_i^* whose boundary braid is b_i. We define S over $N(y_i)$ by this multiple cone. Now we have already constructed S over ∂E. The closed braid in $D_1^2 \times \partial E$ is the closure of the product $b_1 \ldots b_n \beta^{-1}$, which is the trivial element of B_m by the assumption. By Lemma 16.6, we have a braided surface in $D_1^2 \times E$ without branch points such that the boundary fits for the closed braid. In this way we have a desired pointed braided surface S. ∎

We will denote by $S(\overrightarrow{b}, \mathcal{A}, \ell)$ a pointed braided surface S satisfying the condition of Lemma 17.9.

LEMMA 17.10. *Let* $S(\overrightarrow{b}, \mathcal{A}, \ell)$ *be a pointed braided surface as above.*

(1) *The isomorphism class in the strong sense of* $S(\overrightarrow{b}, \mathcal{A}, \ell)$ *is uniquely determined.*

(2) *The isomorphism class in the weak sense of* $S(\overrightarrow{b}, \mathcal{A}, \ell)$ *does not depend on* ℓ.

(3) *The equivalence class in the strong sense of* $S(\overrightarrow{b}, \mathcal{A}, \ell)$ *does not depend on* $\mathcal{A}$.

(4) *The equivalence class in the weak sense of* $S(\overrightarrow{b}, \mathcal{A}, \ell)$ *does not depend on* $\mathcal{A}$ *and* ℓ.

Proof. (1) Let S' be another pointed braided surface satisfying the condition. Modify S' up to isomorphism in the strong sense so that $pr_1(S' \cap pr_2^{-1}(y)) = Q_m$ for any point y of a_i^*. The restriction of S' to $D_1^2 \times \partial N(y_i)$ is the closed m-braid such that the m-braid obtained from it by cutting along $pr_2^{-1}(y_i^*)$ is b_i, since the braid monodromy does not change under isomorphism in the strong/weak sense. Thus we can modify S' up to isomorphism in the strong sense, so that the restriction of S' to $D_1^2 \times N(y_i)$ is the same with that of S (Lemma 16.11). Now the difference of S and S' is contained in the interior of $D_1^2 \times E$. By Lemma 16.7, we can deform S' into S. (3) Let $S' = S(\overrightarrow{b}, \mathcal{A}', \ell)$. There is an ambient isotopy $\{g_u\}_{u \in [0,1]}$ of D_2^2 carrying $\mathcal{A}$ to $\mathcal{A}'$ keeping ∂D_2^2 fixed. Let $\{h_u\}_{u \in [0,1]}$ be a fiber-preserving isotopy of $D_1^2 \times D_2^2$ such that $h_u = \text{id} \times g_u : D_1^2 \times D_2^2 \to D_1^2 \times D_2^2$ for $u \in [0,1]$. This carries S to $h_1(S)$, which is another $S(\overrightarrow{b}, \mathcal{A}', \ell)$. By the previous assertion (1), the latter is isomorphic to S' in the strong sense. Therefore S and S' are equivalent in the strong sense. The assertions (2) and (4) are obtained from (1) and (3) by using a collar neighborhood of ∂D_2^2 in D_2^2. ∎

Let y_0, β, ℓ be as before.

PROPOSITION 17.11 (Characterization of a braid system). *An n-tuple of m-braids $\overrightarrow{b} = (b_1, \dots, b_n)$ is a braid system of a pointed braided surface with n branch points whose boundary is ℓ if and only if $\overrightarrow{b} \in P_\beta^n(A_m)$. In particular, $\overrightarrow{b}$ is a braid system of a surface braid with n branch points if and only if $\overrightarrow{b} \in P_1^n(A_m)$.*

Proof. It follows from Lemmas 17.7 and 17.9. ∎

PROPOSITION 17.12 (Characterization of a braid system). *An n-tuple of m-braids $\overrightarrow{b} = (b_1, \dots, b_n)$ is a braid system of a pointed* simple *braided surface with n branch points whose boundary is ℓ if and only if $\overrightarrow{b} \in P_\beta^n(SA_m)$. In particular, $\overrightarrow{b}$ is a braid system of a* simple *surface braid with n branch points if and only if $\overrightarrow{b} \in P_1^n(SA_m)$.*

Proof. It follows from Lemmas 17.8 and 17.9. ∎

17.4. Braid Monodromy Principal, I

THEOREM 17.13. *Let S and S' be pointed braided surfaces with the same base point y_0, and let ρ_S and $\rho_{S'}$ be their braid monodromies.*

(1) *S and S' are isomorphic in the strong sense if and only if $\partial S = \partial S'$, $\Sigma(S) = \Sigma(S')$, and their braid monodromies are the same.*

(2) *S and S' are isomorphic in the weak sense if and only if $\Sigma(S) = \Sigma(S')$, and their braid monodromies are the same.*

(3) *S and S' are equivalent in the strong sense if and only if $\partial S = \partial S'$ and their braid monodromies are equivalent.*

(4) *S and S' are equivalent in the weak sense if and only if their braid monodromies are equivalent.*

Proof. The only if parts are Exercises 17.4 and 17.5. Since a braid system with a Hurwitz arc system determines a braid monodromy and since a replacement of a Hurwitz arc system induces an equivalence of braid monodromies, we see the if parts. The details are left to the reader as an exercise. ∎

As a special case, we have the following.

THEOREM 17.14. *Let S and S' be surface braids, and let ρ_S and $\rho_{S'}$ be their braid monodromies with the same base point y_0.*

(1) *S and S' are isomorphic if and only if $\Sigma(S) = \Sigma(S')$ and $\rho_S = \rho_{S'}$.*

(2) *S and S' are equivalent if and only if ρ_S and $\rho_{S'}$ are equivalent.*

17.5. G-Monodromy and G-System

Let G be a group. Let y_0 be a point of the boundary of a 2-disk D^2. Let Σ be a set of n interior points of D^2.

A *G-monodromy* on (D^2, Σ, y_0) is a homomorphism

$$\rho : \pi_1(D^2 \setminus \Sigma, y_0) \to G.$$

Two *G-monodromies* $\rho : \pi_1(D^2 \setminus \Sigma, y_0) \to G$ and $\rho' : \pi_1(D^2 \setminus \Sigma', y_0) \to G$ are *equivalent* if there exists a homeomorphism $g : D^2 \to D^2$ with $g(\Sigma) = \Sigma'$ and $g|_{\partial D^2} = \mathrm{id}$ such that $\rho' \circ g_* = \rho$, where $g_* : \pi_1(D^2 \setminus \Sigma, y_0) \to \pi_1(D^2 \setminus \Sigma', y_0)$ is the isomorphism induced by g.

Let $\mathcal{A} = (a_1, \ldots, a_n)$ be a Hurwitz arc system for Σ with base point y_0, and let $\eta_1, \ldots, \eta_n$ be the generator system of $\pi_1(D^2 \setminus \Sigma, y_0)$ associated with $\mathcal{A}$.

The *G-system for a G-monodromy ρ associated with $\mathcal{A}$* is an n-tuple

$$(\rho(\eta_1), \ldots, \rho(\eta_n)).$$

LEMMA 17.15. *Let $\overrightarrow{h} = (h_1, \ldots, h_n)$ be the G-system of a G-monodromy $\rho : \pi_1(D^2 \setminus \Sigma) \to G$ associated with $\mathcal{A}$, and let $\overrightarrow{h'} = (h'_1, \ldots, h'_n)$ be the braid system of a G-monodromy $\rho' : \pi_1(D^2 \setminus \Sigma') \to G$ associated with $\mathcal{A}'$. Then ρ and ρ' are equivalent if and only if $\overrightarrow{h}$ and $\overrightarrow{h'}$ are slide equivalent.*

Proof. There exists a homeomorphism $g : D^2 \to D^2$ carrying $\mathcal{A}$ onto $\mathcal{A}'$ with $g|_{\partial D^2} = \mathrm{id}$. Consider a *G-monodromy* $\rho' \circ g_* : \pi_1(D^2 \setminus \Sigma) \to G$. This is equivalent to ρ' by definition, and the *G-system* associated with $\mathcal{A}$ is $\overrightarrow{h'}$. Therefore, without loss of generality, we may assume that $\mathcal{A} = \mathcal{A}'$ and hence $\Sigma = \Sigma'$. Let $\eta_1, \ldots, \eta_n$ be the generator of $\pi_1(D^2 \setminus \Sigma, y_0)$ associated with $\mathcal{A}$. Suppose that $\overrightarrow{h'} = \mathrm{slide}(\gamma)(\overrightarrow{h})$ for some $\gamma \in B_n$. Let $f : \pi_1(D^2 \setminus \Sigma) \to \pi_1(D^2 \setminus \Sigma)$ be an isomorphism which is $\mathrm{Artin}(\gamma)$ with respect to $\mathcal{A}$. By Lemma 2.14,

$$\mathrm{slide}(\gamma)(\eta_1, \ldots, \eta_n) = P^n(f)(\eta_1, \ldots, \eta_n).$$

Therefore

$$
\begin{aligned}
P^n(\rho')(\eta_1, \ldots, \eta_n) &= \overrightarrow{h'} \\
&= \mathrm{slide}(\gamma)(\overrightarrow{h}) \\
&= \mathrm{slide}(\gamma) \circ P^n(\rho)(\eta_1, \ldots, \eta_n) \\
&= P^n(\rho) \circ \mathrm{slide}(\gamma)(\eta_1, \ldots, \eta_n) \\
&= P^n(\rho) \circ P^n(f)(\eta_1, \ldots, \eta_n) \\
&= P^n(\rho \circ f)(\eta_1, \ldots, \eta_n).
\end{aligned}
$$

Since $\eta_1, \ldots, \eta_n$ generate $\pi_1(D^2 \setminus \Sigma(S))$, we have $\rho' = \rho \circ f$. Since f is induced from a homeomorphism $g : (D^2, \Sigma) \to (D^2, \Sigma)$ with $g|_{\partial D^2} = \mathrm{id}$ (Lemma 2.14), ρ and ρ' are equivalent. Conversely suppose that there is a homeomorphism $g : (D^2, \Sigma) \to (D^2, \Sigma)$ with $g|_{\partial D^2} = \mathrm{id}$ such that $\rho' = \rho \circ g_*$. By Lemma 2.14, there is an element $\gamma \in B_n$ such that $\mathrm{slide}(\gamma)(\eta_1, \ldots, \eta_n) = P^n(g_*)(\eta_1, \ldots, \eta_n)$. Then $\overrightarrow{h'} = \mathrm{slide}(\gamma)\overrightarrow{h}$.

17.6. Braid Monodromy Principal, II

THEOREM 17.16. *Let S and S' be pointed braided surfaces with n branch points and let $\overrightarrow{b}$ and $\overrightarrow{b}'$ be their braid systems, which may not be associated with the same Hurwitz arc system.*

(1) *S and S' are equivalent in the strong sense if and only if $\partial S = \partial S'$ and $\overrightarrow{b}$ and $\overrightarrow{b}'$ are slide equivalent.*

(2) *S and S' are equivalent in the weak sense if and only if $\overrightarrow{b}$ and $\overrightarrow{b}'$ are slide equivalent.*

Proof. (1) follows from Theorem 17.13 and Lemma 17.15. The assertion (2) is proved using a collar neighborhood of ∂D_2^2 in D_2^2. ∎

As a special case, we have the following theorem.

THEOREM 17.17. *Let S and S' be surface braids with n branch points and let $\overrightarrow{b}$ and $\overrightarrow{b}'$ be their braid systems, which may not be associated with the same Hurwitz arc system. S and S' are equivalent if and only if $\overrightarrow{b}$ and $\overrightarrow{b}'$ are slide equivalent.*

PROPOSITION 17.18. *Two (simple) surface braids of degree 2 are equivalent if and only if they have the same number of branch points.*

Proof. Let $\overrightarrow{b} = (b_1, \ldots, b_n)$ be a braid system of a simple surface braid of degree 2 with n branch points. Each b_i is an element of $SA_2 = \{\sigma_1, \sigma_1^{-1}\}$ with $b_1 \cdots b_n = 1$ (Proposition 17.12). If $b_i = \sigma_1^{-1}$ and $b_{i+1} = \sigma_1$ for some i, apply slide(σ_i) and we have $b_i = \sigma_1$ and $b_{i+1} = \sigma_1^{-1}$. Thus $\overrightarrow{b}$ is slide equivalent to $(\sigma_1, \ldots, \sigma_1, \sigma_1^{-1}, \ldots, \sigma_1^{-1})$. The equation $b_1 \cdots b_n = 1$ implies that the number of σ_1s is equal to that of σ_1^{-1}s. ∎

CHAPTER 18

Chart Descriptions

18.1. Introduction

Let S_1 and S_2 be surface braids which have the motion pictures that are depicted in Fig. 14.1(A) and (B). Are they equivalent? When two surface braids are given in the motion picture method, it is difficult to know whether they are equivalent or not. For the reader who is familiar with some techniques in 2-knot theory (the motion picture method or surface link diagram, etc.), it may not be difficult to prove that S_1 and S_2 are ambient isotopic in the 4-disk $D_1^2 \times D_2^2$ keeping the boundary fixed. However this does not imply that they are equivalent as surface braids (recall Sect. 14.5).

The motion picture method is very useful to describe surface braids and to deform them locally. However it is somewhat inconvenient when we want to deform surface braids more globally.

In this section we introduce another method to describe surface braids, called the chart description. It enables us to describe and visualize surface braids much more easily than the motion picture method and to deform them globally and quickly without being fettered by a bi-parametrization of D_2^2. Moreover, the chart description has a lot of applications to 2-knot theory.

For example, the motion pictures illustrated in Fig. 18.1 and in Fig. 18.12 look very different. But they describe equivalent surface braids. By use of the chart description, this will be easily seen (Example 18.23).

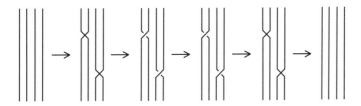

FIGURE 18.1

18.2. BWTS Charts

Before discussing the chart description of a surface braid, we start with the chart description of a classical braid and a braid word transformation sequence.

Recall that any braid is presented by a word (*braid word*) of the standard generators $\sigma_1, \ldots, \sigma_{m-1}$. Let w be a braid word. Take points on the interval $I = [0, 1]$ with the same number as the length of w and put a label σ_i^ϵ to the kth point if the kth letter of w is σ_i^ϵ. We denote by Λ_w the set of points with the labels. For simplicity, we shall abbreviate a label σ_i^ϵ to ϵi. Then labels are in $\{-(m-1), \ldots, -1, +1, \ldots, +(m-1)\}$; see Fig. 18.2.

DEFINITION 18.1. A *braid word chart* for a braid word w is a set Λ_w of points in the interval I with labels.

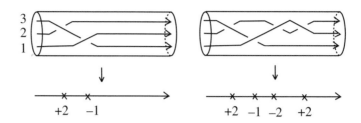

FIGURE 18.2

Two word expressions of the same braid are related by a finite sequence of the following transformations (Theorem 1.17):

(i) Insertion of the word $\sigma_i \sigma_i^{-1}$ or $\sigma_i^{-1} \sigma_i$,

(ii) Deletion of the word $\sigma_i \sigma_i^{-1}$ or $\sigma_i^{-1} \sigma_i$,

(iii) Substitution of $\sigma_i \sigma_j$ for $\sigma_j \sigma_i$ where $|i - j| > 1$,

(iv) Substitution of $\sigma_i \sigma_j \sigma_i$ for $\sigma_j \sigma_i \sigma_j$ where $|i - j| = 1$.

We call these *elementary braid word transformations*.

(iii') Substitution of $\sigma_i^\epsilon \sigma_j^\delta$ for $\sigma_j^\delta \sigma_i^\epsilon$ where $|i - j| > 1$ and $\epsilon, \delta \in \{\pm 1\}$,

(iv') Substitution of $\sigma_i^{\epsilon_1} \sigma_j^{\epsilon_2} \sigma_i^{\epsilon_3}$ for $\sigma_j^{\epsilon_3} \sigma_i^{\epsilon_2} \sigma_j^{\epsilon_1}$ where $|i - j| = 1$ and $\epsilon_1, \epsilon_2, \epsilon_3 \in \{\pm 1\}$ such that $\epsilon_1 = \epsilon_2$ or $\epsilon_2 = \epsilon_3$.

The transformations (iii') and (iv') are consequences of (i) $\sim$ (iv). We call transformations (i) $\sim$ (iv), (iii') and (iv') *braid word transformations*.

DEFINITION 18.2. A *braid word transformation sequence* is a finite sequence $w = w_0 \to w_1 \to \cdots \to w_p = w'$ of braid word transformations, where $w_{k-1} \to w_k$ means a braid word transformation. The *degree* of a braid word transformation sequence is the degree of each braid word.

Let w and w' be braid words related by a braid word transformation. We consider a diagram $\Gamma = \Gamma(w \to w')$ in a 2-disk parameterized by $I \times J$ where $I = [0, 1]$ and $J = [j, j']$ as follows: Take a braid word chart Λ_w for w on the edge $I \times \{j\}$, and take a braid word chart $\Lambda_{w'}$ for w' on the opposite edge $I \times \{j'\}$. The words w and w' are identical except for a few consecutive letters involving the braid word transformation. Thus, except for these points, we can connect the points of Λ_w on $I \times \{j\}$ to the corresponding points of $\Lambda_{w'}$ on $I \times \{j'\}$ by mutually disjoint simple arcs. If an edge connects a point of Λ_w and a point of $\Lambda_{w'}$ whose labels are $+i$ (or $-i$, resp.), then we give an orientation to this edge from Λ_w to $\Lambda_{w'}$ (resp. from $\Lambda_{w'}$ to Λ_w) and give a label i.

If the transformation is (i), then $\Lambda_{w'}$ has two additional points which are labeled $+i$ and $-i$. Connect these points by an oriented arc as in Fig. 18.3(A) or (B) with label i.

If the transformation is (ii), then Λ_w has two additional points which are labeled $+i$ and $-i$. Connect these points by an oriented arc as in Fig. 18.3(C) or (D) with label i.

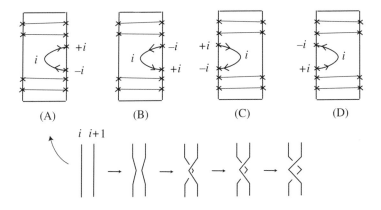

FIGURE 18.3

If the transformation is (iii) or (iii'), then $\Lambda_{w'}$ is obtained from Λ_w by replacing two points. Connect the corresponding points by oriented and labeled arcs as in Fig. 18.4.

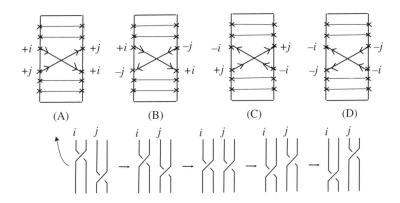

FIGURE 18.4

If the transformation is (iv) or (iv'), then $\Lambda_{w'}$ is obtained from Λ_w by changing three points. Make a vertex, say v, and connect the three points of Λ_w with the vertex v by three arcs and also connect the three points of $\Lambda_{w'}$ with v. Thus v is a vertex of valency six. The six arcs adjacent to v are oriented as in Fig. 18.5, and we give the six edges the labels of Λ_w and $\Lambda_{w'}$ after ignoring their signs. (We usually denote the vertex v by a small circle as in the figure.)

Let $w = w_0 \to w_1 \to \cdots \to w_p = w'$ be a braid word transformation sequence. We construct a diagram, $\Gamma = \Gamma(w = w_0 \to w_1 \to \cdots \to w_p = w')$, in a 2-disk parameterized by $I_3 \times I_4$ as follows: Divide the interval $I_4 = [0, 1]$ into p subintervals, say $J_1 = [t_0, t_1], J_2 = [t_1, t_2], \ldots, J_p = [t_{p-1}, t_p]$, where $t_0 = 0$ and

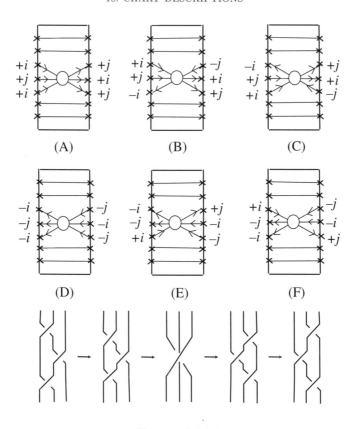

$$\text{FIGURE } 18.5$$

$t_1 = 1$. For each $k \in \{0, 1, \ldots, p\}$, consider a braid word chart Λ_k for the word w_k and put it on the interval $I_3 \times \{t_k\} \subset I_3 \times I_4$. For each $k \in \{1, \ldots, p\}$, construct a diagram $\Gamma(w_{k-1} \to w_k)$ in the rectangle $I_3 \times J_k$ as before. The union of the diagrams $\Gamma(w_{k-1} \to w_k)$ for $k = 1, \ldots, p$ is a diagram in $I_3 \times I_4$. We denote this diagram by $\Gamma = \Gamma(w = w_0 \to w_1 \to \cdots \to w_p = w')$.

DEFINITION 18.3. A *BWTS chart* (or simply, a *chart*) for a braid word transformation sequence $w = w_0 \to w_1 \to \cdots \to w_p = w'$ is a diagram $\Gamma = \Gamma(w = w_0 \to w_1 \to \cdots \to w_p = w')$.

Fig. 18.6 is a chart for a braid word transformation sequence of degree four. The sequence is $w = w_0 \to \cdots \to w_3 = w'$ with

$$
\begin{aligned}
w_0 &= \sigma_3 \sigma_2 \sigma_1^{-1} \\
w_1 &= \sigma_3 \sigma_2 \sigma_1^{-1} \sigma_2^{-1} \sigma_2 \\
w_2 &= \sigma_3 \sigma_1^{-1} \sigma_2^{-1} \sigma_1 \sigma_2 \\
w_3 &= \sigma_1^{-1} \sigma_3 \sigma_2^{-1} \sigma_1 \sigma_2.
\end{aligned}
$$

18.3. Enlarged BWTS Charts

We enlarge the notion of a chart of a braid word transformation sequence (a BWTS chart) so that it involves a simple string recombination of braids.

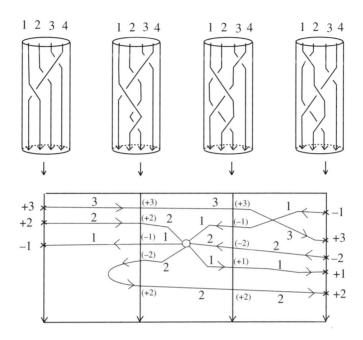

FIGURE 18.6. A BWTS chart

DEFINITION 18.4. A *simple string recombination* is a one-parameter family $\{b_t\}$ of braids as in Fig. 14.4.

In braid word description, a simple string recombination is as follows:

(v) Insertion of a letter σ_i or σ_i^{-1},
(vi) Deletion of a letter σ_i or σ_i^{-1}.

We call transformations (i) $\sim$ (iv), (iii$'$), (iv$'$), (v), and (vi) *enlarged braid word transformations*.

DEFINITION 18.5. An *enlarged braid word transformation sequence* is a finite sequence $w = w_0 \to w_1 \to \cdots \to w_p = w'$ of enlarged braid word transformations, where $w_{k-1} \to w_k$ means an enlarged braid word transformation.

For braid words w and w' related by a transformation of a simple string recombination, consider a diagram $\Gamma = \Gamma(w \to w')$ in a 2-disk parameterized by $I \times J$ where $I = [0, 1]$ and $J = [j, j']$ as in Fig. 18.7.

Let $w = w_0 \to w_1 \to \cdots \to w_p = w'$ be an enlarged braid word transformation sequence. We construct a diagram, $\Gamma = \Gamma(w = w_0 \to w_1 \to \cdots \to w_p = w')$, in a 2-disk parameterized by $I_3 \times I_4$ in the same way as before.

DEFINITION 18.6. An *enlarged BWTS chart* (or simply, a *chart*) for an enlarged braid word transformation sequence $w = w_0 \to w_1 \to \cdots \to w_p = w'$ is a diagram $\Gamma = \Gamma(w = w_0 \to w_1 \to \cdots \to w_p = w')$.

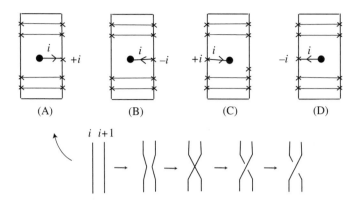

FIGURE 18.7

Fig. 18.8 is a chart for an enlarged braid word transformation sequence. The sequence is $w = w_3 \to \cdots \to w_6 = w'$ with

$$
\begin{aligned}
w_3 &= \sigma_1^{-1}\sigma_3\sigma_2^{-1}\sigma_1\sigma_2 \\
w_4 &= \sigma_1^{-1}\sigma_3\sigma_1\sigma_2\sigma_1^{-1} \\
w_5 &= \sigma_1^{-1}\sigma_3\sigma_1\sigma_1^{-1} \\
w_6 &= \sigma_1^{-1}\sigma_3.
\end{aligned}
$$

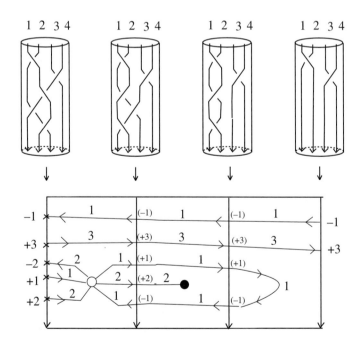

FIGURE 18.8. An enlarged BWTS chart

18.4. Surface Braid Charts

We assume that a graph may be empty or have closed edges without vertices called *hoops* or *simple loops*.

DEFINITION 18.7. A *surface braid chart* (or simply, *chart*) of degree m is a finite graph Γ in D_2^2 whose edges are oriented and labeled satisfying the following conditions:

(0) The graph is disjoint from the boundary of D_2^2.
(1) Every vertex has degree one, four or six.
(2) The labels of edges are integers in $\{1, 2, \ldots, m-1\}$.
(3) For each degree-six vertex, three consecutive edges are oriented inward and the others are oriented outward, and the six edges are labeled i and $i+1$ alternately for some i. See Fig. 18.9(4).
(4) For each degree-four vertex, diagonal edges have the same label and are oriented coherently, and the labels i and j of the diagonals satisfy $|i-j| > 1$. See Fig. 18.9(3).

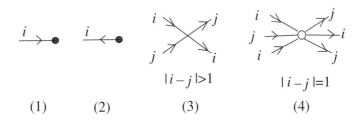

(1)   (2)   (3)   (4)

FIGURE 18.9. Vertices of a surface braid chart

We call a degree-one vertex a *black vertex* and a degree-six vertex a *white vertex*. A *middle* edge of a white vertex means the middle edge of the three consecutive edges oriented inward or that of the other three edges.

Fig. 18.10 is an example of a surface braid chart of degree 4.

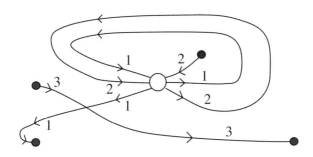

FIGURE 18.10

18.5. From Charts to Surface Braids, I

We construct a surface braid from a surface braid chart.
Fix a bi-parametrization $D_2^2 \cong I_3 \times I_4$ of D_2^2.

DEFINITION 18.8. A surface braid chart Γ is *in general position with respect to the I_4-direction* if there is a division of I_4, say $J_1 = [j_0, j_1], J_2 = [j_1, j_2], \ldots, J_p = [j_{p-1}, j_p]$ for some p, such that the restriction of Γ to the rectangle $I_3 \times J_k$ for each $k \in \{1, \ldots, p\}$ is a diagram as in Figs. 18.3, 18.4, 18.5, and 18.7.

Let Γ be a surface braid chart of degree m which is in general position with respect to the I_4-direction. So there exists a division of I_4, say $J_1, J_2, \ldots, J_p$, such that the restriction of Γ to each $I_3 \times J_k$ is a diagram as in Figs. 18.3, 18.4, 18.5 and 18.7. This gives an enlarged braid word transformation sequence. Consider a motion picture which corresponds to the enlarged braid word transformation sequence. This motion picture describes a surface braid of degree m.

DEFINITION 18.9. A *surface braid associated with* Γ is a surface braid whose motion picture is given as above.

EXAMPLE 18.10. Fig. 18.11 is a chart of degree 4 which is obtained from the chart illustrated in Fig. 18.10 by an ambient isotopy of D_2^2. It is in general position with respect to the I_4-direction. We have an enlarged braid word transformation sequence $w = w_0 \to \cdots \to w_{11} = w'$ with

$$
\begin{aligned}
w_0 &= e \\
w_1 &= \sigma_1^{-1} \sigma_1 \\
w_2 &= \sigma_1^{-1} \\
w_3 &= \sigma_3 \sigma_1^{-1} \\
w_4 &= \sigma_2^{-1} \sigma_2 \sigma_3 \sigma_1^{-1} \\
w_5 &= \sigma_2^{-1} \sigma_1^{-1} \sigma_1 \sigma_2 \sigma_3 \sigma_1^{-1} \\
w_6 &= \sigma_2^{-1} \sigma_1^{-1} \sigma_1 \sigma_2 \sigma_1^{-1} \sigma_3 \\
w_7 &= \sigma_2^{-1} \sigma_1^{-1} \sigma_2^{-1} \sigma_1 \sigma_2 \sigma_3 \\
w_8 &= \sigma_2^{-1} \sigma_1^{-1} \sigma_1 \sigma_2 \sigma_3 \\
w_9 &= \sigma_2^{-1} \sigma_2 \sigma_3 \\
w_{10} &= \sigma_3 \\
w_{11} &= e.
\end{aligned}
$$

A surface braid associated with the chart is described by a motion picture corresponding to the enlarged braid word transformation sequence; see Fig. 18.12.

18.6. From Surface Braids to Charts

Fix a bi-parametrization of D_2^2 with $I_3 \times I_4$.

THEOREM 18.11. *Let S be a simple surface braid. There exists a surface braid chart Γ which is in general position with respect to the I_4-direction such that S is equivalent to a surface braid associated with Γ.*

Proof. Let $\{b_t\}_{t \in [0,1]}$ be the motion picture of S along I_4-direction. Modifying S up to equivalence, we may assume that the I_4-coordinates of the branch points of S are all distinct. So each singular braid in the motion picture has exactly one singular point. Let $t_1, t_2, \ldots, t_n \in I_4$ be the I_4-coordinates of the branch points with $0 < t_1 < t_2 < \cdots < t_n < 1$. For convenience, we put $t_0 = 0$ and $t_{n+1} = 1$.

Let ε be a sufficiently small positive number. For each $k = 1, \cdots, n$, we may assume that the restriction of S to the rectangle $I_3 \times [t_k - \varepsilon, t_k + \varepsilon]$ is a simple string recombination. For each $k \in \{1, \ldots, n\}$, let w_k and w_k' be the braid words representing the geometric braids $b_{t_k - \varepsilon}$ and $b_{t_k + \varepsilon}$ respectively. w_k' is obtained from

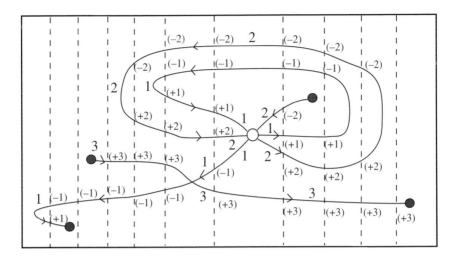

FIGURE 18.11. A chart in general position with respect to the I_4-direction

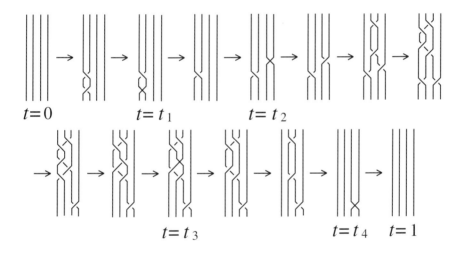

FIGURE 18.12

w_k by deletion or insertion of σ_i or σ_i^{-1} for some i. Let Λ_k and Λ'_k be the braid word charts of w_k and w'_k, and put them on $I_3 \times \{t_k - \varepsilon\} \subset I_3 \times I_4$ and $I_3 \times \{t_k + \varepsilon\}$, respectively. We define Γ over the rectangle $I_3 \times [t_k - \varepsilon, t_k + \varepsilon]$ by a diagram as in Fig. 18.7.

For each $k \in \{1, \ldots, n+1\}$, there are no branch points in the rectangle $I_3 \times [t_{k-1} + \varepsilon, t_k - \varepsilon]$. Thus $b_{t_{k-1}+\varepsilon}$ and $b_{t_k-\varepsilon}$ are equivalent as a braid, and there is a braid word transformation sequence from w'_{k-1} to w_k. (Here we assume w'_0 and w_{n+1} are empty words.) We define Γ over the rectangle $I_3 \times [t_{k-1} + \varepsilon, t_k - \varepsilon]$ by the BWTS chart of this sequence.

Now we have a surface braid chart Γ. Let S' be a surface braid associated with Γ. S and S' are the same over the rectangle $I_3 \times [t_k - \varepsilon, t_k + \varepsilon]$ for each

$k \in \{1, \ldots, n\}$. By Lemma 16.7, we can deform S' up to isomorphism such that S and S' are the same. ∎

18.7. From Charts to Surface Braids, II

We introduce an alternative method to construct a surface braid from a chart. In this method we do not need to fix a bi-parametrization $D_2^2 \cong I_3 \times I_4$, nor do we need to modify the chart by an ambient isotopy of D_2^2. It is easily seen that a surface braid obtained by this method is unique up to equivalence (in fact, up to isomorphism). It will turn out that a surface braid constructed here is equivalent to a surface braid constructed in Sect. 18.5.

Let Γ be a surface braid chart of degree m. We construct a surface braid $S = S(\Gamma)$ of degree m as follows: Let $V(\Gamma)$ be the set of vertices of Γ and $E(\Gamma)$ the set of edges of Γ. Here hoops (closed edges without vertices) are assumed to be edges. For each vertex $v \in V(\Gamma)$, let $N(v)$ be a regular neighborhood of v in D_2^2, and put

$$N(V) = \cup_{v \in V(\Gamma)} N(v), \quad \text{and} \quad X(V) = \text{cl}(D_2^2 \setminus N(V)).$$

For each edge $e \in E(\Gamma)$, we denote by $e|_X$ the restriction of e to $X(V)$, which is a properly embedded arc or a loop in $X(V)$. Let $N(e|_X)$ be a regular neighborhood of $e|_X$ in $X(V)$, and put

$$N(E|_X) = \cup_{e \in E(\Gamma)} N(e|_X).$$

Then the union $N(V) \cup N(E|_X)$ is a regular neighborhood $N(\Gamma)$ of Γ in D_2^2.

Recall that $pr_i : D_1^2 \times D_2^2 \to D_i^2$ ($i = 1, 2$) is the ith factor projection. For each $j \in \{1, \ldots, m-1\}$, we fix a path $g_j : ([-1, 1], \{-1, 1\}) \to (C_m, Q_m)$ in the configuration space C_m which represents the standard generator σ_j of the braid group $B_m = \pi_1(C_m, Q_m)$.

Step 1. Define S over $\text{cl}(D_2^2 \setminus N(\Gamma))$ by

$$pr_1(S \cap pr_2^{-1}(y)) = Q_m \quad \text{for } y \in \text{cl}(D_2^2 \setminus N(\Gamma)).$$

Step 2. In this step we define S over $N(E|_X)$. For each $e \in E(\Gamma)$, identify the regular neighborhood $N(e|_X)$ with $(e|_X) \times [-1, 1]$ such that when we walk along an interval $\{y\} \times [-1, 1]$ ($y \in e|_X$) from -1 to 1, the orientation of the edge e is from right to left. This identification is called a bi-collaring of $N(e|_X)$. If the label given to the edge e is j, then we define S over $N(e|_X)$ by

$$pr_1(S \cap pr_2^{-1}(z)) = g_j(t) \quad \text{for } z = (y, t) \in e|_X \times [-1, 1] = N(e|_X),$$

where $g_j : ([-1, 1], \{-1, 1\}) \to (C_m, Q_m)$ is the path which represents the standard generator σ_j.

Step 3. We define S over $N(V)$. Note that we have already defined S over $\partial N(V)$. For each $v \in V(\Gamma)$, the restriction of S to $D_1^2 \times \partial N(v)$ is a closed geometric braid of degree m, say ℓ_v. If v is a black vertex and if the edge attached to v is labeled j and oriented inward (resp. outward), then ℓ_v is the closure of a geometric braid σ_j (resp. σ_j^{-1}). In this case, we define S over $N(v)$ by the multiple cone of ℓ_v in $D_1^2 \times N(v)$. If v is a degree four vertex and if the four edges are labeled i and j with $|i - j| > 1$, then ℓ_v is the closure of a geometric braid $\sigma_i \sigma_j \sigma_i^{-1} \sigma_j^{-1}$. If v is a white vertex and if the six edges are labeled i and j with $|i - j| = 1$ (assume that two of the edges labeled i are oriented inward), then ℓ_v is the closure of a geometric braid $\sigma_i \sigma_j \sigma_i \sigma_j^{-1} \sigma_i^{-1} \sigma_j^{-1}$. In these two cases, ℓ_v is a closed trivial braid and we can extend it over $N(v)$ by a braided surface without branch points (Lemma 16.6).

In this way we construct a simple surface braid $S = S(\Gamma)$ of degree m such that branch points of S correspond to black vertices of Γ.

LEMMA 18.12. *The surface braid $S(\Gamma)$ constructed above is unique up to equivalence.*

Proof. Let S and S' be surface braids constructed as above. Since any ambient isotopy of D_2^2 rel ∂D_2^2 induces a fiber-preserving isotopy of $D_1^2 \times D_2^2$ rel $D_1^2 \times \partial D_2^2$, using the uniqueness of a regular neighborhood, we may assume that S and S' are constructed by use of the same regular neighborhood $N(\Gamma) = N(V) \cup N(E|_X)$. Then by construction, S and S' are the same over $\mathrm{cl}(D_2^2 \setminus N(\Gamma))$. Note that S and S' are identical over $\partial N(e|_X)$ for each $e \in E(\Gamma)$. By Lemma 16.7, we can deform S' up to isomorphism such that it coincides with S over $N(e|_X)$. Note that S and S' are identical over $\partial N(v)$ for each $v \in V(\Gamma)$. By Lemma 16.11, we can deform S' such that it coincides with S over $N(v)$. ∎

DEFINITION 18.13. A *surface braid associated with* a chart Γ is a surface braid $S(\Gamma)$ constructed as above.

When Γ is in general position with respect to the I_4-direction, this definition is compatible with Definition 18.9.

PROPOSITION 18.14. *Let Γ_1 and Γ_2 be charts of degree m. If they are ambient isotopic in D_2^2, then surface braids S_1 and S_2 associated with them are equivalent.*

EXERCISE 18.15. Prove Proposition 18.14. [*Hint*: It is seen by the same reasoning as the proof of Lemma 18.12. Recall Lemma 14.11.]

EXAMPLE 18.16. Two surface braids illustrated in Fig. 14.1(A) and (B) are associated with charts of degree 2 illustrated in Figs. 18.13(A) and (B). Since these charts are ambient isotopic in D_2^2, the two surface braids are equivalent.

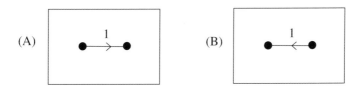

FIGURE 18.13

18.8. A Chart as the Singularity of a Projection

Let Γ be a chart in D_2^2 and let $S = S(\Gamma)$ be a surface braid in $D_1^2 \times D_2^2$ associated with Γ. Take bi-parameterizations $D_1^2 \cong I_1 \times I_2$ and $D_3^2 \cong I_1 \times I_4$. Consider a motion picture $\{b_t\}$ of S along the I_4-direction. Let $\pi : D_1^2 \times I \to I_1 \times I$ and $\pi : D_1^2 \times D_2^2 \to I_1 \times D_2^2$ be projections along I_2. For each $t \in I_4$ such that there are no vertices of Γ in $I_3 \times \{t\}$, we assume that the map $\pi : D_1^2 \times I \to I_1 \times I$ is a generic braid projection of b_t. Then $\pi : D_1^2 \times D_2^2 \to I_1 \times D_2^2$ is a generic projection of S. Let Σ be the singularity set (consisting of double points, triple points and branch points). The image of Σ projected in D_2^2 is exactly the same as the chart Γ. A triple point corresponds to a white vertex, a branch point corresponds to a black vertex, and double curves are mapped to edges of Γ.

REMARK 18.17. In the above correspondence between the singularity set Σ and the chart Γ, a degree-4 vertex of Γ corresponds to just a crossing of the images in D_2^2 of two double curves. There is no meaning in Σ as to the singularity of $\pi(S)$ unless we consider the secondary projection $I_1 \times D_2^2 \to D_2^2$. For this reason, in the original definition of a chart stated in [**312**], degree-4 vertices were not considered to be vertices. In some situations, it is more convenient to consider a chart to be an immersed graph in D_2^2 and to consider degree-4 vertices to be crossings of edges.

18.9. From Charts to Braid Monodromies: Intersection Braid Word

Let Γ be a chart in D_2^2 of degree m. We say that a path $\alpha : [0, 1] \to D_2^2$ is *in general position with respect to* Γ if the following conditions are satisfied.

(1) The image of α is disjoint from the vertex set $V(\Gamma)$ of vertices of Γ.
(2) $\alpha^{-1}(\Gamma)$ is empty or consists of a finite number of interior points of $[0, 1]$, say $t_1, \ldots, t_s$.
(3) For each $j \in \{1, \ldots, s\}$, there is an open regular neighborhood $U(t_j)$ of t_j such that the restriction of α to $U(t_j)$ is an immersion and intersects an edge of Γ transversely at $\alpha(t_i)$.

In the above situation, if $\alpha|_{U(t_j)}$ intersects an edge of Γ labeled i, then we assign the intersection a letter σ_i^ϵ, where $\epsilon = 1$ (resp. $\epsilon = -1$) if the edge of Γ is oriented from right to left (resp. left to right) when one walks along $\alpha|_{U(t_j)}$. Let $w_\Gamma(\alpha)$ be a braid word obtained by reading the letters along α. We call $w_\Gamma(\alpha)$ the *intersection braid word* along α with respect to Γ.

Let S be a surface braid of degree m and $\alpha : [0, 1] \to D_2^2$ a path such that the image is disjoint from the branch point set $\Sigma(S)$ of S. The path α defines a path

$$\rho_S(\alpha) : [0, 1] \to C_m$$

by

$$\rho_S(\alpha)(t) = pr_1(S \cap pr_2^{-1}(\alpha(t))),$$

where C_m is the configuration space $C_m(\text{int}D_1^2)$ and $pr_i : D_1^2 \times D_2^2 \to D_i^2$ is the projection. If $\rho_S(\alpha)(0) = \rho_S(\alpha)(1) = Q_m$, then it represents a geometric braid. We call this the *geometric braid obtained from S by restriction along α* and denote it by $b_S(\alpha)$.

LEMMA 18.18. *Let $S = S(\Gamma)$ be a surface braid associated with Γ. Let $\alpha : [0, 1] \to D_2^2$ be a path in general position with respect to Γ such that $\alpha(0)$, $\alpha(1) \in \partial D_2^2$. Then $w_\Gamma(\alpha)$ is a braid word presentation of $b_S(\alpha)$.*

Proof. From the construction of $S(\Gamma)$, the proof is obvious. ∎

This lemma gives a characterization of $S(\Gamma)$: Let S be a surface braid. Suppose that for any path $\alpha : [0, 1] \to D_2^2$ in general position with respect to Γ with $\alpha(0)$, $\alpha(1) \in \partial D_2^2$, $w_\Gamma(\alpha)$ is a braid word presentation of $b_S(\alpha)$. Then S is isomorphic to $S(\Gamma)$.

If one prefers the construction in Sect. 18.5 to Sect. 18.7, then consider a straight path $\alpha_t : [0, 1] \to D_2^2 = I_3 \times I_4, s \mapsto (s, t)$ for $t \in I_4$. In Fig. 18.11, we parameterize so that the vertical downward direction is the I_3-direction and the horizontal left-to-right direction is the I_4-direction. Thus α_t is a straight path along a vertical line at t-level oriented downward. Recall that, in the figure, there are 10 dotted vertical lines corresponding to $t = j_1, \ldots, t = j_{10}$. The enlarged braid word transformation sequence $w = w_0 \to \cdots \to w_{11} = w'$ given there is exactly the intersection braid words along $\alpha_{j_0}, \ldots, \alpha_{j_{10}}$.

How to obtain a braid monodromy. Let $S = S(\Gamma)$ be a surface braid associated with Γ. The braid monodromy $\rho_S : \pi_1(D_2^2 \setminus \Sigma(S), y_0) \to B_m$ is calculated as follows: Let $\alpha : (I, \partial I) \to (D_2^2 \setminus \Sigma(S), y_0)$ be a loop. Deform it up to homotopy so that it is in general position with respect to Γ. Then $\rho_S([\alpha])$ is a braid expressed by a braid word $w_\Gamma(\alpha)$.

18.10. From Charts to Braid Systems

By use of intersection braid words defined in Sect. 18.9, a braid system is easily obtained from a chart description.

How to obtain a braid system. Let $S = S(\Gamma)$ be a surface braid associated with Γ. Let $\mathcal{A} = (a_1, \ldots, a_n)$ be a Hurwitz arc system with starting point set $\Sigma(S)$ and terminal point y_0. The braid system $(b_1, \ldots, b_n)$ of S is obtained as follows: Let $c_1 \ldots, c_n$ be the loop in $(D_2^2 \setminus \Sigma(S), y_0)$ associated with $\mathcal{A}$ as in Sect. 2.6, which represents the Hurwitz generator system $\eta_1, \ldots, \eta_n$ of $\pi_1(D_2^2 \setminus \Sigma(S), y_0)$. Then b_i is a braid expressed by a braid word $w_\Gamma(c_i)$.

Let $N(y_i)$ be a regular neighborhood of a branch point y_i and let a_i^* be an arc obtained from a_i by removing $a_i \cap \mathrm{int} N(y_i)$. Then

$$w_\Gamma(c_i) = w_\Gamma(a_i^*)^{-1} \, w_\Gamma(\partial N(y_i)) \, w_\Gamma(a_i^*).$$

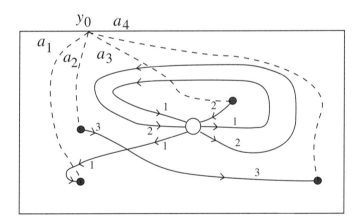

FIGURE 18.14

EXAMPLE 18.19. Let $\mathcal{A} = (a_1, a_2, a_3, a_4)$ be a Hurwitz arc system as in Fig. 18.14 whose starting point set is $\Sigma(S) = \{y_1, y_2, y_3, y_4\}$. Then

$$\begin{aligned}
w_\Gamma(a_1^*) &= \sigma_1 & w_\Gamma(\partial N(y_1)) &= \sigma_1 \\
w_\Gamma(a_2^*) &= 1 & w_\Gamma(\partial N(y_2)) &= \sigma_3^{-1} \\
w_\Gamma(a_3^*) &= \sigma_1 \sigma_2 & w_\Gamma(\partial N(y_3)) &= \sigma_2^{-1} \\
w_\Gamma(a_4^*) &= 1 & w_\Gamma(\partial N(y_4)) &= \sigma_3.
\end{aligned}$$

Thus $w_\Gamma(c_1) = \sigma_1$, $w_\Gamma(c_2) = \sigma_3^{-1}$, $w_\Gamma(c_3) = \sigma_2^{-1}\sigma_1^{-1}\sigma_2^{-1}\sigma_1\sigma_2 = \sigma_2^{-1}$, $w_\Gamma(c_4) = \sigma_3$, and the braid system of the chart is $(\sigma_1, \sigma_3^{-1}, \sigma_2^{-1}, \sigma_3)$.

18.11. Chart Moves

Let Γ and Γ' be charts in D_2^2 of the same degree. Suppose that there is a 2-disk E in D_2^2 such that the loop ∂E is in general position with respect to Γ and Γ', the restriction of Γ to the outside of E is identical with that of Γ', and one of the following conditions, (CI), (CII) and (CIII), is satisfied. Then we say that Γ' is obtained from Γ by a *chart move*, or a *C-move*, of type I, type II or type III. A C-move is also called a CI-move, CII-move or CIII-move.

(CI) There are no black vertices (degree-one vertices) in $\Gamma \cap E$ and $\Gamma' \cap E$.
(CII) $\Gamma \cap E$ and $\Gamma' \cap E$ are as in Fig. 18.15, where $|i - j| > 1$.
(CIII) $\Gamma \cap E$ and $\Gamma' \cap E$ are as in Fig. 18.16, where $|i - j| = 1$.

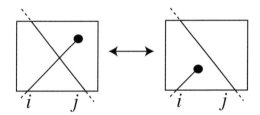

FIGURE 18.15. CII-move

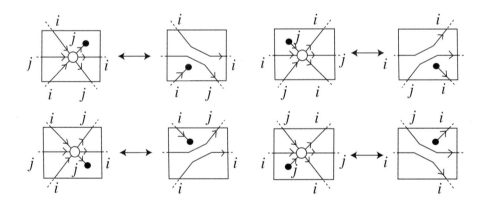

FIGURE 18.16. CIII-move

Some examples of CI-moves are given in Fig. 18.17, where edges in (4) − (7) are oriented in any way such that the conditions on degree-four and degree-six vertices in the definition of a chart are satisfied. It is known that any CI-move is a consequence of the moves illustrated in Fig. 18.17; cf. [**89**].

THEOREM 18.20. *Two charts of degree m present the same, up to equivalence, simple surface braid if and only if they are related by a finite sequence of C-moves (and ambient isotopies of D_2^2).*

Proof. Let S and S' be surface braids $S(\Gamma)$ and $S(\Gamma')$ associated with Γ and Γ'.

First we prove the if part. It is sufficient to prove this in a case where Γ' is obtained from Γ by a C-move. Let E be a 2-disk where the C-move is applied.

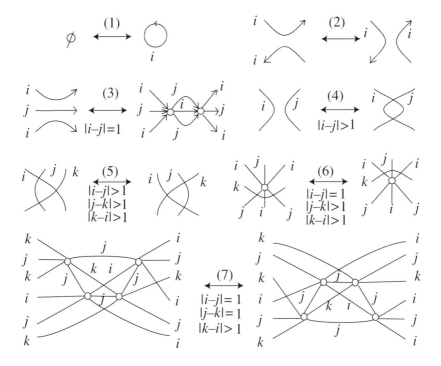

FIGURE 18.17. Some CI-moves

Consider a Hurwitz arc system $\mathcal{A} = (a_1, \ldots, a_n)$ for $\Sigma(S) = \{y_1, \ldots, y_n\}$ in D_2^2, and let $c_1, \ldots, c_n$ be loops as in Sect. 18.10. Recall that a braid system $(b_1, \ldots, b_n)$ of S is calculated by $b_k = w_\Gamma(c_k) = w_\Gamma(a_k^*)^{-1} w_\Gamma(\partial N(y_k)) w_\Gamma(a_k^*)$. If the C-move is of type I, we may assume that each arc a_k is disjoint from E. Then the C-move does not change the intersection braid word $w_\Gamma(c_k)$. If the move is of type II or type III, then E contains a branch point (a black vertex), say y_k. Deforming a_k near E, we may assume that $w_\Gamma(c_k)$ and $w_{\Gamma'}(c_k)$ are related by one of the following:

$$w_\Gamma(c_k) = w^{-1} \sigma_j^{-\epsilon} \sigma_i^\delta \sigma_j^\epsilon w \qquad \longleftrightarrow \qquad w_{\Gamma'}(c_k) = w^{-1} \sigma_i^\delta w$$
$$w_\Gamma(c_k) = w^{-1} \sigma_j^{-\epsilon} \sigma_i^{-\epsilon} \sigma_j^\delta \sigma_i^\epsilon \sigma_j^\epsilon w \qquad \longleftrightarrow \qquad w_{\Gamma'}(c_k) = w^{-1} \sigma_i^\delta w.$$

Thus $w_\Gamma(c_k)$ and $w_{\Gamma'}(c_k)$ represent the same m-braid. The other $w_\Gamma(c_1), \ldots, w_\Gamma(c_n)$ are the same as $w_{\Gamma'}(c_1), \ldots, w_{\Gamma'}(c_n)$. Thus Γ and Γ' have the same braid system. Therefore S and S' are equivalent.

Now we prove the only if part. Let $\{h_u\}_{u \in [0,1]}$ be a fiber-preserving ambient isotopy of $D_1^2 \times D_2^2$ which carries S to S' and let $\{\underline{h}_u\}$ be the ambient isotopy of D_2^2 induced by $\{h_u\}$. Put $\Gamma'' = \underline{h}_1(\Gamma')$. Consider a fiber-preserving ambient isotopy $\{g_u\}$ of $D_1^2 \times D_2^2$ such that $g_u = h_u \circ (\mathrm{id} \times \underline{h}_u^{-1})$. By this isotopy, $S = S(\Gamma)$ and $S'' = S(\Gamma'')$ are isomorphic. Note that Γ'' is ambient isotopic to Γ' in D_2^2. Therefore, without loss of generality, we may assume that S and S' are isomorphic.

Let $\mathcal{A} = (a_1, \ldots, a_n)$ be a Hurwitz arc system for $\Sigma(S) = \Sigma(S')$, and let $c_1, \ldots, c_n$ be loops as before. We assume that each a_k is in general position with respect to Γ and Γ'.

Since S and S' are isomorphic, for each $k \in \{1, \ldots, n\}$, $w_\Gamma(c_k)$ and $w_{\Gamma'}(c_k)$ represent the same m-braid.

We introduce the notion of a symmetric braid word.

DEFINITION 18.21. A braid word is a *symmetric braid word* if it is $x^{-1}\sigma_i^\epsilon x$ for some braid word x. Two symmetric braid words are *symmetrically equivalent* if one is transformed into the other by a finite sequence of the following transformations:

(T1) $x^{-1}\sigma_j^\epsilon x \leftrightarrow y^{-1}\sigma_j^\epsilon y$, where x, y are braid words which represent the same m-braid, and $\epsilon \in \{1, -1\}$;

(T2) $x^{-1}\sigma_i^{-\epsilon}\sigma_j^\delta\sigma_i^\epsilon x \leftrightarrow x^{-1}\sigma_j^\delta x$, where x is a braid word, $|i - j| \neq 1$, and $\epsilon, \delta \in \{1, -1\}$;

(T3) $x^{-1}\sigma_i^{-\epsilon}\sigma_j^\delta\sigma_i^\epsilon x \leftrightarrow x^{-1}\sigma_j^\epsilon\sigma_i^\delta\sigma_j^{-\epsilon}x$, where x is a braid word, $|i - j| = 1$, and $\epsilon, \delta \in \{1, -1\}$.

LEMMA 18.22. *Two symmetric braid words represent the same m-braid if and only if they are symmetrically equivalent.*

For a proof of this lemma, refer to [**321**]. Assuming this lemma, we continue the proof of the theorem. Recall that $w_\Gamma(c_k) = w_\Gamma(a_k^*)^{-1}w_\Gamma(\partial N(y_k))w_\Gamma(a_k^*)$. Thus $w_\Gamma(c_k)$ is a symmetric braid word, and so is $w_{\Gamma'}(c_k)$. Since they represent the same m-braid, there is a finite sequence of symmetric braid words from $w_\Gamma(c_k)$ to $w_{\Gamma'}(c_k)$ related by transformations (T1) $\sim$ (T3). According to this sequence, we apply C-moves and transform Γ so that $w_\Gamma(a_k^*) = w_{\Gamma'}(a_k^*)$ and $w_\Gamma(\partial N(y_k)) = w_\Gamma(\partial N(y_k))$.

Fig. 18.18(1) depicts an insertion of $\sigma_i\sigma_i^{-1}$ in $w_\Gamma(a_k^*)$, where a dotted line stands for a_k. By reversing the orientation of the small loop, we can also insert $\sigma_i^{-1}\sigma_i$. Fig. 18.18(2) depicts a deletion of $\sigma_i\sigma_i^{-1}$ in $w_\Gamma(a_k^*)$. A deletion of $\sigma_i^{-1}\sigma_i$ is realized similarly. Fig. 18.18(3) depicts a replacement of $\sigma_i\sigma_j$ by $\sigma_j\sigma_i$ in $w_\Gamma(a_k^*)$. Fig. 18.18(4) depicts a replacement of $\sigma_i\sigma_j\sigma_i$ by $\sigma_j\sigma_i\sigma_j$ in $w_\Gamma(a_k^*)$. Applying these C-moves, we can realize a transformation (T1) on a symmetric braid word $w_\Gamma(c_k)$ by a sequence of CI-moves. For a transformation (T2), apply a C1-move as in Fig. 18.19(1) if $i = j$, or apply a CII-move as in Fig. 18.19(2). For a transformation (T3), apply a C1-move followed by a CIII-move as in Fig 18.19(3).

Therefore, for each $k \in \{1, \ldots, n\}$, we can transform Γ so that $w_\Gamma(a_k^*) = w_{\Gamma'}(a_k^*)$ and $w_\Gamma(\partial N(y_k)) = w_{\Gamma'}(\partial N(y_k))$. The restriction of Γ to $N(y_k) \cup a_k^*$ is the same as that of Γ'. Let E be a 2-disk obtained from $D_2^2 \cup_{k=1}^n \mathrm{int} N(y_k)$ by cutting along $a_1^*, \ldots, a_n^*$. It contains no black vertices of Γ and Γ'. We can change Γ into Γ' by a CI-move on E. This completes the proof of Theorem 18.20. ∎

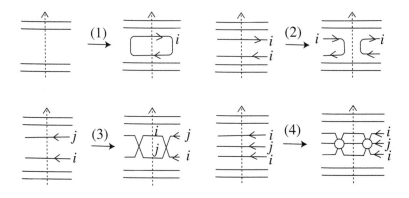

FIGURE 18.18. Realization of T1

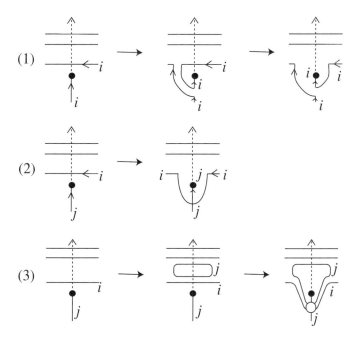

FIGURE 18.19. Realization of T2 and T3

EXAMPLE 18.23. Fig. 18.20 is a deformation by chart moves. A surface braid whose motion picture is as in Fig. 18.12 and which is associated with a chart in Fig. 18.20(1) is equivalent to a surface braid associated with a chart Fig. 18.20(4) whose motion picture is Fig. 18.1.

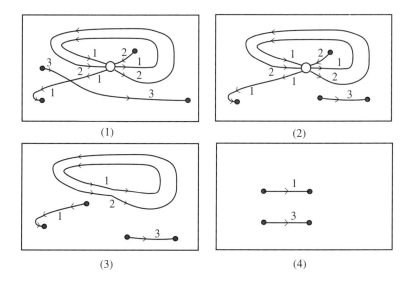

FIGURE 18.20. Deformation by chart moves

18.12. Further Examples of Chart Moves

LEMMA 18.24. *If $|i - j| = 1$, then the operations in* Fig. 18.21 *are realized by chart moves.*

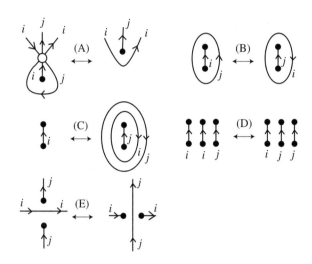

FIGURE 18.21

Proof. See Fig. 18.22. ∎

EXERCISE 18.25. Fig. 18.23 is a chart of degree 4 without black vertices. By definition, this chart is eliminated by a single CI-move. Show that it can be eliminated by a combination of the CI-moves listed in Fig. 18.17.

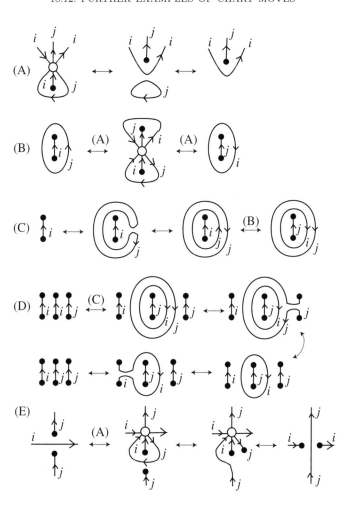

FIGURE 18.22

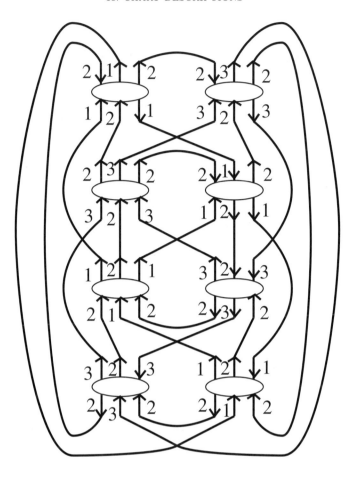

FIGURE 18.23. A chart without black vertices

Non-simple Surface Braids

This chapter is devoted to research on non-simple surface braids (cf. [**320**]). It is written for researchers. The reader who wants to read quickly may skip over this chapter.

19.1. Singular Points in a Motion Picture

Let S be a surface braid of degree m in $D_1^2 \times D_2^2$. Fix an identification of D_2^2 and $I_3 \times I_4$ and let $\{b_t\}$ $(t \in I_4)$ be the motion picture of S. Recall that if $x \in b_t$ is a singular point of the geometric singular m-braid b_t for some exceptional value $t \in I_4$, then it is a singular point of the surface braid S. Conversely, if $x \in S$ is a singular point of S, then it is a singular point of b_t for some $t \in I_4$. The degree $\deg(b_t; x)$ of $x \in b_t$ as a singular point of the singular m-braid b_t equals the local degree $\deg(S; x)$ of the surface braid S at x. For example, see Fig. 14.2. The point x_3 is a singular point of degree 3.

Let x be a singular point of S and p the branch point of S with $p = pr_2(x)$, where $pr_2 : D_1^2 \times D_2^2 \to D_2^2$. Let $p = (p_3, p_4)$ under the bi-parametrization $D_2^2 = I_3 \times I_4$. Let $\Delta = J_3 \times J_4$ be a regular neighborhood of p in $D_2^2 = I_3 \times I_4$, where J_3 and J_4 are subintervals of I_3 and I_4. The restriction of S to $pr_2^{-1}(\Delta) = D_1^2 \times \Delta$ is a braided surface with a single branch point. Let $\{\beta_t\}$ $(t \in J_4)$ be the motion picture of $S \cap (D_1^2 \times \Delta)$ with the time direction J_4. Let $J_4 = [t', t'']$.

LEMMA 19.1. *Deforming S up to isomorphism, we can assume that $\partial \beta_t = Q_m \times \partial J_3$ for each $t \in J_4$ and*

(1) $\beta_{t'} = Q_m \times I_3$, *or*

(2) $\beta_{t''} = Q_m \times I_3$.

Moreover, if we assume (1), the m-braid $\beta_{t''}$ is the inverse of a local braid monodromy at p, and if we assume (2), $\beta_{t'}$ is a local braid monodromy at p. In particular $\beta_{t'}$ and $\beta_{t''}$ are elements of A_m.

EXERCISE 19.2. Prove this lemma. [*Hint*: Recall Sect. 16.6.]

The second row of Fig. 19.1 is an example of a local motion picture $\{\beta_t\}$ at a branch point whose branch type is $(3, 2, 1)$. In the dotted box, three (or two, resp.) strands are attached such that there is a geometric 3-braid (resp. 2-braid) whose closure is an unknot in the 3-sphere. The first row is such an example.

19.2. Reduction of the Singular Index

Let $m \geq 3$ and suppose that an m-braid b is obtained from an $(m - 1)$-braid b' by a (positive/negative) stabilization, i.e., $b = b'\sigma_{m-1}^\epsilon$. Note that $b \in A_m$ if and only if $b' \in A_{m-1}$. Assume $b \in A_m$ and consider a braided surface $S = C(\ell)$ in $D_1^2 \times D_2^2$ with a single branch point whose braid system is (b), where ℓ is a complete

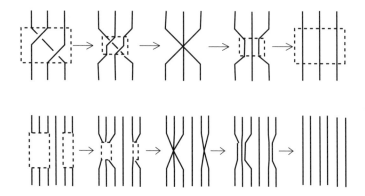

FIGURE 19.1

split closed braid in $D_1^2 \times \partial D_2^2$ and $C(\ell)$ is a multiple cone as in Sect. 16.5. Let $S' = S(b', \sigma_{m-1}^\epsilon)$ be a braided surface with two branch points whose braid system is $(b', \sigma_{m-1}^\epsilon)$ and $\partial S = \partial S' = \ell$. In Fig. 19.2, (1) and (2) are motion pictures of S and S' provided ℓ is connected.

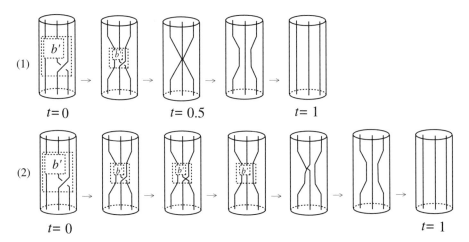

FIGURE 19.2. Reduction of singular index

LEMMA 19.3. *In the above situation, if $\ell = \partial S$ is connected, then $S = C(b)$ is braid ambient isotopic to $S' = S(b', \sigma_{m-1}^\epsilon)$.*

Proof. Let $\{\ell_t\}_{t \in [0,1]}$ be a one-parameter family of (singular) closed braids such that $\ell_0 = \ell$, $\ell_1 = k_1 \cup k_2$ which is the split union of a closed $(m-1)$-braid k_1 corresponding to b' and a trivial loop k_2, and a string recombination as in Fig. 14.4(C) or (D) at $t = 0.5$ occurs. Let N_1 and N_2 be disjoint 2-disks in D_1^2 such that $pr_1(k_1) \subset \mathrm{int}(N_1)$ and $pr_1(k_2) \subset \mathrm{int}(N_2)$. Assume $D_2^2 = \cup_{r \in [0,1]} S_r^1$, where S_r^1 is a circle with radius r in $\mathbb{R}^2$ and $O = S_0^1$ is the origin. Let z_0 be a point of N_1 and $p = (z_0, O) \in D_1^2 \times D_2^2$. We assume that the braided surface $S = C(\ell)$ is the cone $p * \ell$. For each $r \in [0,1]$, let $\lambda_r : D_1^2 \times S^1 \to D_1^2 \times S^1$ be a map defined by

$$\lambda_r(z, w) + ((1 - r)z_0 + rz, w) \quad \text{for } (z, w) \in D_1^2 \times S^1.$$

Then $S \cap (D_1^2 \times S_r^1) = \lambda_r(\ell)$ for $r \in [0, 1]$. Define a braid isotopy $\{S_t\}_{t \in [0,1]}$ with $S_0 = S$ by

$$S_t \cap (D_1^2 \times S_r^1) = \begin{cases} \lambda_r(k_1) \cup \lambda_{t/2}(k_2) & \text{for } r \in [0, t/2] \\ \lambda_r(\ell_{2(t-r)/t}) & \text{for } r \in (t/2.t) \\ \lambda_r(\ell) & \text{for } r \in [t, 1]. \end{cases}$$

Since S_1 has two branch point and a braid system $(b', \sigma_{m-1}^\epsilon)$, it is equivalent to S'. The trace $\cup_{t \in [0,1]} S_t \times \{t\}$ in $D_1^2 \times D_2^2 \times [0, 1]$ is locally flat, and by the isotopy extension theorem (cf. [**771**]) we have that S and S' are braid ambient isotopic. ∎

EXERCISE 19.4. The assertion of Lemma 19.3 holds without the assumption that ℓ is connected. Prove this by using Lemma 19.3.

THEOREM 19.5. *Any non-simple surface braid of degree 3 is braid ambient isotopic to a simple surface braid.*

Proof. Let S be a surface braid of degree 3. Suppose that there is a non-simple branch point y. The branch type of y is (3). For a regular neighborhood Δ of y in D_2^2, the restriction $S \cap D_1^2 \times \Delta$ is a braided surface with a single branch point y. The boundary, say ℓ, must be a trivial knot in the 3-sphere $\partial(D_1^2 \times \Delta)$. It is known that such a closed 3-braid ℓ is obtained from a closed 2-braid by a stabilization, [**686**]. Applying Lemma 19.3, the branch point is changed into two simple branch points. ∎

19.3. Fission/Fusion of Branch Points

LEMMA 19.6. *Let S and S' be surface braids of degree m. The following conditions are mutually equivalent.*

(1) *S and S' are equivalent.*
(2) *S and S' are braid ambient isotopic by a braid ambient isotopy $\{h_t\}_{t \in [0,1]}$ of $D_1^2 \times D_2^2$ such that for all $t \in [0, 1]$, $h_t(S)$ has n branch points.*
(3) *S and S' are braid isotopic by a braid isotopy $\{S_t\}_{t \in [0,1]}$ such that for all $t \in [0, 1]$, $h_t(S)$ has n branch points.*

Proof. (1) $\Rightarrow$ (2) $\Rightarrow$ (3) is trivial. We prove that (3) $\Rightarrow$ (1). Since the one-parameter family $\{S_t\}$ has a constant number of branch points, there exist n paths $\xi_1, \ldots, \xi_n$ in D_2^2 such that the branch point set $\Sigma(S_t)$ of S_t is $\{\xi_1(t), \ldots, \xi_n(t)\}$, for each t (cf. [**36**] for example). For each $t \in [0, 1]$, let $N(\xi_i(t))$ be a regular neighborhood of $\xi_i(t)$ in D_2^2. There exists an open regular neighborhood O_t of t in $[0, 1]$ with $\xi_i(O_t) \subset N(\xi_i(t))$ for every $i \in \{1, \ldots, n\}$. Note that for any $t' \in O_t$, S_t and $S_{t'}$ are equivalent. This is seen as follows: Let $\mathcal{A} = (a_1, \ldots, a_n)$ and $\mathcal{A}' = (a_1', \ldots, a_n')$ be Hurwitz arc systems for $\Sigma(S_t)$ and $\Sigma(S_{t'})$ respectively such that for each $i \in \{1, \ldots, n\}$, $a_i \setminus \text{int} N(\xi_i(t)) = a_i' \setminus \text{int} N(\xi_i(t))$. The braid systems of S_t and $S_{t'}$ associated with $\mathcal{A}$ and $\mathcal{A}'$ are the same. Therefore S_t and $S_{t'}$ are equivalent (Theorem 17.17). Using compactness of $[0, 1]$, we see that S_0 and S_1 are equivalent. ∎

LEMMA 19.7. *Let S and S' be braid ambient isotopic or braid isotopic. There is a finite sequence $S = S_0, S_1, \ldots, S_s = S'$ such that for each $i \in \{1, \ldots, s\}$, S_i is equivalent to S_{i-1} or obtained from S_{i-1} by a branch point fission/fusion.*

Proof. Let $\{S_t\}_{t \in [0,1]}$ be a braid isotopy between S and S'. We call $t \in [0, 1]$ a *regular value* if there is a regular neighborhood N_t in $[0, 1]$ such that for any

$t' \in N_t$, S_t and $S_{t'}$ are equivalent; otherwise it is an *exceptional value*. For each exceptional value $t \in [0, 1]$, let N_t be a regular neighborhood of t. Deform $\{S_t\}_{t \in [0,1]}$ such that $S_{t'} = S_t$ for all $t' \in N_t$. If some branch point fusions and fissions occur simultaneously for an exceptional value, then deform the isotopy locally such that they occur at distinct values. So we may assume that a single branch point fission/fusion occurs at each exceptional value. By the previous lemma, we have the result. ∎

For an m-braid $b \in B_m$, we define the *singular index*, $\tau(b)$, by $m - c(\ell)$, where $c(\ell)$ is the number of components of a closed m-braid ℓ obtained from b.

DEFINITION 19.8. Let $\overrightarrow{b} = (b_1, \ldots, b_n) \in P^n(A_m)$ and $\overrightarrow{b}' = (b'_1, \ldots, b'_{n+q}) \in P^n(A_m)$ for some $q > 0$. $\overrightarrow{b}'$ is obtained from $\overrightarrow{b}$ by an *Euler fission* if $b_i = b'_i$ for i with $1 \le i \le n-1$, $b_n = b'_n b'_{n+1} \cdots b'_{n+q}$ in the braid group B_m and $\tau(b_n) = \sum_{i=0}^{q} \tau(b'_{n+i})$. The inverse is an *Euler fusion*.

LEMMA 19.9. *If S' is obtained from S by a branch point fission, then there exist braid systems $\overrightarrow{b}$ and $\overrightarrow{b}'$ of S and S' such that $\overrightarrow{b}'$ is obtained from $\overrightarrow{b}$ by an Euler fission.*

EXERCISE 19.10. Prove this lemma. [*Hint*: For a braid system $(b_1, \ldots, b_n)$ of S, by the Riemann-Hurwitz formula, the Euler characteristic $\chi(S)$ of S is given by $m - \sum_{i=1}^{n} \tau(b_i)$.]

PROPOSITION 19.11. *Let $\overrightarrow{b}$ and $\overrightarrow{b}'$ be braid systems of S and S'. If S and S' are braid ambient isotopic or braid isotopic, then there is a finite sequence $\overrightarrow{b} = \overrightarrow{b}_0, \overrightarrow{b}_1, \ldots, \overrightarrow{b}_s = \overrightarrow{b}'$ such that for each $i \in \{1, \ldots, s\}$, $\overrightarrow{b}_i$ is slide equivalent to $\overrightarrow{b}_{i-1}$ or obtained from $\overrightarrow{b}_{i-1}$ by an Euler fission/fusion.*

Proof. This follows from the above two lemmas and Theorem 17.17. ∎

EXERCISE 19.12. Let S and S' be surface braids of degree m. The following conditions are mutually equivalent.

(2) S and S' are equivalent.

(3*) S and S' are braid ambient isotopic by a braid ambient isotopy $\{h_u\}_{u \in [0,1]}$ of $D_1^2 \times D_2^2$ such that fissions/fusions of branch points do not occur.

(3*') S and S' are braid ambient isotopic by a braid ambient isotopy $\{h_u\}_{u \in [0,1]}$ of $D_1^2 \times D_2^2$ such that for all $u \in [0, 1]$, $h_u(S)$ has the same number of branch points.

(3**) S and S' are braid ambient isotopic by a braid ambient isotopy $\{h_u\}_{u \in [0,1]}$ of $D_1^2 \times D_2^2$ such that fissions/fusions of singular points do not occur.

(4*) S and S' are braid isotopic by a braid isotopy $\{S_u\}_{u \in [0,1]}$ such that fissions/fusions of branch points do not occur.

(4*') S and S' are braid isotopic by a braid isotopy $\{S_u\}_{u \in [0,1]}$ such that for all $u \in [0, 1]$, $h_u(S)$ has the same number of branch points.

(4**) S and S' are braid isotopic by a braid isotopy $\{S_u\}_{u \in [0,1]}$ such that fissions/fusions of singular points do not occur.

19.4. Stable Non-simple Surface Braids

THEOREM 19.13 ([**320**]). *For a positive integer $m \ge 4$ and a positive even number n, there is a non-simple surface braid S of degree m with n branch points* *such that any surface braid which is braid isotopic to S is equivalent to S. In particular, S is never braid isotopic to a simple surface braid.*

Proof. Let b be the 4-braid $\sigma_3^{-2}\sigma_2\sigma_3^{-1}\sigma_2\sigma_1^3\sigma_2^{-1}\sigma_1\sigma_2^{-1}$. The closed 4-braid in $D_1^2 \times \partial D_2^2$ obtained from b is connected, and it is an unknot in $\partial(D_1^2 \times D_2^2)$. This is an element of A_4 with singular index $\tau(b) = 3$. Let S_0 be a braided surface of degree 4 with a single branch point y whose braid monodromy is (b). We first assert that no branch point fission is applicable to S. Suppose that S' is obtained from S by a branch point fission and let $\overrightarrow{b}' = (b_1', \ldots, b_s')$ be obtained from (b) by an Euler fission. Since $\tau(b) = 3$, there are two possibilities; (1) $s = 3$ and $\tau(b_1') = \tau(b_2') = \tau(b_3') = 1$ or (2) $s = 2$ and $\{\tau(b_1'), \tau(b_2')\} = \{1, 2\}$.

Suppose that case (1) occurs. Each b_i' is an element of SA_4, which is a conjugate of σ_1 or σ_1^{-1}. Consider the abelianization map $e : B_4 \to \mathbb{Z}, \sigma_i \mapsto 1$. Since $e(b) = 1$, two of b_i' $(i = 1, 2, 3)$ are conjugations of σ_1 and the other is a conjugation of σ_1^{-1}. Thus b is conjugate to a 4-braid b' of the form $w_1^{-1}\sigma_1 w_1 w_2^{-1}\sigma_1 w_2\sigma_1^{-1}$ for some $w_1, w_2 \in B_4$. Let $\phi : B_4 \to SL(2; \mathbb{Z})$ be the representation defined by

$$\phi(\sigma_1) = \phi(\sigma_3) = \begin{pmatrix} 1 & 1 \\ 0 & 1 \end{pmatrix} \quad \text{and} \quad \phi(\sigma_2) = \begin{pmatrix} 1 & 0 \\ -1 & 1 \end{pmatrix}.$$

Then

$$\phi(b) = \begin{pmatrix} 73 & 46 \\ -27 & -17 \end{pmatrix} \quad \text{and} \quad \mathrm{Tr}(\phi(b)) = 56 \equiv 1 \pmod 5.$$

On the other hand, put

$$\phi(w_1) = \begin{pmatrix} a & b \\ c & d \end{pmatrix} \quad \text{and} \quad \phi(w_2) = \begin{pmatrix} e & f \\ g & h \end{pmatrix}.$$

Then $\mathrm{Tr}(\phi(b')) = 2 + c^2 + g^2 - cdg^2 - d^2g^2 + c^2gh + 2cdgh - c^2h^2$. Since b and b' are conjugate, $\mathrm{Tr}(\phi(b)) = \mathrm{Tr}(\phi(b'))$. This holds only if $c \equiv d \equiv 0 \pmod 5$ or $g \equiv h \equiv 0 \pmod 5$, which contradicts that $\phi(w_1), \phi(w_2) \in SL(2; \mathbb{Z})$.

Suppose that case (2) occurs. If S' has three singular points, then they are all simple singular points; i.e. the local degrees are two. Two of them are in the same fiber $D_1^2 \times \{y_1\}$ and the other is in another fiber $D_1^2 \times \{y_2\}$ for some $y_1, y_2 \in D_2^2$. Deform S' by a braid isotopy so that all singular points are in distinct fibers. Now the situation is reduced to case (1). Thus this never happens. If S' has two singular points, then one of them is a simple singular point and the other is a non-simple singular point of local degree three. By the same reasoning as Theorem 19.5, the latter one can be changed into two simple singular points. By the same reason, this yields a contradiction.

Therefore we see that no Euler fission can be applied to (b) and hence no branch point fission can be applied to S_0.

For $m \geq 4$, we regard the above b as an m-braid. It is an element of A_m with $\tau(b) = m - 3$. Let S be a surface braid of degree m whose braid system is $(b, \ldots, b, b^{-1}, \ldots, b^{-1})$, where the first $n/2$ are $b \in B_m$ and the others are $b^{-1} \in B_m$. Suppose that we can apply a branch point fusion to S. Let s_+ and s_- be the numbers of branch points whose monodromies are b and b^{-1} respectively such that they are fused into a single branch point by the fusion. Then the surface braid S' has a braid system $(b, \ldots, b, b^{-1}, \ldots, b^{-1}, b^{s_+ - s_-})$, where the first $n/2 - s_+$ are b and the second $n/2 - s_-$ are b^{-1}. (Note that b and b^{-1} commute with each other.) By the condition of an Euler fission, $\tau(b^{s_+ - s_-}) = s_+\tau(b) + s_-\tau(b^{-1}) = 3(s_+ + s_-) > 3$.

On the other hand, since $b^{s_+ - s_-}$ is the split union of a 4-braid and a trivial $(m-4)$-braid, the singular index $\tau(b^{s_+ - s_-}) \leq 3$. This is a contradiction. Thus we cannot apply a branch point fusion to S.

Suppose that we can apply a branch point fission to S. Let y be a branch point of S which is decomposed, and let Δ be a regular neighborhood of y. The restriction of S to $D_1^2 \times \Delta$ is a braided surface of degree m which is the split union of a braided surface S_0 of degree 4 (or its mirror image) and a trivial braided surface of degree $m - 4$. If a branch point fission happens, then it is for S_0 (or its mirror image). This is impossible. ∎

In [**320**] it is shown that there exist infinitely many non-simple surface braids S as in Theorem 19.13.

REMARK 19.14. Recall that the set of simple branched coverings is open and dense in the whole space of branched covering maps (cf. [**36**, **37**] for example). This fact implies that the set of simple surface braids of degree m is open in the whole space of surface braids of degree m. If $m \leq 3$, the set of simple surface braids of degree m is dense by Theorem 19.5. If $m \geq 4$, it is not dense by Theorem 19.13.

CHAPTER 20

1-Handle Surgery on Surface Braids

20.1. Nice 1-Handles

In Sect. 11.1, we introduced the notions of a 1-handle and a 1-handle surgery for surface links in $\mathbb{R}^4$. Similarly, they are defined for surfaces embedded in 4-manifolds. In this section we consider 1-handles for surface braids in $D_1^2 \times D_2^2$.

Let S be a surface braid in $D_1^2 \times D_2^2$ of degree $m \geq 2$.

DEFINITION 20.1. A 1-handle B attaching to S is a *nice 1-handle* if a core α of B is in a fiber $pr_2^{-1}(y) = D_1^2 \times \{y\}$ for a point $y \in \mathrm{int}D_2^2$, B misses the singular points of S, and the surgery result $\mathrm{h}(S; B)$ is a surface braid.

LEMMA 20.2. *Let S_0 be the trivial surface braid $Q_2 \times D_2^2$ of degree 2, and let α_0 be a straight segment in $pr_2^{-1}(y)$ for a point $y \in \mathrm{int}D_2^2$ connecting the two points of $S \cap pr_2^{-1}(y)$. There is a nice 1-handle B whose core is α_0. Moreover, for any nice 1-handle B attaching to S_0, $\mathrm{h}(S_0; B)$ is unique up to equivalence.*

Proof. The 1-handle illustrated in Fig. 20.1 (via the motion picture method) is the desired one. The core α_0 is the bold segment in the figure, and the surgery result $\mathrm{h}(S_0; B)$ is a surface braid as in Fig. 14.1(A). The latter assertion follows from Proposition 17.18; a surface braid of degree 2 is uniquely determined up to equivalence from the number of branch points. (Since $\chi(\mathrm{h}(S_0; B)) = 0$, $\mathrm{h}(S_0; B)$ has two branch points.) ∎

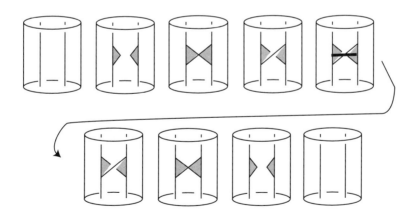

FIGURE 20.1. A nice 1-handle

PROPOSITION 20.3. *Let S be a surface braid and let α be a simple arc in a fiber $pr_2^{-1}(y)$ for a point $y \in \mathrm{int}D_2^2$. If α misses the singular points of S, then there is a* *nice 1-handle attaching to S with core α, say B. Moreover the equivalence class of* $\mathrm{h}(S; B)$ *depends on α but not on B.*

Proof. Modifying S and α by a fiber-preserving ambient isotopy of $D_1^2 \times D_2^2$, we may assume that the restriction of S to a regular neighborhood of α is S_0 and α is α_0 as in the previous lemma. A desired 1-handle is obtained from a 1-handle in the lemma by the inverse ambient isotopy. Uniqueness follows from the uniqueness in the lemma. ∎

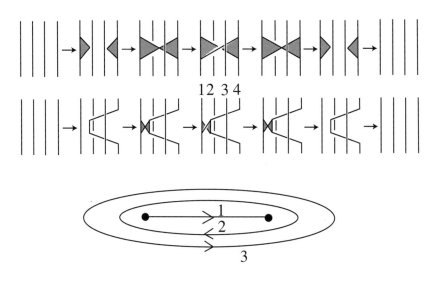

FIGURE 20.2. A nice 1-handle

Recall that any oriented 1-handle is determined up to local ambient isotopy from its core. So any nice 1-handle can be regarded as the image of a 1-handle as in Fig. 20.1 by a fiber-preserving ambient isotopy of $D_1^2 \times D_2^2$.

EXAMPLE 20.4. See the top row of Fig. 20.2. It is a motion picture of a nice 1-handle. It is equivalent to the 1-handle that is in the second row of the picture. The surgery result is described locally by a chart as in the third row.

20.2. Free Edges and Oval Nests

DEFINITION 20.5. A *free edge* of a chart is an edge whose endpoints are black vertices. An *oval nest* is a free edge together with some concentric simple loops.

Let S be a surface braid and let Γ be its chart description. If Γ' is a chart which is obtained from Γ by inserting a free edge with label i, the surface braid S' described by Γ' is obtained from S by surgery along a nice 1-handle as in Fig. 20.1. If Γ' is a chart which is obtained from Γ by inserting an oval nest, then the surface braid S' described by Γ' is obtained from S by surgery along a nice 1-handle; see Fig. 20.2 for example. The converse is also true.

Part 4

Braid Presentation of Surface Links

CHAPTER 21

The Normal Braid Presentation

21.1. Simple Bands

DEFINITION 21.1. A *simple band* attaching to a geometric m-braid $b \subset D^2 \times I$ is a band which is contained in a subcylinder $D' \times I' \subset D^2 \times I$ as in Fig. 21.1. A *birth-type* simple band is (A) or (B) and a *death-type* simple band is (C) or (D). (In the figure, $(-)$ stands for an attaching arc of a band and $(+)$ is an arc appearing after surgery.)

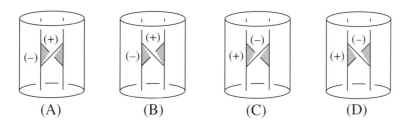

(A)   (B)   (C)   (D)

FIGURE 21.1. Simple bands

DEFINITION 21.2. A band is a *pre-simple band* if it is ambient isotopic to a simple band by a fiber-preserving ambient isotopy of $D^2 \times I$.

A core of a pre-simple band is an arc in a fiber $D^2 \times \{y\}$ for a point $y \in I$.

Note that, even if it is of death-type, any pre-simple band is ambient isotopic to a *birth-type* simple band by a fiber-preserving ambient isotopy of $D^2 \times I$. See Fig. 21.2.

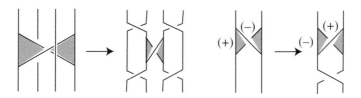

FIGURE 21.2. Pre-simple bands

When we use braid word expression, surgery along a simple band induces an insertion or deletion of a single letter σ_i or σ_i^{-1} for some i. Surgery along a pre-simple band induces an insertion or deletion of a conjugate of a single letter σ_i or σ_i^{-1}, i.e., an element of SA_m.

Fig. 21.3 shows the motion pictures of surgery along simple bands. The surface described by the motion picture as in (A) is ambient isotopic to the surface described in (A) of Fig. 14.4. Similarly (B), (C) and (D) of Fig. 21.3 are ambient isotopic to (B), (C) and (D) of Fig. 14.4.

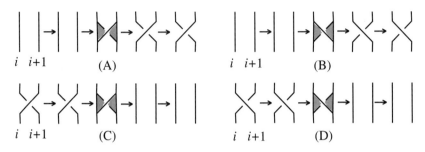

FIGURE 21.3. Surgery along simple bands

Let S be a simple surface braid in $D_1^2 \times D_2^2$ and let $\{b_t\}$ $(t \in I_4)$ be a motion picture of S. We may assume that the local motion picture around each singular point is one of (A), (B), (C), and (D) of Fig. 14.4. By an ambient isotopy of $D_1^2 \times D_2^2$, the surface braid S is deformed such that the neighborhood of each singular point is one of (A), (B), (C), and (D) of Fig. 21.3. S is no longer a surface braid. However by reversing this process, we may regard S as a simple surface braid. It is often convenient to describe a simple surface braid in such a form using simple bands (or pre-simple bands).

For example, the surface in $D_1^2 \times D_2^2$ illustrated in Fig. 21.4 is ambient isotopic to the surface braid illustrated in Fig. 14.3.

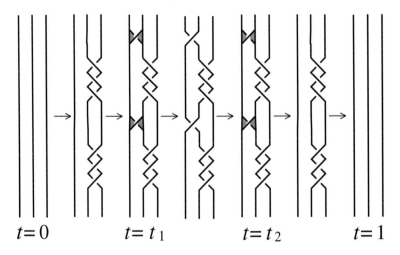

$t = 0$ $t = t_1$ $t = t_2$ $t = 1$

FIGURE 21.4

21.2. The Normal Braid Form of a Surface Link

Let us identify $\mathbb{R}^3 \backslash z$-axis with $\mathbb{R}_+^2 \times S^1$ in the usual way, where $\mathbb{R}_+^2$ is the half-plane $\{(x, 0, z) \in \mathbb{R}^3 | x > 0\}$.

DEFINITION 21.3. A surface link is in a *normal braid form* of degree m if there exist closed braids L_- and L_+ in $\mathbb{R}^3$ about the z-axis and a band set $\mathcal{B}$ attaching to L_- such that

1. each band of $\mathcal{B}$ is a birth-type simple band, and L_+ is obtained from L_- by surgery along $\mathcal{B}$,
2. the images of the bands of $\mathcal{B}$ under the projection $\mathbb{R}_+^2 \times S^1 \to S^1$ are mutually disjoint arcs in S^1,
3. L_- and L_+ are trivial closed braids, and
4. F is a closed realizing surface of $L_- \to L_+$.

A surface link in a normal braid form of degree m looks as in Fig. 21.5 for some n. L_- and L_+ are closed trivial m-braids. $\mathcal{B}$ is a union of n birth-type simple bands which correspond to $\sigma_{n_i}^{\epsilon_i}$ for some $n_i \in \{1, \dots, m-1\}$ and $\epsilon_i \in \{\pm 1\}$. Since L_- and L_+ are trivial closed m-braids, the following conditions are satisfied:

1. $b_1 \dots b_n = 1 \in B_m$,
2. $\sigma_{n_1}^{\epsilon_1} b_1 \dots \sigma_{n_n}^{\epsilon_n} b_n = 1 \in B_m$.

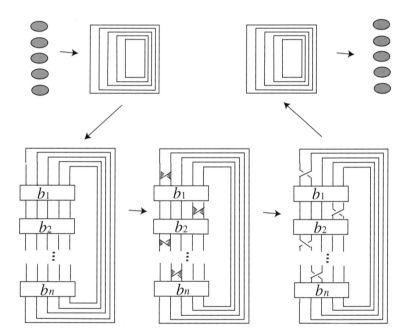

FIGURE 21.5. Normal braid form

21.3. The Normal Braid Form Theorem

THEOREM 21.4 (Normal Braid Form Theorem). *Any oriented surface link in* $\mathbb{R}^4$ *is ambient isotopic to a surface link in a normal braid form.*

The following proof is a refined version of the proof in [**316**]. The next two sections, 21.4 and 21.5, will help the reader understand this braiding procedure.

Proof. Let F be an oriented surface link. By Theorem 9.14, we may assume that it is a closed realizing surface of a hyperbolic transformation $L_- \to L_+$ along a band set $\mathcal{B} = B_1 \cup \cdots \cup B_n$, where L_- and L_+ are trivial links. Let $N(B_i)$ be a regular neighborhood of a band B_i. The intersection $N(B_i) \cap L_-$ is a pair of trivial arcs. By an ambient isotopy of $\mathbb{R}^3$, deform L_- and $N(B_i)$ such that the intersection $N(B_i) \cap L_-$ and the band B_i are as in Fig. 21.1. See Fig. 21.6 for this process. (For a death-type simple band, change it into a birth-type as in Fig. 21.2.) Furthermore we may assume that the second condition of the normal braid form is satisfied. The remainder $L_- \cap (\mathbb{R}^3 \backslash (N(B_1) \cup \cdots \cup N(B_n)))$ is the union of some arcs or loops. Applying the argument in the proof of Alexander's theorem (Sect. 4.2), without moving $N(B_1) \cup \cdots \cup N(B_n)$, we can deform L_- into a closed m-braid around z-axis for some m. Then L_+ is also a closed m-braid.

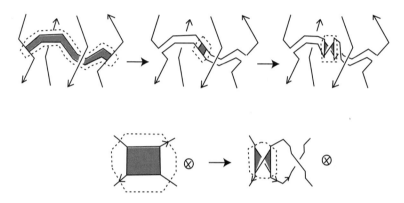

FIGURE 21.6. Change a band into a simple band

If both L_- and L_+ are trivial closed braids, then we have the theorem. Suppose that L_+ is not a trivial closed braid. By Markov's theorem, there exists a positive integer p and a finite sequence of closed braids

$$L_+ = L_+^0 \to L_+^1 \to L_+^2 \to \cdots \to L_+^{2p+1}$$

such that

(1) for each odd i, $L_+^{i-1} \to L_+^i$ is an isotopic transformation of type $\mathcal{R}$ (see below),

(2) for each even i, $L_+^{i-1} \to L_+^i$ is an isotopic transformation by an ambient isotopy of $\mathbb{R}^3$ realizing a stabilization or a destabilization,

(3) L_+^{2p+1} is a trivial closed braid.

We call such a link transformation sequence an *unbraiding sequence*. (Here an ambient isotopy $\{h_t\}$ of $\mathbb{R}^3$ is *of type $\mathcal{R}$* if each h_t preserves $\mathbb{R}_+^2 \times \{\theta\} \subset \mathbb{R}_+^2 \times S^1 = \mathbb{R}^3 \backslash z$-axis setwise for each $\theta \in S^1$. An isotopic transformation *of type $\mathcal{R}$* is an isotopic transformation by an ambient isotopy of type $\mathcal{R}$.)

Let C be a sufficiently small loop around the z-axis and let D be a small 2-disk bounded by C. We consider two cases as follows:

(Case 1) $L_+^1 \to L_+^2$ is a stabilization.
(Case 2) $L_+^1 \to L_+^2$ is a destabilization.

Modifying L_+^1 and L_+^2 by ambient isotopies of $\mathbb{R}^3$ of type $\mathcal{R}$, we assume that there exists a birth-type simple band B attaching to $L_+^1 \cup C$ with $L_+^2 = \mathrm{h}(L_+^1 \cup C; B)$ (see Fig. 21.7) in Case 1, or there exists a death-type simple band B attaching to L_+^1

with $L_+^2 \cup C = h(L_+^1; B)$ in Case 2. Moreover the image of B under the projection $\mathbb{R}_+^2 \times S^1 \to S^1$ is disjoint from the image of the band set $\mathcal{B}$ used in $L_- \to L_+$.

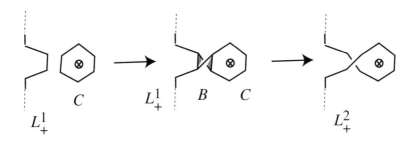

FIGURE 21.7

We may assume that $D \cup B$ is an embedded 2-disk and that the isotopic transformation $L_+^1 \to L_+^2$ is an ambient isotopy induced by the cellular move along $D \cup B$. By Lemma 9.13(2), we may assume that F is a closure of a realizing surface $F(L_- \to L_+ = L_+^0 \to L_+^1 \to L_+^2)_{[-1,3]} = F(L_- \to L_+)_{[-1,1]} \cup F(L_+^0 \to L_+^1)_{[1,2]} \cup F(L_+^1 \to L_+^2)_{[2,3]}$ of $L_- \to L_+ = L_+^0 \to L_+^1 \to L_+^2$. Since $L_+^1 \to L_+^2$ is an isotopic transformation induced from a cellular move along $D \cup B$, we can deform F by an ambient isotopy such that

$$\pi(F \cap \mathbb{R}^3[t]) = \begin{cases} L_+^1 & \text{for } t \in [2, 2.5), \\ L_+^1 \cup D \cup B & \text{for } t = 2.5, \\ L_+^2 & \text{for } t \in (2.5, 3], \end{cases}$$

and the other part is identical with the original one.

Let $\{h_t\}_{t \in [0,1]}$ be an ambient isotopy of $\mathbb{R}^3$ of type $\mathcal{R}$ which realizes $L_+^0 \to L_+^1$ so that

$$\pi(F \cap \mathbb{R}^3[t]) = h_{t-1}(L_+^0) \quad \text{for } t \in [1, 2].$$

Let B', D' and C' be the preimages of B, D and C by the ambient isotopy $\{h_t\}$, i.e., $h_1(B') = B$, $h_1(D') = D$ and $h_1(C') = C$. Then B' is a pre-simple band attaching to a closed braid $L_+^0 \cup C'$ in Case 1, or attaching to a closed braid L_+^0 in Case 2. Since the image of B under the projection $\mathbb{R}_+^2 \times S^1 \to S^1$ is disjoint from the image of $\mathcal{B}$, the band B' is disjoint from the bands $\mathcal{B}$. Thus we can regard B' as a pre-simple band attaching to $L_- \cup C'$ in Case 1, or attaching to L_- in Case 2. We denote by L_+' the closed braid obtained from $L_- \cup C'$ (in Case 1) or L_- (in Case 2) by surgery along B'. Then $h_1(L_+') = L_+^2$ (in Case 1) or $h_1(L_+') = L_+^2 \cup C$ (in Case 2).

For Case 1, consider a 3-ball M such that

$$\pi(M \cap \mathbb{R}^3[t]) = \begin{cases} D' & \text{for } t \in [-1 - \epsilon, 0), \\ D' \cup B' & \text{for } t = [0, 1], \\ h_{t-1}(D' \cup B') & \text{for } t \in (1, 2), \\ D \cup B & \text{for } t \in [2, 2.5], \\ \emptyset & \text{otherwise,} \end{cases}$$

where ϵ is a sufficiently small positive number. The intersection of $F \cap M$ is a 2-disk in ∂M, and we can apply a cellular move to F along M. The result is a closed realizing surface of

$$L_- \cup C' \to L_+{}' \to L_+^2,$$

where $L_- \cup C' \to L_+{}'$ is a hyperbolic transformation along a band set $\mathcal{B} \cup \mathcal{B}'$, and $L_+{}' \to L_+^2$ is an isotopic transformation by $\{h_t\}$. Put $\widetilde{L}_- = L_- \cup C'$, $\widetilde{L}_+ = L_+{}'$ and $\widetilde{\mathcal{B}} = \mathcal{B} \cup \mathcal{B}'$.

For Case 2, consider 3-balls M_1 and M_2 such that

$$\pi(M_1 \cap \mathbb{R}^3[t]) = \begin{cases} B' & \text{for} \quad t \in [0,1], \\ h_{t-1}(B') & \text{for} \quad t \in (1,2), \\ B & \text{for} \quad t \in [2, 2.5], \\ \emptyset & \text{otherwise}, \end{cases}$$

and

$$\pi(M_2 \cap \mathbb{R}^3[t]) = \begin{cases} D & \text{for} \quad t = [2.5, 3 + \epsilon], \\ \emptyset & \text{otherwise}, \end{cases}$$

where ϵ is a sufficiently small positive number. Apply cellular moves to F along M_1 and M_2. The result is a closed realizing surface of

$$L_- \to L_+{}' \to L_+^2 \cup C,$$

where $L_- \to L_+{}'$ is a hyperbolic transformation along $\mathcal{B} \cup \mathcal{B}'$, and $L_+{}' \to L_+^2 \cup C$ is an isotopic transformation by $\{h_t\}$. Put $\widetilde{L}_- = L_-$, $\widetilde{L}_+ = L_+{}'$, and $\widetilde{\mathcal{B}} = \mathcal{B} \cup \mathcal{B}'$.

In both cases, $\widetilde{L}_-$, $\widetilde{L}_+$ and $\widetilde{\mathcal{B}}$ satisfy the conditions of (1), (2) and (4) of the normal braid form. (Here B' is a pre-simple band. Deforming $\widetilde{L}_-$, $\widetilde{L}_+$ and $\widetilde{\mathcal{B}}$ by an ambient isotopy of $\mathbb{R}^3$ of type $\mathcal{R}$, we may assume that B' is a simple band.) Note that $\widetilde{L}_+$ has an unbraiding sequence whose length is $p-1$ and $\widetilde{L}_-$ has an unbraiding sequence with the same length as L_-. Continue this procedure until L_+ is a closed trivial braid. Then apply the same argument to L_-. $\blacksquare$

21.4. The 2-Twist Spun Trefoil

Let F be the 2-twist spun trefoil in $\mathbb{R}^4$. Fig. 21.8(1) is a middle cross section (the $(t = 0)$-level) of F such that F is a closed realizing surface of a hyperbolic transformation $L_- \to L_+$. Deform the middle section into a braid form as in the figure. Fig. 21.8(4) is in a braid form of degree 5. Fig. 21.9 is obtained from Fig. 21.8(4) by cutting along a half plane indicated by the dotted line.

Braid word descriptions of L_+ and L_- are

$$\begin{bmatrix} L_+ \\ L_- \end{bmatrix} = \overline{1}\ \overline{1}\ \overline{1}\ \overline{2}\ \overline{1}\ 2\ \overline{3}\ \overline{2}\ 3 \begin{bmatrix} \overline{4} \\ e \end{bmatrix} 1\ 3\ \overline{2}\ \overline{3}\ 2\ \overline{1}\ \overline{2}\ 1\ 1\ 1\ 4 \begin{bmatrix} e \\ 2 \end{bmatrix}.$$

L_+ and L_- are not trivial closed braids. (In fact, L_+ and L_- are 2-component links. Thus they are not trivial as closed 5-braids.)

We have to transform them into trivial closed braids. This can be done by the method given in the proof of the normal braid form theorem. However, if we use the chart description method to do this, then we can obtain a chart for F much more quickly. Let us demonstrate this method.

A chart describing the hyperbolic transformation of $L_- \to L_+$ is illustrated in Fig. 21.10. We construct a chart above the $(t = 1)$-level as follows. See Fig. 21.11. First we try to simplify the braid word of L_+ appearing at the $(t = 1)$-level and reduce the number of σ_4's and σ_4^{-1}'s in the braid word (or the number of arcs labeled

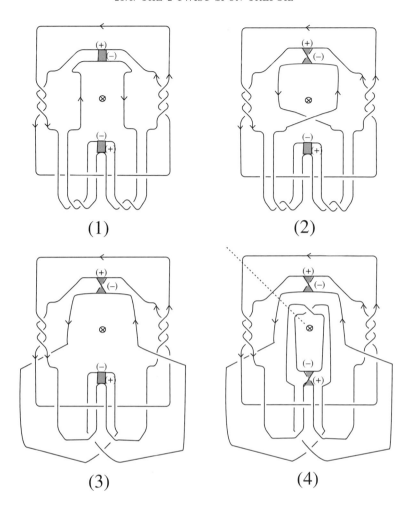

(1)   (2)

(3)   (4)

FIGURE 21.8. Braiding a 2-twist spun trefoil

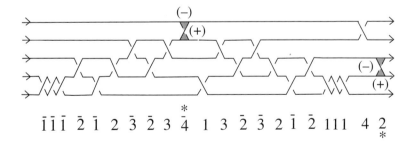

$$\bar{1}\bar{1}\bar{1} \ \bar{2}\ \bar{1} \ 2 \ \bar{3} \ \bar{2} \ 3 \ \overset{*}{\bar{4}} \ 1 \ 3 \ \bar{2} \ \bar{3} \ 2 \ \bar{1} \ \bar{2} \ 111 \ 4 \ \underset{*}{2}$$

FIGURE 21.9

4 at the $(t = 1)$-level in the chart) by a braid word transformation sequence. Draw a BWTS chart of the sequence between $t = 1$ and $t = 2$. At the $(t = 2)$-level, we have only one arc with label 4 (since the orientation of the arc is downward, the

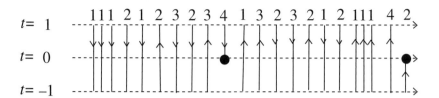

FIGURE 21.10. Middle part

label is -4, to be precise.) This implies that the closed 5-braid appearing at the $(t = 2)$-level can be destabilized into a closed 4-braid. Instead of a destabilization, we attach a death-type simple band and do surgery to obtain a closed 5-braid whose last string is trivial. See Fig. 21.3(D). The new closed braid appears at the $(t = 3)$-level. This process is described by a chart between $t = 2$ and $t = 3$. Recall that a black vertex corresponds to a hyperbolic transformation along a simple band.

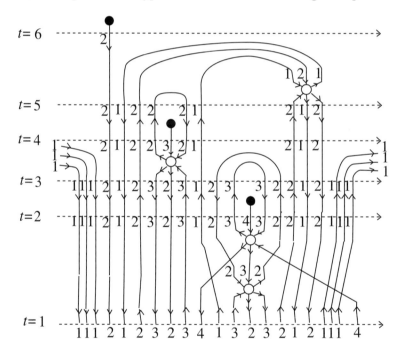

FIGURE 21.11. Upper part

Next, we try to simplify the braid word, which is a word in $\{1, 2, 3\}$, and reduce the number of 3's in the braid word (or the number of edges with label 3 at the $(t = 3)$-level in the chart). See the chart between $t = 3$ and $t = 4$. Three arcs labeled 1 run toward the right side and appear from the left side. This implies that we applied conjugations by σ_1 three times. Since we are considering a braid word chart for a closed braid, we may assume that the right side and the left side are identified. Thus the region between $t = 3$ and $t = 4$ is regarded as an annulus.

The closed braid appearing at $t = 4$ has a braid word description with only one 3 (this is -3 to be precise). Attach a band as in Fig. 21.3(D) again and do surgery in order to change the fourth string to be trivial. This process is described by the chart between $t = 4$ and $t = 5$. Now the fourth and the fifth strings are trivial.

We try to simplify and reduce the number of 2's in the braid word. By a braid word transformation sequence described by the chart between $t = 5$ and $t = 6$ in Fig. 21.11, we can deform the closed braid appearing at $t = 5$ into the closed braid appearing at $t = 6$. Attach a band as in Fig. 21.3(D) again and do surgery in order to make the third string trivial. The result is a trivial closed 5-braid.

Note that, in the whole process above, we used ambient isotopic deformations in 3-space and hyperbolic transformations, each of which splits a component. Thus the surface embedded in $\mathbb{R}^3[1, \infty)$ described by the chart in Fig. 21.11 and the restriction to $\mathbb{R}^3[1, \infty)$ of the original F given in Fig. 21.8(4) are ambient isotopic without moving the hyperplane $\mathbb{R}^3[1]$.

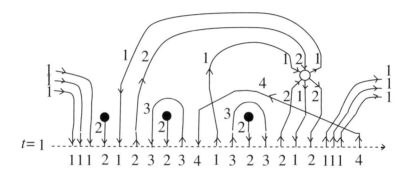

FIGURE 21.12. Upper part (after chart moves)

We may use chart moves to simplify the chart. Fig. 21.12 is the chart obtained from Fig. 21.11 by chart moves.

Apply the same argument to the part that is below $t = -1$. Combine the upper part and the lower part into a whole chart as in Fig. 21.13. This is a chart description of degree 5 of the 2-twist spun trefoil F. Using chart moves, we can simplify the chart as in Fig. 21.14.

Furthermore, change the chart as in Fig. 21.15. Here we have deleted an arc labeled 4 and regard the chart as a chart of degree 4. This deletion is not a chart move, and the charts in Fig. 21.14 and in Fig. 21.15 do not describe the same surface braids. In fact their degrees are different. However, they are ambient isotopic as a surface in $\mathbb{R}^4$. This modification is called a *destabilization of a closed surface braid* and is discussed in Sect. 25.2. Now we have the chart of degree 4 illustrated in Fig. 21.15 which describes a 2-twist spun trefoil.

Since we are considering a chart of a closed surface braid, we may deform it by an ambient isotopy of the 2-sphere. Deform the chart into the chart depicted in Fig. 21.16. The middle cross section is illustrated in Fig. 21.17. The braid words description of L_+ and L_- are

$$\begin{bmatrix} L_+ \\ L_- \end{bmatrix} = \overline{3}\,\overline{1} \begin{bmatrix} \overline{2} \\ e \end{bmatrix} \overline{1}\,\overline{1}\,2\,\overline{1} \begin{bmatrix} 1 \\ e \end{bmatrix} \begin{bmatrix} \overline{3} \\ e \end{bmatrix} \overline{2}\,3\,2\,2 \begin{bmatrix} 3 \\ e \end{bmatrix} \begin{bmatrix} \overline{1} \\ e \end{bmatrix} \overline{2}\,\overline{2}\,1 \begin{bmatrix} 2 \\ e \end{bmatrix} 1\,1.$$

L_+ and L_- are trivial closed 4-braids. (We do not need to check this. There are no black vertices above the $(t = 1)$-level and below the $(t = -1)$-level in Fig. 21.16.

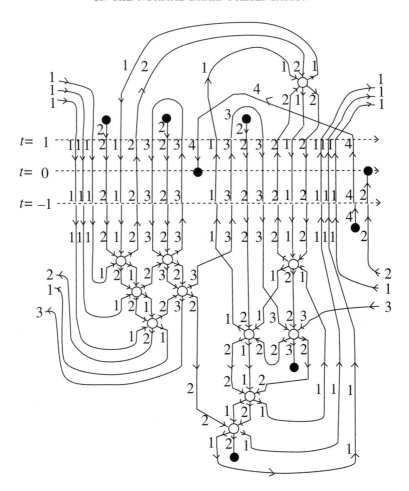

FIGURE 21.13. A chart of degree 5 describing a 2-twist spun trefoil

Thus the closed braids L_+ and L_- appearing at these levels must be trivial.) So we have a 2-twist spun trefoil which is in a normal braid form of degree 4.

21.5. Strategy for Braiding

The argument in Sect. 21.4 is a little simpler than that of the proof of the normal braid form theorem given in Sect. 21.3. One reason is that, in the process of transforming L_+ into a trivial closed braid, we did not move the bands (corresponding to black vertices) into the middle cross section (the $(t = 0)$-level). Another reason is that the particular closed braids L_+ and L_- in Sect. 21.4 have a feature that they can be transformed into trivial closed braids without stabilizations. In general one cannot expect this, and the argument will be longer.

Here we summarize the process performed in Sect. 21.4 in a general situation.

Step I: Middle part

(I-1) Describe a surface link F by use of L_- and $\mathcal{B}$, and deform them by an isotopy of $\mathbb{R}^3$ into a closed braid and a set of simple bands.

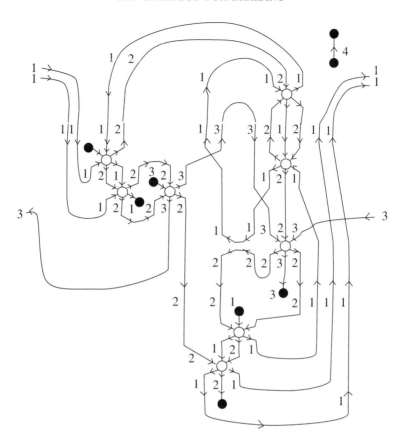

FIGURE 21.14. A chart of degree 5 describing a 2-twist spun trefoil (after chart moves in S^2)

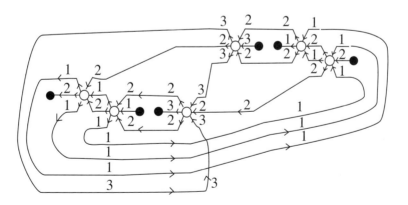

FIGURE 21.15. A chart of degree 4 describing a 2-twist spun trefoil

(I-2) Draw a chart between $t = -1$ and $t = 1$ which describes the hyperbolic transformation $L_- \to L_+$ along $\mathcal{B}$. Each band corresponds to a black vertex.

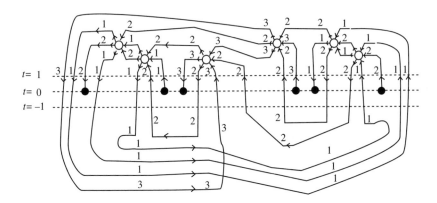

FIGURE 21.16. A chart which gives a normal braid form

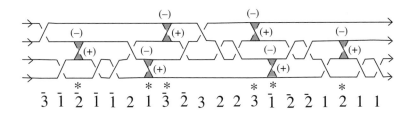

$$\bar{3}\ \bar{1}\ \bar{2}\ \bar{1}\ \bar{1}\ 2\ 1\ \bar{3}\ \bar{2}\ 3\ 2\ 2\ 3\ \bar{1}\ \bar{2}\ \bar{2}\ 1\ 2\ 1\ 1$$

FIGURE 21.17. A middle cross section of a 2-twist spun trefoil in a normal braid form

Step II: Upper part

(II-1) Try to simplify the braid word describing L_+ and reduce the number of the letter m (and $-m$) by a braid word transformation sequence, where m is the maximum letter in the word. If we can deform L_+ so that there is only one letter m, then add a death-type simple band as in Fig. 21.3(C) or (D) in order to make the $(m+1)$st string trivial. In the chart description, a black vertex labeled m is introduced. Continue this process until we have an empty word; i.e. L_+ is a trivial closed braid.

This procedure (II-1) *is possible if L_+ can be transformed into a trivial closed braid without stabilizations.*

(II-2) If (II-1) is impossible or the process fails, then find a transformation sequence (an unbrading sequence) from L_+ to a trivial closed braid which involves not only destabilizations but also stabilizations. For each destabilization, introduce a black vertex in the same way as (II-1). For each stabilization, let m be the degree of L_+. We introduce a trivial component to L_+ (so that the degree becomes $m+1$) and a birth-type simple band connecting them and do surgery as in Fig. 21.3(A) or (B). In the chart description, this is introducing a black vertex labeled m, and we consider that the chart is of degree $m+1$. Finally L_+ is changed into a trivial closed braid, and we obtain a chart above the ($t=1$)-level.

In this procedure (II-2), *the degree of L_+ increases. The difference from the original degree is equal to the number of stabilizations.*

Step III: Lower part

Before starting Step III, note that the degree of L_-, which equals the degree of L_+, has increased if the procedure (II-2) was applied.

(III-1) and (III-2) are similar to (II-1) and (II-2).

Now we have a chart description of F. If you want, do the next step.

Step IV: Simplification

(IV-1) Simplify the chart by chart moves. Since we are considering a closed surface braid (see Chapter 23), we may assume that the chart is in a 2-sphere. Take advantage of isotopic deformation in the sphere.

(IV-2) If the degree is m and there is only one edge with label $m - 1$, then we can eliminate this (free) edge by regarding the chart as a chart of degree $m - 1$. This is a 2-dimensional destabilization (Sect. 25.2). Go back to (IV-1).

CHAPTER 22

Braiding Ribbon Surface Links

In Chapter 21, a method to obtain a braid presentation (called a normal braid form) of a surface link was introduced. In this chapter, we consider a braid presentation of a ribbon surface link, called a normal ribbon braid form.

22.1. The Braid Form of a Ribbon Surface Link

DEFINITION 22.1. A surface link is in a *normal ribbon braid form* of degree m if there exist closed braids L_0 and L_+ in $\mathbb{R}^3$ about the z-axis and a band set $\mathcal{B}$ attaching to L_0 such that

(1) each band of $\mathcal{B}$ is a death-type simple band, and L_+ is obtained from L_0 by surgery along $\mathcal{B}$,
(2) the images of the bands of $\mathcal{B}$ under the projection $\mathbb{R}_+^2 \times S^1 \to S^1$ are mutually disjoint arcs in S^1,
(3) L_+ is a trivial closed braid, and
(4) F is a closed realizing surface of $L_+ \to L_0 \to L_+$, where $L_+ \to L_0$ is the inverse of a hyperbolic transformation $L_0 \to L_+$ along $\mathcal{B}$.

It is not required that L_0 be a trivial link or a trivial closed braid.

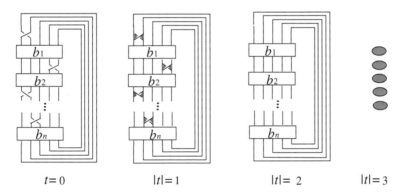

$$t = 0 \qquad |t| = 1 \qquad |t| = 2 \qquad |t| = 3$$

FIGURE 22.1. Normal ribbon braid form

THEOREM 22.2. *Any oriented ribbon surface link in $\mathbb{R}^4$ is ambient isotopic to a surface link in a normal ribbon braid form.*

This theorem is due to Rudolph [**778, 783**]. The proof given below is a modified version of that of Theorem 21.4, cf. [**312**].

Proof. Let F be an oriented ribbon surface link. By Theorem 11.4, we may assume that it is a closed realizing surface of $L_+ \to L_0 \to L_+$, where $L_0 \to L_+$ is a hyperbolic transformation along a band set $\mathcal{B}$ and $L_+ \to L_0$ is the inverse. L_+ is a trivial link. By the same argument as the proof of Theorem 21.4, we may assume that the conditions of a normal ribbon braid form are satisfied except for the third condition.

Suppose that L_+ is not a trivial closed braid. Consider an unbraiding sequence

$$L_+ = L_+^0 \to L_+^1 \to L_+^2 \to \cdots \to L_+^{2p+1}$$

of L_+; that is,

(1) for each odd i, $L_+^{i-1} \to L_+^i$ is an isotopic transformation of type $\mathcal{R}$,

(2) for each even i, $L_+^{i-1} \to L_+^i$ is an isotopic transformation by an ambient isotopy of $\mathbb{R}^3$ realizing a stabilization or a destabilization,

(3) L_+^{2p+1} is a trivial closed braid.

Let C be a sufficiently small loop around the z-axis and let D be a small 2-disk bounded by C. We consider two cases as follows:

(Case 1) $L_+^1 \to L_+^2$ is a stabilization.

(Case 2) $L_+^1 \to L_+^2$ is a destabilization.

Modifying L_+^1 and L_+^2 by ambient isotopies of $\mathbb{R}^3$ of type $\mathcal{R}$, we assume that there exists a birth-type simple band B attaching to $L_+^1 \cup C$ with $L_+^2 = \mathrm{h}(L_+^1 \cup C; B)$ (see Fig. 21.7) in Case 1, or there exists a death-type simple band B attaching to L_+^1 with $L_+^2 \cup C = \mathrm{h}(L_+^1; B)$ in Case 2. Moreover the image of B under the projection $\mathbb{R}_+^2 \times S^1 \to S^1$ is disjoint from the image of the band set $\mathcal{B}$ used in $L_- \to L_+$.

We may assume that $D \cup B$ is an embedded 2-disk and that the isotopic transformation $L_+^1 \to L_+^2$ is an ambient isotopy induced by the cellular move along $D \cup B$.

By Lemma 9.13(2), we assume that F is a closure of a realizing surface $F(L_+^2 \to L_+^1 \to L_+ \to L_0 \to L_+ \to L_+^1 \to L_+^2)_{[3,3]} = F(L_+^2 \to L_+^1)_{[-3,-2]} \cup F(L_+^1 \to L_+^0)_{[-2,-1]} \cup F(L_+ \to L_0 \to L_+)_{[-1,1]} \cup F(L_+^0 \to L_+^1)_{[1,2]} \cup F(L_+^1 \to L_+^2)_{[2,3]}$. We assume the saddle bands of $F(L_+ \to L_0 \to L_+)_{[-1,1]}$ are at the $(t = -0.5)$-level and the $(t = 0.5)$-level. Since $L_+^1 \to L_+^2$ is an isotopic transformation induced from a cellular move along $D \cup B$, we can deform F by an ambient isotopy such that

$$\pi(F \cap \mathbb{R}^3[t]) = \begin{cases} L_+^1 & \text{for } t \in [2, 2.5), \\ L_+^1 \cup D \cup B & \text{for } t = 2.5, \\ L_+^2 & \text{for } t \in (2.5, 3], \end{cases}$$

and the other part in $\mathbb{R}^3[0, \infty)$ is identical with the original one and the part in $\mathbb{R}^3(-\infty, 0]$ is symmetric to the part in $\mathbb{R}^3[0, \infty)$.

Let $\{h_t\}$ be an ambient isotopy of $\mathbb{R}^3$ of type $\mathcal{R}$ which realizes $L_+^0 \to L_+^1$ so that

$$\pi(F \cap \mathbb{R}^3[t]) = h_{t-1}(L_+^0) \qquad \text{for } t \in [1, 2].$$

Let B', D' and C' be the preimages of B, D and C by the ambient isotopy $\{h_t\}$, i.e., $h_1(B') = B$, $h_1(D') = D$ and $h_1(C') = C$. Then B' is a pre-simple band attaching to a closed braid $L_+^0 \cup C'$ in Case 1, or attaching to a closed braid L_+^0 in Case 2. As before, we can regard B' as a pre-simple band attaching to $L_0 \cup C'$ in Case 1, or attaching to L_0 in Case 2.

For Case 1, let $L_0{}'$ and $L_+{}'$ be closed braids obtained from $L_0 \cup C'$ and $L_+ \cup C'$ by surgery along B', respectively.

For Case 2, let L_0' and L_+' be closed braids which are $h(L_0; B') \setminus C'$ and $h(L_+; B') \setminus C'$; in other words L_0' and L_+' are obtained from L_0 and L_+ by destabilizations along $B' \cup D'$.

In both cases, consider a 3-ball M in $\mathbb{R}^4$ such that

$$\pi(M \cap \mathbb{R}^3[t]) = \begin{cases} D' \cup B' & \text{for } t \text{ with } |t| \in [0,1], \\ h_{|t|-1}(D' \cup B') & \text{for } t \text{ with } |t| \in (1,2), \\ D \cup B & \text{for } t \text{ with } |t| \in [2,2.5], \\ \emptyset & \text{otherwise.} \end{cases}$$

The intersection of $F \cap M$ is a 2-disk in ∂M, and we can apply a cellular move to F along M. The result is a closed realizing surface of

$$L_+^2 \to L_+' \to L_0' \to L_+' \to L_+^2,$$

where $L_0' \to L_+'$ is a hyperbolic transformation along $\mathcal{B}$, $L_+' \to L_+^2$ is an isotopic transformation by $\{h_t\}$, and $L_+' \to L_0'$ and $L_+^2 \to L_+'$ are their inverses. Put $\widetilde{L}_0 = L_0'$, $\widetilde{L}_+ = L_+'$ and $\widetilde{\mathcal{B}} = \mathcal{B}$. They satisfy the conditions of (1), (2) and (4) of the normal ribbon braid form. (Here B' is a pre-simple band. Deforming $\widetilde{L}_0$, $\widetilde{L}_+$ and $\widetilde{\mathcal{B}}$ by an ambient isotopy of $\mathbb{R}^3$ of type $\mathcal{R}$, we may assume that B' is a death-type simple band.) Note that $\widetilde{L}_+$ has an unbraiding sequence whose length is $p-1$. Continue this procedure until L_+ is a closed trivial braid. ∎

22.2. The Spun Trefoil

Let F be a closed realizing surface of $L_+ \to L_0 \to L_+$ where $L_0 \to L_+$ is a hyperbolic transformation and $L_+ \to L_0$ is the inverse such that $L_0 \cup \mathcal{B}$ is as in Fig. 22.2(1). F is a spun trefoil. Deform $L_0 \cup \mathcal{B}$ into a braid form as in the figure. Fig. 22.3 is obtained from Fig. 21.8(2) by cutting along a half plane indicated by a dotted line. Fig. 22.4(1) is a whole chart for F. It is isotopic to Fig. 22.4(2). Fig. 14.3 corresponds to Fig. 22.4(2).

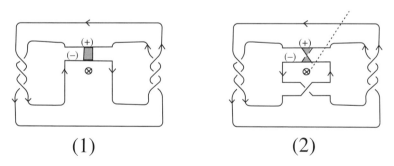

(1)   (2)

FIGURE 22.2. Braiding a spun trefoil

In general, a surface link which is obtained from a knotted curve as in the left hand side of Fig. 22.5 has a chart description of degree 3 as in the right hand side. (If s is odd, it is a 2-knot. If s is even, it is a surface link of $S^2 \cup T^2$.)

A surface link which is obtained from a 3-braid link by spinning (Fig. 22.6) has a chart description of degree 5 as in Fig. 22.8. A surface link which is obtained from a 4-braid link by spinning (Fig. 22.6) has a chart description of degree 7 as in Fig. 22.9.

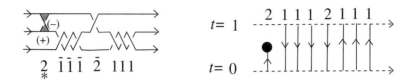

FIGURE 22.3

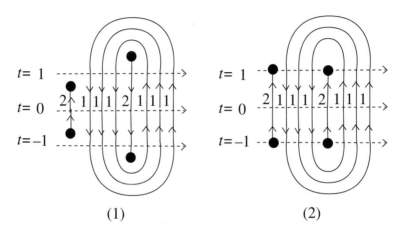

(1)   (2)

FIGURE 22.4

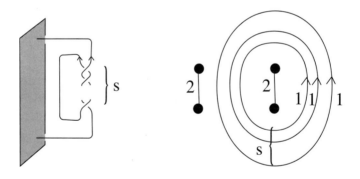

FIGURE 22.5. A chart of degree 3 for a spun 2-braid link

EXERCISE 22.3. Construct a chart description of a spun 2-knot/link of a k-braid knot/link for $k = 5, 6, \ldots$.

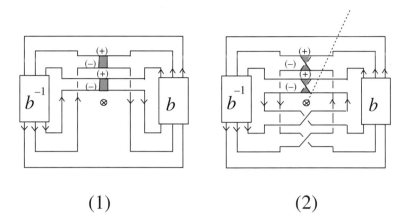

(1)   (2)

FIGURE 22.6

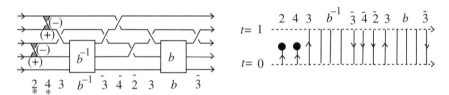

FIGURE 22.7

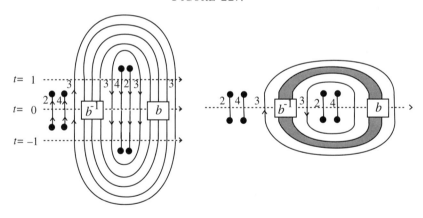

FIGURE 22.8. A chart of degree 5 for a spun 3-braid link

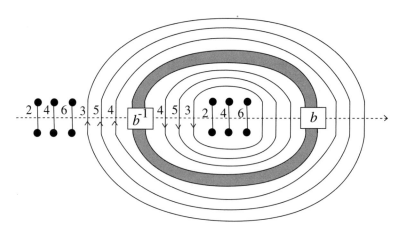

FIGURE 22.9. A chart of degree 7 for a spun 4-braid link

CHAPTER 23

Alexander's Theorem in Dimension Four

23.1. Closed Surface Braids in $D^2 \times S^2$

DEFINITION 23.1. A *closed surface braid in* $D^2 \times S^2$ is a closed oriented surface S embedded locally flatly in $D^2 \times S^2$ such that the restriction map $pr_2|_S : S \to S^2$ of the projection $pr_2 : D^2 \times S^2 \to S^2$ is a branched covering map. It is a *simple closed surface braid* if the branched covering is simple.

Two closed surface braids S and S' in $D^2 \times S^2$ are *equivalent* if there is a fiber-preserving ambient isotopy of $D^2 \times S^2$ which carries S and S'. Here we regard $D^2 \times S^2$ as a D^2-bundle over S^2.

Let S be a surface braid in $D_1^2 \times D_2^2$ of degree m. Identify a 2-sphere S^2 with the union of D_2^2 and $\overline{D}_2^2$, where $\overline{D}_2^2$ is a 2-disk and ∂D_2^2 and $\partial \overline{D}_2^2$ are identified. Extending S to a closed surface $\widehat{S}$ in $D_1^2 \times S^2 = D_1^2 \times (D_2^2 \cup \overline{D}_2^2)$ such that

$$pr_1(\widehat{S} \cap pr_2^{-1}(y)) = Q_m \quad \text{for } y \in \overline{D}_2^2,$$

we have a closed surface braid $\widehat{S}$.

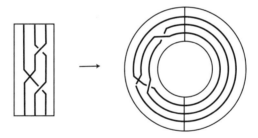

FIGURE 23.1. Closure of a surface braid

DEFINITION 23.2. The *closure* in $D^2 \times S^2$ of a surface braid S is the closed surface braid $\widehat{S}$.

EXERCISE 23.3. (1) The equivalence class of a closure in $D^2 \times S^2$ of a surface braid S does not depend on the identification $S^2 \cong D_2^2 \cup \overline{D}_2^2$. (2) If S and S' are equivalent, then their closures in $D^2 \times S^2$ are equivalent.

Let F be a closed surface braid of degree m in $D_1^2 \times S^2$. Let y be a point of $S^2 \setminus \Sigma(F)$, where $\Sigma(F)$ is the branch point set. $pr_1(F \cap pr_2^{-1}(y))$ consists of m interior points of D_1^2. Deform F up to isomorphism (the definition of isomorphism is similar to Definition 14.10) so that $pr_1(F \cap pr_2^{-1}(z)) = Q_m$ for $z \in N(y)$. Then the restriction of F to $D_1^2 \times \mathrm{cl}\,(S^2 \setminus N(y))$ is a surface braid of degree m. We call it a *cut* of F along $pr_2^{-1}(y)$. A cut is uniquely determined up to "2-dimensional conjugation" defined later. Moreover it turns out that the conjugacy class does not depend on a choice of $y \in S^2 \setminus \Sigma(F)$.

23.2. Closed Surface Braids in $\mathbb{R}^4$

Let S^2 be a standard 2-sphere in $\mathbb{R}^4$ and let $N(S^2)$ be a regular neighborhood of S^2, which is identified with $D^2 \times S^2$.

DEFINITION 23.4. A *closed surface braid* in $\mathbb{R}^4$ is an oriented surface link in $\mathbb{R}^4$ which is contained in $N(S^2) = D^2 \times S^2$ as a closed surface braid. The *closure* in $\mathbb{R}^4$ of a surface braid S is the closed surface braid $\widehat{S}$ in $N(S^2) = D^2 \times S^2 \subset \mathbb{R}^4$.

Obviously, if surface braids S and S' are equivalent, then their closures in $\mathbb{R}^4$ are ambient isotopic in $\mathbb{R}^4$.

When we use the motion picture method, a closure in $\mathbb{R}^4$ is obtained as follows: Let $\{b_t\}_{t \in [0,1]}$ be a motion picture of a surface braid of S degree m. For each $t \in [0,1]$, put the braid $b_t \subset D^2 \times I$ in the hyperplane $\mathbb{R}^3[t]$ of $\mathbb{R}^4 = \mathbb{R}^3(-\infty, \infty)$ and take a closure of b_t in the 3-space, say $\widehat{b_t}$. The union of the closed braids for $t \in [0,1]$ forms an embedded surface in $\mathbb{R}^3[0,1]$. The restriction to $\mathbb{R}^3[0]$ is a trivial closed braid. Thus we can cap off by a trivial disk system in $\mathbb{R}^3(-\infty, 0]$. Similarly, we can cap off the opposite side by a trivial disk system in $\mathbb{R}^3[1, \infty)$. See Fig. 23.2. In this way, we have a closure of S in $\mathbb{R}^4$.

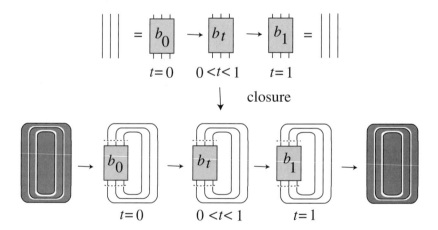

FIGURE 23.2. Closure of the motion picture of a surface braid

DEFINITION 23.5. The *closure* of the motion picture $\{b_t\}_{t \in [0,1]}$ is the motion picture of the closure of S in $\mathbb{R}^4$ obtained as above.

Fig. 23.3 is a closure of the motion picture illustrated in Fig. 14.3.

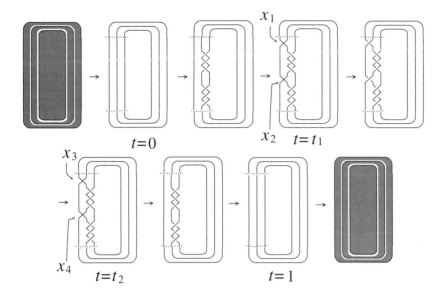

$$FIGURE\ 23.3$$

23.3. Alexander's Theorem in Dimension Four

The Alexander theorem states that any oriented link in $\mathbb{R}^3$ is ambient isotopic to a closed braid. An analogous result holds for surface links in $\mathbb{R}^4$.

THEOREM 23.6 (Alexander's Theorem in Dimension Four). *Any oriented surface link is ambient isotopic to a (simple) closed surface braid in $\mathbb{R}^4$.*

Proof. Any oriented surface link in $\mathbb{R}^4$ is ambient isotopic in a normal braid form (Theorem 21.4). Deform each simple band into a saddle point as in Fig. 14.4. Then we have a simple closed surface braid. ∎

THEOREM 23.7. *Any oriented ribbon surface link is ambient isotopic to a (simple) closed surface braid in $\mathbb{R}^4$ which is symmetric with respect to the hyperplane $\mathbb{R}^3[0]$.*

Proof. It follows from Theorem 22.2. ∎

23.4. The Chart Description of a Surface Link

DEFINITION 23.8. A (surface braid) chart Γ is a *chart description* of an oriented surface link F if the closure in $\mathbb{R}^4$ of the surface braid associated with Γ is ambient isotopic to F.

Since any oriented surface link can be presented as a closure of a simple surface braid (Theorem 23.6), it has a chart description. When we handle a closed surface braid (or a closure of a surface braid), a chart may be regarded as a chart in S^2 rather than in D_2^2.

23.5. The Braid Index of a Surface Link

DEFINITION 23.9. The *braid index* of an oriented surface link F is the minimum degree of simple closed surface braids in $\mathbb{R}^4$ that are equivalent to F. The index is denoted by $\mathrm{Braid}(F)$.

It is obvious that $\mathrm{Braid}(F) = 1$ if and only if F is a trivial 2-sphere.

REMARK 23.10. For an oriented surface link F, let $\mathrm{Braid}^*(F)$ be the minimum degree of closed surface braids in $\mathbb{R}^4$ (possibly non-simple) that are equivalent to F. By definition, we have $\mathrm{Braid}^*(F) \leq \mathrm{Braid}(F)$. It is unknown whether there exists an oriented surface link F with $\mathrm{Braid}^*(F) < \mathrm{Braid}(F)$.

23.6. Another Kind of Braid Presentation

Recall that any oriented surface link F is ambient isotopic to a closed realizing surface of a hyperbolic transformation $L_- \to L_+$ along a band set $\mathcal{B}$ such that L_- and L_+ are trivial links in $\mathbb{R}^3$. The following braid form was introduced by F. González-Acuña [206].

THEOREM 23.11. *We can choose* L_-, L_+ *and* $\mathcal{B}$ *as in* Fig. 23.4.

Fig. 23.4 is a picture of L_-, $L_- \cup \mathcal{B}$ and L_+, where L_- and L_+ are closed m-braids for some m, $\mathcal{B}$ is a union of n simple bands of birth type for some n which correspond to $\sigma_{2i-1}^{\epsilon_i}$ for $i = 1, \dots, n$ and $\epsilon_i \in \{\pm 1\}$.

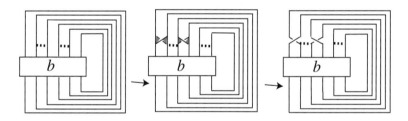

FIGURE 23.4

23.7. Notes

Theorem 23.6 (without the condition of being simple) was announced by Viro [928], and a proof was given in [316] as a corollary of Theorem 21.4. Theorem 23.7 was announced in [312]. However this is due to Rudolph [778], since it is obtained from his result that a ribbon surface in $\mathbb{R}^3[0, \infty)$ is deformed into a braided surface. In [316] a characterization of a 2-knot group was given by use of Theorem 23.6, which is restated in Sect. 27.3. González-Acuña [206] independently gave another kind of characterization of a 2-knot group by use of his braid form (Theorem 23.11). Further results based on the relation between surface braids and surface links will be discussed in Part 5.

CHAPTER 24

Split Union and Connected Sum

24.1. Natural Injection

Let p and q be non-negative integers.

For a (geometric) m-braid b, we denote by $\iota_p^q(b)$ an $(m+p+q)$-braid which is b with p trivial strings in front of b and q trivial strings behind b. Then we have a monomorphism

$$\iota_p^q : B_m \to B_{m+p+q}$$

of the m-braid group to the $(m+p+q)$-braid group. In other words, $\iota_p^q : B_m \to B_{m+p+q}$ is a homomorphism sending a standard generator σ_i of B_m to a standard generator σ_{i+p} of B_{m+p+q}.

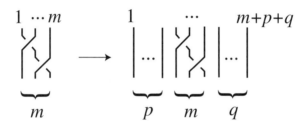

FIGURE 24.1. Natural injection $\iota_p^q : B_m \to B_{m+p+q}$

Similarly, for a surface braid S of degree m, we denote by $\iota_p^q(S)$ a surface braid of degree $(m+p+q)$ which is S with p trivial sheets in front of S and q trivial sheets behind S. If $\{b_t\}$ is a motion picture of S, then $\{\iota_p^q(b_t)\}$ is a motion picture of $\iota_p^q(S)$ (see Fig. 24.2). In terms of a braid system, if $(b_1, \ldots, b_n)$ is a braid system of S, then $(\iota_p^q(b_1), \ldots, \iota_p^q(b_n))$ is a braid system of $\iota_p^q(S)$. We have a monomorphism

$$\iota_p^q : \mathcal{B}_m \to \mathcal{B}_{m+p+q}$$

of the surface braid monoid of degree m to that of degree $m + p + q$.

Let Λ be a (surface braid) chart of degree m. We denote by $\iota_p^q(\Lambda)$ a chart of degree $m + p + q$ obtained from Λ by shifting all labels on the edges by p and by regarding it as a chart of degree $m + p + q$. It is obvious that $\iota_p^q(\Lambda)$ is a chart description of a surface braid $\iota_p^q(S)$.

24.2. Piling

Let S_1 and S_2 be surface braids of degree m_1 and m_2, respectively.

DEFINITION 24.1. The *pile product* (or *piling*) of surface braids S_1 and S_2 is a surface braid $\iota_0^{m_2}(S_1) \cdot \iota_{m_1}^0(S_2)$.

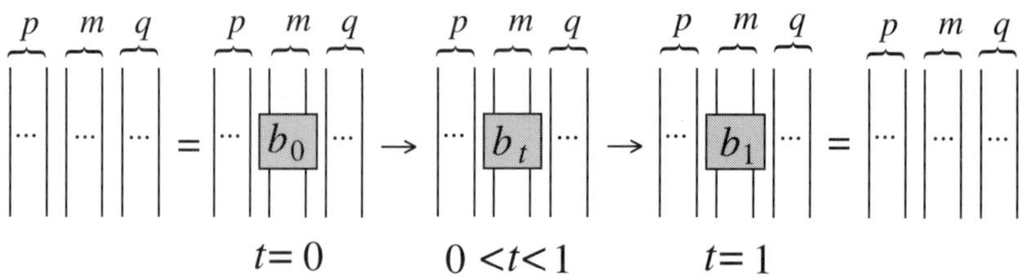

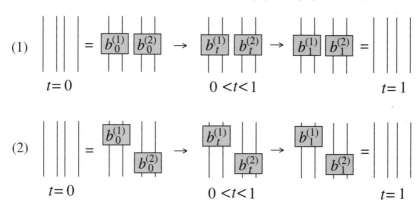

FIGURE 24.2. Natural injection $\iota_p^q : \mathcal{B}_m \to \mathcal{B}_{m+p+q}$

Let $\{b_t^{(1)}\}$ and $\{b_t^{(2)}\}$ be motion pictures of S_1 and S_2. Then a motion picture $\{\iota_0^{m_2}(b_t^{(1)}) \cdot \iota_{m_1}^0(b_t^{(2)})\}$ is a motion picture of the pile product $\iota_0^{m_2}(S_1) \cdot \iota_{m_1}^0(S_2)$. This motion picture is equivalent to a motion picture $\{b_t^{(1)} \amalg b_t^{(2)}\}$; see Fig. 24.3.

FIGURE 24.3. Piling

PROPOSITION 24.2. *A closure of the pile product $\iota_0^{m_2}(S_1) \cdot \iota_{m_1}^0(S_2)$ in $\mathbb{R}^4$ is a split union of closures of S_1 and S_2.*

Proof. See the left side of Fig. 24.4. In each hyperplane $\mathbb{R}^3[t]$ for $t \in [0,1]$ the dotted curve means a separating 2-sphere. The trace of the 2-spheres for $t \in [0,1]$ forms a 3-manifold homeomorphic to $S^2 \times [0,1]$. Attach 3-balls in $\mathbb{R}^3(-\infty, 0]$ and in $\mathbb{R}^3[0, \infty)$ to make a separating 3-sphere. ∎

PROPOSITION 24.3. *Let F be a split union of surface links F_1 and F_2; then* $\mathrm{Braid}(F) = \mathrm{Braid}(F_1) + \mathrm{Braid}(F_2)$.

Proof. By Proposition 24.2, we have $\mathrm{Braid}(F) \leq \mathrm{Braid}(F_1) + \mathrm{Braid}(F_2)$. Suppose that F is a simple closed surface braid of degree m. The restriction of F to the part corresponding to F_i $(i = 1, 2)$ is also a simple closed surface braid whose degree is greater than or equal to $\mathrm{Braid}(F_i)$. Thus m is greater than or equal to $\mathrm{Braid}(F_1) + \mathrm{Braid}(F_2)$. ∎

24.3. Connected Sum

DEFINITION 24.4. The *connected sum* of S_1 and S_2 is a surface braid
$$\iota_0^{m_2-1}(S_1) \cdot \iota_{m_1-1}^0(S_2).$$

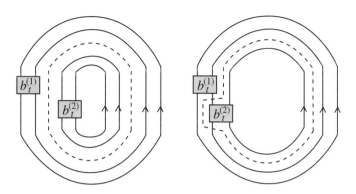

FIGURE 24.4. Split union (the left) and connected sum (the right)

Let $\{b_t^{(1)}\}$ and $\{b_t^{(2)}\}$ be motion pictures of two surface braids S_1 and S_2 of degree m_1 and m_2.

DEFINITION 24.5. The *connected sum* of $\{b_t^{(1)}\}$ and $\{b_t^{(2)}\}$ is a motion picture $\{\beta_t\}_{t\in[0,1]}$ such that

$$\beta_t = \iota_0^{m_2-1}(b_t^{(1)}) \cdot \iota_{m_1-1}^0(b_t^{(2)}).$$

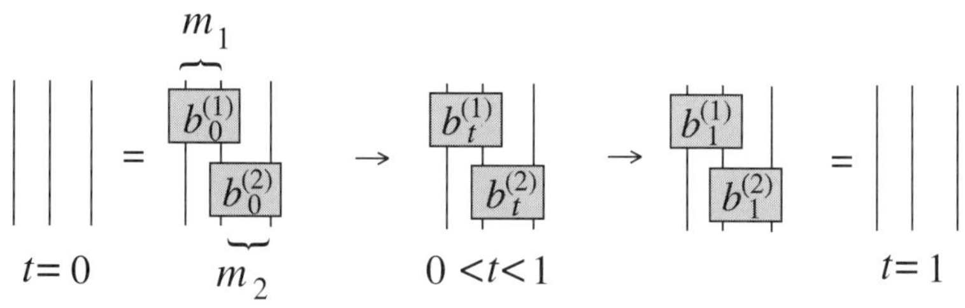

FIGURE 24.5. Connected sum

PROPOSITION 24.6. *A closure of the connected sum* $\iota_0^{m_2-1}(S_1) \cdot \iota_{m_1-1}^0(S_2)$ *in* $\mathbb{R}^4$ *is a connected sum of closures of S_1 and S_2.*

Proof. See the right side of Fig. 24.4. In each hyperplane $\mathbb{R}^3[t]$ for $t \in [0,1]$ the dotted curve means a decomposing 2-sphere which intersects the link at two points. The trace of the 2-spheres for $t \in [0,1]$ forms a 3-manifold homeomorphic to $S^2 \times [0,1]$. Attach 3-balls in $\mathbb{R}^3(-\infty,0]$ and in $\mathbb{R}^3[0,\infty)$ to make a decomposing 3-sphere. ∎

COROLLARY 24.7. *Let F be a connected sum of surface links F_1 and F_2; then* $\mathrm{Braid}(F) \leq \mathrm{Braid}(F_1) + \mathrm{Braid}(F_2) - 1.$

24.4. Charts of Piling and Connected Sum

For two charts Λ_1 and Λ_2 of the same degree, we denote by $\Lambda_1 \cdot \Lambda_2$ a chart which is a split union in D_2^2 of Λ_1 and Λ_2. If S_1 and S_2 are surface braids described by Λ_1 and Λ_2, then the product $S_1 \cdot S_2$ is described by $\Lambda_1 \cdot \Lambda_2$.

Recall that if S is a surface braid described by Λ, then $\iota_a^b(S)$ is a surface braid described by $\iota_a^b(\Lambda)$.

Using this notation, we define a pile product and a connected sum of charts as follows:

DEFINITION 24.8. (1) A *pile product* of charts Λ_1 and Λ_2 of degree m_1 and m_2 is a chart $\iota_0^{m_2}(\Lambda_1) \cdot \iota_{m_1}^0(\Lambda_2)$.
(2) A *connected sum* of charts Λ_1 and Λ_2 of degree m_1 and m_2 is a chart $\iota_0^{m_2-1}(\Lambda_1) \cdot \iota_{m_1-1}^0(\Lambda_2)$.

This definition is compatible with a pile product and a connected sum of surface braids.

Do not confuse a connected sum of charts Λ_1 and Λ_2 with a "connected sum" (in a usual sense in knot theory) of 1-manifolds or graphs embedded in a 2-disk.

CHAPTER 25

Markov's Theorem in Dimension Four

25.1. 2-Dimensional Conjugation

Let S be a surface braid of degree m and let b be an m-braid.

DEFINITION 25.1. A surface braid S' is obtained from S by a *conjugation* (or a 2-*dimensional conjugation*) associated with $b \in B_m$ if S' has a braid system $(b^{-1}b_1 b, \cdots, b^{-1}b_n b)$ where $(b_1, \cdots, b_n)$ is a braid system for S.

Let t_0 and t_1 be numbers in $[0, 1]$ which are sufficiently close to 0 and 1, respectively. Let $\{b_t\}_{t \in [0,1]}$ be a motion picture of S. Consider a motion picture $\{\beta_t\}_{t \in [0,1]}$ such that

$$\beta_t = b^{-1} b_{(t-t_0)/(t_1-t_0)} b \quad \text{for } t \in [t_0, t_1]$$

and $\{\beta_t\}_{t \in [0,t_0]}$ and $\{\beta_t\}_{t \in [t_1,1]}$ have no singular points (Fig. 25.1). Then $\{\beta_t\}_{t \in [0,1]}$ is a motion picture of a surface braid obtained from S by a conjugation associated with b.

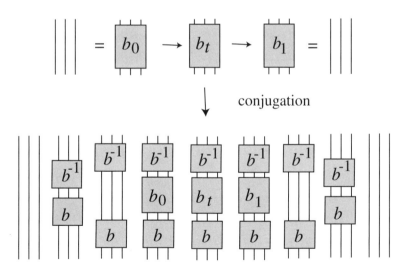

FIGURE 25.1. Conjugation

Conjugation may be defined as follows: Let $N(\partial D_2^2) = \partial D_2^2 \times I$ be a collar neighborhood of ∂D_2^2 in D_2^2 with $\partial D_2^2 \times \{1\} = \partial D_2^2$, where $I = [0, 1]$. Let b be a geometric m-braid with $\partial b = Q_m \times \partial I$ and $\{h_t\}_{t \in I}$ an isotopy of D_1^2 associated with b; i.e., $h_0 = \text{id}$ and for each $t \in I$, $h_t(Q_m) = pr_1(b \cap (D_1^2 \times \{t\}))$ and $h_t|_{\partial D_1^2} = \text{id}$. Let

187

H be a fiber-preserving homeomorphism of $D_1^2 \times D_2^2$ which fixes $pr_2^{-1}(y) = D_1^2 \times \{y\}$ for $y \in D_2^2 \setminus N(\partial D_2^2)$ and acts on $D_1^2 \times \{y\}$ by h_t for $y = (z, t) \in \partial D_2^2 \times I = N(\partial D_2^2)$. For a surface braid S, the image $H(S)$ is a surface braid obtained from S by a conjugation associated with b.

It is obvious that if S and S' are related by a conjugation, then their closures in $D_1^2 \times S^2$ are equivalent as closed surface braids in $D_1^2 \times S^2$. Therefore their closures in $\mathbb{R}^4$ are ambient isotopic in $\mathbb{R}^4$ as surface links.

EXERCISE 25.2. Let F be a closed surface braid in $D_1^2 \times D_2^2$ (or in $\mathbb{R}^4$). Prove that the conjugacy class of a cut of F along $pr_2^{-1}(y)$ for $y \in S^2 \setminus \Sigma(F)$ does not depend on a choice of y.

EXERCISE 25.3. Show that there is a bijection between the set of conjugacy classes of surface braids in $D_1^2 \times D_2^2$ of degree m and the set of equivalence classes of closed surface braids in $D_1^2 \times S^2$ of degree m.

EXERCISE 25.4. (**Conjugation in Chart Description**) Let S be a simple surface braid of degree m described by a chart Γ of degree m. Let Γ' be a chart of degree m which is obtained from Γ by introducing some concentric simple loops (hoops) surrounding Γ. Then the surface braid described by Γ' is a conjugate of S by some m-braid b. Observe a relationship between b and the hoops.

25.2. 2-Dimensional Stabilization

Let S be a surface braid of degree m.

DEFINITION 25.5. A surface braid S' of degree $m + 1$ is obtained from S by a *stabilization* (or a *2-dimensional stabilization*) if S' has a braid system $(b_1, \cdots, b_n, \sigma_m^{-1}, \sigma_m)$ of degree $m + 1$ where $(b_1, \cdots, b_n)$ is a braid system of degree m for S. A *destabilization* (or a *2-dimensional destabilization*) is the inverse of a stabilization.

Let t_0 and t_1 be numbers in $[0, 1]$ which are sufficiently close to 0 and 1, respectively. Let $\{b_t\}_{t \in [0,1]}$ be a motion picture of S. Consider a motion picture $\{\beta_t\}_{t \in [0,1]}$ of degree $m + 1$ such that

$$\beta_t = b_{(t-t_1)/(t_1-t_0)} \sigma_m \quad \text{for } t \in [t_0, t_1]$$

and $\{\beta_t\}_{t \in [0,t_0]}$ has a simple singular point which introduces σ_m, and $\{\beta_t\}_{t \in [t_1,1]}$ has a simple singular point which removes σ_m (Fig. 25.2). Then $\{\beta_t\}_{t \in [0,1]}$ is a motion picture of a surface braid obtained from S by a stabilization.

EXERCISE 25.6. Verify that if a surface braid S' is obtained from a surface braid S by a stabilization, then their closures in $\mathbb{R}^4$ are ambient isotopic.

Stabilization can be defined as follows: Recall a map ι_p^q from Sect. 24.1. Let S be a surface braid of degree m. Then $\iota_0^1(S)$ is a surface braid of degree $m+1$ which is S together with a trivial disk. Let S^* be a surface braid of degree 2 with two branch points, which is uniquely determined up to equivalence as the generator of the monoid $\mathcal{B}_2 = S\mathcal{B}_2$. A surface braid obtained from S by a *stabilization* is the product $\iota_0^1(S) \cdot \iota_{m-1}^0(S^*)$, which is a connected sum of S and S^*.

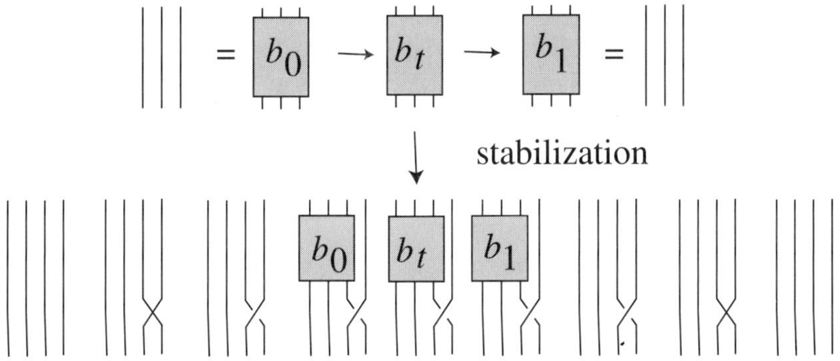

FIGURE 25.2. Stabilization

EXERCISE 25.7. (**Stabilization in Chart Description**) Let S be a simple surface braid of degree m described by a chart Γ of degree m. Let Γ' be a chart of degree $m + 1$ which is obtained from Γ (precisely speaking $\iota_0^1(\Gamma)$) by introducing a free edge with label m separated from Γ. Then the surface braid described by Γ' is obtained from S by a stabilization.

25.3. Stabilization for Closed Surface Braids

For a closed surface braid F in $D_1^2 \times S^2$, we denote by $\iota_0^1(F) = F \amalg G$ a closed surface braid which is a split union of F and a trivial closed surface braid G of degree 1; namely, $\iota_0^1(F) = F \cup (\{z'\} \times S^2)$, where z' is a point of D_1^2 which is separated from the image $pr_1(F)$. Let α be a simple arc connecting a point of F and a point of G such that α is in a fiber $pr_2^{-1}(y)$ for some regular point $y \in S^2 \setminus \Sigma(F)$.

LEMMA 25.8. *In the above situation, there is a nice 1-handle in $D_1^2 \times S^2$ attaching to $F \cup G$ with core α. Furthermore the surgery result, which we denote by $F(\alpha)$, is uniquely determined up to equivalence from F and α.*

Proof. See Proposition 20.3. ∎

DEFINITION 25.9. A closed surface braid F' is obtained from F by a *stabilization* if F' is obtained from $\iota_0^1(F) = F \amalg G$ by surgery along a nice 1-handle whose core α is in a fiber $pr_2^{-1}(y)$ for a point $y \in S^2 \setminus \Sigma(F)$ and one of the endpoints of α is on G.

LEMMA 25.10. *A closed surface braid F' is obtained from F by a stabilization (up to equivalence) if and only if there exist cuts S and S' of F and F' such that S' is obtained from S by a stabilization.*

Proof. Suppose that F' is equivalent to $F(\alpha)$. Without loss of generality we may assume that $F \cap (D_1^2 \times N(p)) = Q_m \times N(p)$ and α is a straight segment connecting z_m and z_{m+1} in $D_1^2 \times \{p\}$. Then $F(\alpha) \cap D_1^2 \times (S^2 \setminus \mathrm{int}N(p))$ is $\iota_0^1(S)$ for some cut S of F and $F(\alpha) \cap (D_1^2 \times N(p))$ is $\iota_{m-1}^1(S^*)$. Hence F' has a cut S' which is obtained from S by a stabilization. Conversely suppose that F' has a cut S' which is obtained from a cut S by a stabilization. Since $\iota_0^1(S) \cdot \iota_{m-1}^0(S^*)$ is obtained from $\iota_0^1(S) \cdot \iota_{m-1}^0(S_0)$ by surgery along a nice 1-handle, we have the result. ∎

EXERCISE 25.11. (**Stabilization in Chart Description**) Let F be a simple closed surface braid of degree m in $D_1^2 \times S^2$ described by a chart Γ in S^2 of degree

m. Let Γ' be a chart of degree $m+1$ which is obtained from Γ (precisely speaking $\iota_0^1(\Gamma)$) by introducing a free edge with label m. (We may put the free edge in any region of $S^2 \setminus \Gamma$.) Then the closed surface braid in $D_1^2 \times S^2$ described by Γ' is obtained from S by a stabilization. Observe what happens when we put the free edge in another region of $S^2 \setminus \Gamma$, or when we replace the free edge by an oval nest such that the label of the free edge is m and the labels of the concentric simple loops are less than m.

25.4. Markov's Theorem in Dimension Four

THEOREM 25.12 (Markov's Theorem in Dimension Four). *Let S and S' be surface braids. Their closures in $\mathbb{R}^4$ are ambient isotopic if and only if they are related by a finite sequence of braid ambient isotopies, conjugations, stabilizations and destabilizations.*

This theorem is proved in Chapter 26.

LEMMA 25.13. *Let F and F' be closed surface braids in $D_1^2 \times S^2$ and let S and S' be cuts of them. (1) If F is braid ambient isotopic to F', then S' is obtained from S by braid ambient isotopies and conjugations. (2) If F' is obtained from F by a stabilization, then S' is obtained from S by conjugations and a stabilization.*

Proof. (1) Since F is braid ambient isotopic to F', there is a sequence of closed surface braids $F = F_0, F_1, \cdots, F_t = F'$ such that F_i $(i = 1, \cdots, t)$ is obtained from F_{i-1} by a braid ambient isotopy of $D_1^2 \times S^2$ rel $D_1^2 \times N_i$, where N_i is a small 2-disk in S^2 containing no branch points of $pr_2|_{F_{i-1}}$ and $pr_2|_{F_i}$. Take cuts S_{i-1}^{**} and S_i^* of F_{i-1} and F_i which are braid ambient isotopic in $D_1^2 \times (S^2 \setminus \text{int} N_i)$. Put $S_0^* = S$ and $S_t^{**} = S'$. Since for each $i \in \{0, \cdots, t\}$, S_i^* and S_i^{**} are cuts of the same F_i, they are related by conjugations. So we have (1). By Lemma 25.10, we have (2). ∎

By Lemma 25.13, Theorem 25.12 is equivalent to the following.

THEOREM 25.14. *Two closed surface braids F and F' in $D_1^2 \times S^2$ are ambient isotopic in $\mathbb{R}^4$ if and only if they are related by a finite sequence of braid ambient isotopies, stabilizations and destabilizations.*

REMARK 25.15. Suppose that F and F' are closed surface braids in $D_1^2 \times S^2$ which are ambient isotopic in $\mathbb{R}^4$. By Theorem 25.14, they are related by a finite sequence of braid ambient isotopies, stabilizations and destabilizations. It is proved in [**327**] that we can transform F to F' by a sequence of braid ambient isotopies and stabilizations followed by a sequence of braid ambient isotopies and destabilizations.

CHAPTER 26

Proof of Markov's Theorem in Dimension Four

26.1. Introduction

We have seen that any oriented surface link is ambient isotopic to a closed braid, which is an analogue of Alexander's theorem in classical knot theory. In this chapter we prove Markov's theorem in dimension four.

The argument given here is based on the author's manuscript [**315**] (cf. [**317**]). The proof follows the argument in Birman's book [**42**], which gives a complete proof of Markov's theorem. There are two big differences from the classical dimensional case. One is existence of branch points in surface link diagrams, which makes it difficult to construct a sawtooth and to achieve our key argument "height reduction process". The other is that a surgery result of a triangulated surface link along a sawtooth is no longer triangulated. To avoid these problems, we shall define the notion of a division as a generalization of a triangulation.

The whole argument is completely technical, and the reader who is not interested in the details should skip over or through this chapter.

Throughout this chapter a surface link is oriented.

26.2. Division of a Surface

We denote by ℓ the 4th axis of $\mathbb{R}^4$ and by $\pi : \mathbb{R}^4 \to \mathbb{R}^3$ the projection along it. The distance $d(x, y)$ between $x = (x_i)$ and $y = (y_i) \in \mathbb{R}^4$ is $\max\{|x_i - y_i| \,|\, i = 1, 2, 3, 4\}$. For a subset X of $\mathbb{R}^4$, a point p of X and a positive number ϵ, $N_\epsilon(p; X)$ is the (closed) ϵ-neighborhood of p in X. For subsets X, Y of $\mathbb{R}^4$, the *join* $X * Y$ is $\{\lambda p + \mu q \in \mathbb{R}^4 \,|\, p \in X, q \in Y, \lambda + \mu = 1, \lambda \geq 0, \mu \geq 0\}$. (It is not necessary for each point of $X * Y$ to be uniquely expressed in the form $\lambda p + \mu q$.) For $p, q, r \in \mathbb{R}^4$ in general position, A_{pqr} or $|p, q, r|$ stands for the oriented 2-simplex determined by them. In this chapter a 2-simplex means a linear 2-simplex—i.e. a triangle.

Let F be a surface link in $\mathbb{R}^4$. A *division* of F is a finite collection $K = \{A_1, ..., A_s\}$ of 2-simplices such that $F = \cup_{i=1}^{s} A_i$ and $A_i \cap A_j$ $(i \neq j)$ is an arc, a point or an empty set contained in $\partial A_i \cap \partial A_j$. A division K' of F is called a *subdivision* of a division K if each 2-simplex of K' is a subset of a 2-simplex of K. Let K be a division of F and $A = A_{pqr}$ a 2-simplex of K. For an interior point s of an edge $\overline{qr}$, $(K \setminus \{A_{pqr}\}) \cup \{A_{pqs}, A_{psr}\}$ is a subdivision of K, which is said to be obtained from K by an *elementary dividing operation*. A subdivision K' of K is called a *good subdivision* of K if $K = K'$ or there is a sequence of elementary dividing operations transforming K into K'.

LEMMA 26.1. *For divisions K_1 and K_2 of F, there is a common good subdivision.*

Proof. (Case I) First, consider the case where K_2 is a subdivision of K_1. Let A be a 2-simplex of K_1. We denote by $K_i|_A$ $(i = 1, 2)$ the restriction of K_i to A. $K_1|_A = \{A\}$ and $K_2|_A$ is a division of A. Let $v_1, \cdots, v_m$ be vertices of $K_2|_A$ such that v_1, v_2 and v_3 are vertices of A. Let $v_2 = p_1, p_2, \cdots, p_s = v_3$ be points on $\overline{v_2 v_3}$ appearing in this order such that every vertex of $K_2|_A$ lies on $\overline{v_1 p_i}$ for some i $(i = 1, \cdots, s)$. Divide each component of $K_2|_A$ by $\overline{v_1 p_i}$ $(i = 1, \cdots, s)$. Each component of the result is a 2-simplex or a convex quadralateral. If it is a convex quadralateral, divide it into two 2-simplices by a diagonal line. Let $\mathbb{A}$ be the division of A obtained in this way. We prove that $\mathbb{A}$ is a good subdivision of both $K_1|_A$ and $K_2|_A$. Then applying the same argument for each 2-simplex of A, we have a common good subdivision of K_1 and K_2. First we prove that $\mathbb{A}$ is a good subdivision of $K_1|_A$. Let $\mathbb{A}_1 = \{A_{v_1 p_i p_{i+1}}(i = 1, \cdots, s - 1)\}$, which is a good subdivision of $K_1|_A$. Since $\mathbb{A}$ is a subdivision of $\mathbb{A}_1$, it is sufficient to show that for each i $(i = 1, \cdots, s - 1)$, the restriction of $\mathbb{A}$ to $A_{v_1 p_i p_{i+1}}$ is a good subdivision of $\{A_{v_1 p_i p_{i+1}}\}$. Let $p_i = a_1, a_2, \cdots, a_q = v_1$ and $p_{i+1} = b_1, b_2, \cdots, b_r = v_1$ be vertices of $\mathbb{A}$ on $\overline{p_i v_1}$ and on $\overline{p_{i+1} v_1}$ appearing in this order respectively. Either $\overline{a_1 b_2}$ or $\overline{b_1 a_2}$ is an edge of a 2-simplex of $\mathbb{A}$ which has $\overline{a_1 b_1}$ as an edge. If it is $\overline{a_1 b_2}$ (resp. $\overline{b_1 a_2}$), apply an elementary dividing operation to $A_{v_1 p_i p_{i+1}}$ along it. Suppose that it is $\overline{a_1 b_2}$. Then either $\overline{a_1 b_3}$ or $\overline{b_2 a_2}$ is an edge of a 2-simplex of $\mathbb{A}$ which has $\overline{a_1 b_2}$ as an edge. If it is $\overline{a_1 b_3}$ (resp. $\overline{b_2 a_2}$), then apply an elementary dividing operation to $A_{v_1 a_1 b_2}$ along it. Continuing this argument, we see that the restriction of $\mathbb{A}$ to $A_{v_1 p_i p_{i+1}}$ is a good subdivision of $A_{v_1 p_i p_{i+1}}$. Hence $\mathbb{A}$ is a good subdivision of $K_1|_A$.

Secondly we prove that $\mathbb{A}$ is a good subdivision of $K_2|_A$. Since $\mathbb{A}$ is a subdivision of $K_2|_A$, it suffices to show that for each 2-simplex B of $K_2|_A$, the restriction of $\mathbb{A}$ to B is a good subdivision of $\{B\}$. (i) Consider the case where two edges of B are contained in $\cup_{i=1}^s \overline{v_1 p_i}$. Then v_1 is a vertex of B. Divide B into some 2-simplices along $\cup_{i=1}^s \overline{v_1 p_i}$; then the result $\mathbb{B}$ is a good subdivision of B. Since $\mathbb{A}$ is a subdivision of $\mathbb{B}$, it is sufficient to prove that, for each 2-simplex C of $\mathbb{B}$, $\mathbb{A}|_C$ is a good subdivision of C. Note that two edges of C are in $\overline{v_1 p_i} \cup \overline{v_1 p_{i+1}}$ for some i and $C = A_{v_1 a_k b_\ell}$. From the construction of $\mathbb{A}$, we see that $\mathbb{A}|_C$ is a good subdivision of C. (ii) Consider the case where just one edge of B is contained in $\cup_{i=1}^s \overline{v_1 p_i}$. Let this edge be $\overline{a_k a_{k+1}}$ contained in $\overline{v_1 p_i}$. The vertex c of B opposite to the edge is in $\overline{v_1 p_j}$ for some $j \neq i$. Suppose that $j > i$. If $j \neq i + 1$, then the restriction of B to $A_{v_1 p_i p_{i+1}}$ is a convex quadralateral. Let b_ℓ and $b_{\ell+1}$ be vertices of the quadralateral which are in $\overline{v_1 p_{i+1}}$. By the definition of $\mathbb{A}$, there are two 2-simplices of $\mathbb{A}$ contained in this quadralateral, which are $\{A_{a_k b_\ell a_{k+1}}, A_{a_{k+1} b_\ell b_{\ell+1}}\}$ or $\{A_{a_k b_\ell b_{\ell+1}}, A_{a_k b_{\ell+1} a_{k+1}}\}$. In the former case, divide B along $\overline{a_{k+1} b_\ell}$ and then along $\overline{b_\ell b_{\ell+1}}$. In the latter case, divide B along $\overline{a_k b_{\ell+1}}$ and then along $\overline{b_\ell b_{\ell+1}}$. These dividings are elementary dividing operations. Apply the same argument to the restriction of B to $A_{v_1 p_{i+1} p_{i+2}}, \cdots, A_{v_1 p_{j-2} p_{j-1}}$ inductively. The restriction of B to $A_{v_1 p_{j-1} p_j}$ is a 2-simplex of $\mathbb{A}$. In this way we have a sequence of elementary dividing operations that make B into $\mathbb{A}|_B$. The case $j = i+1$ is trivial and the case $j < i$ is proved by a similar argument. (iii) Consider the case where no edges of B are contained in $\cup_{i=1}^s \overline{v_1 p_i}$. There is a vertex c of B such that c lies on $\overline{v_1 p_i}$ for some i and $\overline{v_1 p_i}$ divides B into 2-simplices. This is an elementary dividing operation. Since each 2-simplex of the result satisfies the above condition (ii), we see that $\mathbb{A}|_B$ is a good subdivision of B. This completes the proof of Case I.

(Case II) General case. Let K_3 be a common subdivision of K_1, K_2. By Case I, there is a common good subdivision K_4 of K_1, K_3, and a common good subdivision K of K_2, K_4. Then K is a common good subdivision of K_1, K_2. ∎

26.3. General Position with Respect to ℓ

A 2-plane E in $\mathbb{R}^4$ is said to be *in general position with respect to* ℓ if there is no 3-plane in $\mathbb{R}^4$ containing $E \cup \ell$, or equivalently if $\pi(E)$ is a 2-plane in $\mathbb{R}^3$ missing the origin O of $\mathbb{R}^3$. A 2-simplex A in $\mathbb{R}^4$ is said to be *in general position with respect to* ℓ if the 2-plane in $\mathbb{R}^4$ determined by A is in general position with respect to ℓ. Then, in case A is oriented, $\pi(A)$ is an oriented 2-simplex in $\mathbb{R}^3$ and forms together with O an oriented 3-simplex in $\mathbb{R}^3$. If the orientation of the 3-simplex coincides with that of $\mathbb{R}^3$ (or does not, resp.), then we call A a *positive* 2-*simplex* (or *negative* 2-*simplex*, resp.). The *sign* of A is $+1$ (or -1, resp.). This is denoted by $\tau(A)$. In other words, take a ray in $\mathbb{R}^3$ from O through an interior point x of $\pi(A)$. If an orientation of $\mathbb{R}^3$ at x determined by oriented $\pi(A)$ and the ray coincides with (or is opposite to, resp.) the orientation of $\mathbb{R}^3$, then A is positive (or negative, resp.). Unless otherwise stated, we always assume that a 2-simplex A contained in a surface link F in $\mathbb{R}^4$ has an orientation which is induced from F.

A surface link F in $\mathbb{R}^4$ is said to be *in general position with respect to* ℓ if any 2-simplex in F is in general position with respect to ℓ and the map $\pi|_F : F \to \mathbb{R}^3$ is a generic projection.

26.4. $\mathcal{E}$ Operation

Let V be a 3-simplex in $\mathbb{R}^4$ such that 2-faces $A_1, \cdots, A_4$ of V are in general position with respect to ℓ. Put $J_i = \{\lambda p \in \mathbb{R}^3 | p \in \pi(A_i), \lambda \geq 0\}$ for $i \in \{1, 2, 3, 4\}$. One of the following holds: (α) $J_a = J_b \cup J_c \cup J_d$, (β) $J_a \cup J_b = J_c \cup J_d$, or (γ) $J_a \cup J_b \cup J_c \cup J_d = \mathbb{R}^3$, where $\{a, b, c, d\}$ is the set $\{1, 2, 3, 4\}$. According to this, we call V a 3-simplex of *type α, type β* or *type γ*.

Let F and V be a surface link and a 3-simplex in $\mathbb{R}^4$ resp. such that $F \cap V$ is a union of i ($i = 1, 2$ or 3) 2-faces of V. Put $F \cap V = D_1$ and $\mathrm{cl}(\partial V \setminus F \cap V) = D_2$. Recall that a surface link $F' = (F \setminus D_1) \cup D_2$ in $\mathbb{R}^4$ is said to be obtained from F by an elementary move along V.

If ∂V is in general position with respect to ℓ, an elementary move $\mathcal{F}$ is called an $\mathcal{E}_i$ *operation* ($i = 1, 2$ or 3) or an $\mathcal{E}$ *operation*. Suppose that each 2-face of V contained in F (or in F', resp.) has an orientation induced from F (resp. F'). In case $\mathcal{F}$ is an $\mathcal{E}_1$ operation, let A_1 be the 2-face of V in F and A_2, A_3, A_4 2-faces of V in F'. We call $\mathcal{F}$ an $\mathcal{E}_1^j$ *operation* ($j = 1, \cdots, 8$) if signs of $A_1, \cdots, A_4$ are as in Table 26.1. In the case where $\mathcal{F}$ is $\mathcal{E}_2$, let A_1, A_2 be in F and A_3, A_4 in F'. $\mathcal{F}$ is an $\mathcal{E}_2^j$ *operation* ($j = 1, \cdots, 6, -1, -2, -6$) if their signs are as in Table 26.2. An $\mathcal{E}_2^{-j}$ *operation* ($j = 1, 2, 6$) is the inverse of an $\mathcal{E}_2^j$ operation. If $\mathcal{F}$ is $\mathcal{E}_3$, we call it an $\mathcal{E}_3^j$ *operation* ($j = 1, \cdots, 8$) when its inverse is $\mathcal{E}_1^j$.

Let K, K' be divisions of F, F' such that each 2-face of V is a 2-simplex of K or K' and $K|_{\mathrm{cl}(F \setminus D_1)} = K'|_{\mathrm{cl}(F' \setminus D_2)}$. Then we say that K' is obtained from K by an *elementary move* along V.

Let K be a division of a surface link which is in general position with respect to ℓ. We define the *height* of K by the number of negative 2-simplices of K. It is denoted by $h(K)$. If a division K' is obtained from K by an $\mathcal{E}$ operation, then the difference $h(K') - h(K)$ is as in Tables 26.1 and 26.2.

j	$\mathcal{E}_1^j$	$\tau(A_1)$	$\tau(A_2), \tau(A_3), \tau(A_4)$	type of V	$h(K') - h(K)$
1	$\mathcal{E}_1^1$	$+$	$+, +, +$	α	0
2	$\mathcal{E}_1^2$	$+$	$+, -, -$	α	2
3	$\mathcal{E}_1^3$	$-$	$+, +, -$	α	0
4	$\mathcal{E}_1^4$	$-$	$-, -, -$	α	2
5	$\mathcal{E}_1^5$	$+$	$+, +, -$	β	1
6	$\mathcal{E}_1^6$	$-$	$+, -, -$	β	1
7	$\mathcal{E}_1^7$	$+$	$-, -, -$	γ	3
8	$\mathcal{E}_1^8$	$-$	$+, +, +$	γ	-1

TABLE 26.1

j	$\mathcal{E}_2^j$	$\tau(A_1), \tau(A_2)$	$\tau(A_3), \tau(A_4)$	type of V	$h(K') - h(K)$
1	$\mathcal{E}_2^1$	$+, +$	$+, -$	α	1
-1	$\mathcal{E}_2^{-1}$	$+, -$	$+, +$	α	-1
2	$\mathcal{E}_2^2$	$-, -$	$+, -$	α	-1
-2	$\mathcal{E}_2^{-2}$	$+, -$	$-, -$	α	1
3	$\mathcal{E}_2^3$	$+, +$	$+, +$	β	0
4	$\mathcal{E}_2^4$	$+, -$	$+, -$	β	0
5	$\mathcal{E}_2^5$	$-, -$	$-, -$	β	0
6	$\mathcal{E}_2^6$	$+, +$	$-, -$	γ	2
-6	$\mathcal{E}_2^{-6}$	$-, -$	$+, +$	γ	-2

TABLE 26.2

26.5. Deformation Chains

DEFINITION 26.2 ($\mathcal{R}$ Operation). Let F be a surface link and F' a surface obtained from F by an $\mathcal{E}$ operation $\mathcal{F}$ along a 3-simplex V. If all 2-faces of V are positive when we assign them orientations induced from F and F' as before, then $\mathcal{F}$ is called an $\mathcal{R}$ operation. In other words, an $\mathcal{R}$ operation is an $\mathcal{E}_1^1$, $\mathcal{E}_2^3$ or $\mathcal{E}_3^1$ operation.

DEFINITION 26.3 ($\mathcal{W}$ Operation). Let F be a surface link and F_1 the surface obtained from F by an $\mathcal{E}_1^5$ operation $\mathcal{F}_1$ along a 3-simplex V_1. Let $A_1, \cdots, A_4$ be 2-faces of V_1 such that $A_1 \subset F$, $A_2, A_3, A_4 \subset F_1$ and $\tau(A_4) = -1$. Let F_2 be a surface obtained from F_1 by an $\mathcal{E}_1^8$ operation $\mathcal{F}_2$ along a 3-simplex V_2 with $F_1 \cap V_2 = A_4$. If $F \cap V_1 * V_2 = A_1$, then the composition of $\mathcal{F}_1$ and $\mathcal{F}_2$ is called a $\mathcal{W}$ operation.

DEFINITION 26.4 ($\mathcal{U}$ Operation). Let F be a surface link and F_1 a surface obtained from F by an $\mathcal{E}_2^1$ operation $\mathcal{F}_1$ along a 3-simplex V_1. Let $A_1, \cdots, A_4$ be 2-faces of V_1 such that $A_1, A_2 \subset F$, $A_3, A_4 \subset F_1$ and $\tau(A_4) = -1$. Let F_2 be a surface obtained from F_1 by an $\mathcal{E}_1^8$ operation $\mathcal{F}_2$ along V_2 with $F_1 \cap V_2 = A_4$. If $F \cap V_1 * V_2 = A_1 \cup A_2$, then the composition of $\mathcal{F}_1$ and $\mathcal{F}_2$ is called a $\mathcal{U}$ operation.

DEFINITION 26.5 ($\mathcal{T}$ Operation). Let F be a surface link and F_1 a surface obtained from F by an $\mathcal{E}_3^8$ operation $\mathcal{F}_1$ along a 3-simplex V_1. Let $A_1, \cdots, A_4$ be 2-faces of V_1 such that $A_1, A_2, A_3 \subset F$ and $A_4 \subset F_1$. Let F_2 be a surface obtained from F_1 by an $\mathcal{E}_1^8$ operation $\mathcal{F}_2$ along V_2 with $F_1 \cap V_2 = A_4$. The composition of $\mathcal{F}_1$ and $\mathcal{F}_2$ is called a $\mathcal{T}$ *operation*. (We do not require $F \cap V_1 * V_2 = A_1 \cup A_2 \cup A_3$.)

DEFINITION 26.6 (Sawtooth). Let F be a surface link and A a 2-simplex in F which is in general position with respect to ℓ. Let $\mathbb{A} = \{A_1, \cdots, A_s\}$ be a division of A. A family of 3-simplices $\mathbb{V} = \{V_1, \cdots, V_s\}$ in R^4 is called a *sawtooth on A avoiding F with basis* $\mathbb{A}$ if it satisfies the following conditions: (1) for each $i \in \{1, \cdots, s\}$, $F \cap V_i = A_i$ and V_i is of type γ, and (2) $V_i \cap V_j = A_i \cap A_j$ for $i \neq j$. We say that $\mathbb{V}$ *avoids* a subset X of $\mathbb{R}^4$ if $X \cap V_i = X \cap A_i$ for $i \in \{1, \cdots, s\}$.

DEFINITION 26.7 ($\mathcal{S}$ Operation). In the situation of Definition 26.6, if A is a negative 2-simplex, then the composition of $\mathcal{E}_1^8$ operations along $V_1, \cdots, V_s$ is called an $\mathcal{S}$ *operation* along $\mathbb{V}$. It replaces a negative 2-simplex A by $3s$ positive 2-simplices.

The inverse of a $\mathcal{Z}$ operation ($\mathcal{Z}$ is $\mathcal{E}, \mathcal{R}, \mathcal{W}, \mathcal{U}, \mathcal{T}$ or $\mathcal{S}$) is called a $\mathcal{Z}^{-1}$ *operation*. (In the case where $\mathcal{Z}$ is $\mathcal{E}, \mathcal{R}$ or $\mathcal{T}$, it is also a $\mathcal{Z}$ operation.)

DEFINITION 26.8 (Deformation Chain). A sequence $F_0, F_1, \cdots, F_s$ of surfaces in $\mathbb{R}^4$ is a *deformation chain* if each F_i ($i = 0, \cdots, s$) is in general position with respect to ℓ and obtained from F_{i-1} ($i \neq 0$) by an $\mathcal{E}, \mathcal{R}, \mathcal{W}^{\pm 1}, \mathcal{U}^{\pm 1}$ or $\mathcal{S}^{\pm 1}$ operation. (We exclude $\mathcal{T}$ operations.)

DEFINITION 26.9 (Markov Equivalence). Two surfaces F and F' in $\mathbb{R}^4$ are *Markov equivalent* if there is a deformation chain between them, say $F = F_0, F_1, \cdots, F_s = F'$, such that each F_i ($i = 1, \cdots, s$) is obtained from F_{i-1} by an $\mathcal{R}, \mathcal{W}^{\pm 1}$ or $\mathcal{U}^{\pm 1}$ operation.

26.6. Operations at the Division Level

In the situation of Definition 26.2, if a division K' of F' is obtained from a division K of F by an $\mathcal{E}$ operation along V, then K' is said to be obtained from K by an $\mathcal{R}$ operation along V. In the situation of Definition 26.3, 26.4 or 26.5, let K, K_1 and K_2 be divisions of F, F_1 and F_2 respectively. If K_1 is obtained from K by an $\mathcal{E}$ operation along V_1 and K_2 is obtained from K_1 by an $\mathcal{E}$ operation along V_2, then K_2 is said to be obtained from K by a $\mathcal{W}, \mathcal{U}$ or $\mathcal{T}$ operation. In the situation of Definition 26.7, let K, K' be divisions of F, F'. We say that K' is obtained from K by an $\mathcal{S}$ operation along $\mathbb{V}$ if A is a negative 2-simplex of K, each 2-face of V_i ($i = 1, \cdots, s$) except A_i is a 2-simplex of K', and K and K' coincide on what remains.

DEFINITION 26.10 ($\mathcal{A}$ Operation and $\mathcal{B}$ Operation). Let K be a division of F and A a positive (or negative, resp.) 2-simplex of K which is in general position with respect to ℓ. Let K' be a division of F obtained from K by an elementary dividing operation on A. It is said to be obtained from K by an $\mathcal{A}$ (or $\mathcal{B}$, resp.) *operation* on A. If F is in general position with respect to ℓ, then $h(K) = h(K')$ (or $h(K) = h(K') + 1$, resp.).

DEFINITION 26.11 (Deformation Chain in the Division Level). A   sequence $K_0, K_1, \cdots, K_s$ of divisions of surfaces in $\mathbb{R}^4$ is a *deformation chain* if $F_i = |K_i|$ ($i = 0, \cdots, s$) is in general position with respect to ℓ and K_i ($i = 1, \cdots, s$) is obtained from K_{i-1} by an $\mathcal{A}^{\pm 1}, \mathcal{B}^{\pm 1}, \mathcal{E}, \mathcal{R}, \mathcal{W}^{\pm 1}, \mathcal{U}^{\pm 1}$ or $\mathcal{S}^{\pm 1}$ operation. (We exclude $\mathcal{T}$ operations.) The *length* is s and the *maximal height* is $\max\{h(K_i) \,|\, i \in \{0, \cdots, s\}\}$.

DEFINITION 26.12 (Markov Equivalence). Two divisions K and K' of surfaces in $\mathbb{R}^4$ are *Markov equivalent* if there is a deformation chain $K = K_0, K_1, \cdots, K_s = K'$ with constant height; i.e. K_i $(i = 1, \cdots, s)$ is obtained from K_{i-1} by an $\mathcal{A}^{\pm 1}, \mathcal{R}, \mathcal{W}^{\pm 1}$ or $\mathcal{U}^{\pm 1}$ operation.

LEMMA 26.13. *Let K and K' be divisions of a surface link F in general position with respect to ℓ. Let R be a subset of F (possibly $R = F$) such that $K|_R$ and $K'|_R$ consist of 2-simplices of K and K' respectively. If $K|_{\mathrm{cl}(F \backslash R)} = K'|_{\mathrm{cl}(F \backslash R)}$, then there is a deformation chain $K = K_0, K_1, \cdots, K_s = K'$ such that K_i $(i = 1, \cdots, s)$ is obtained from K_{i-1} by an $\mathcal{A}^{\pm 1}$ or $\mathcal{B}^{\pm 1}$ operation. In the case where all 2-simplices contained in R are positive, we may assume that each operation is an $\mathcal{A}^{\pm 1}$ operation.*

Proof. By Lemma 26.1, there is a good subdivision L of both $K|_R$ and $K'|_R$. Let K^* be the division of F such that $K^*|_R = L$ and $K^*|_{\mathrm{cl}(F \backslash R)} = K|_{\mathrm{cl}(F \backslash R)}$. It is a good subdivision of both K and K'. So there are sequences of elementary dividing operations transforming K and K' into K^* respectively. ∎

LEMMA 26.14. *Let F and F' be surface links in general position with respect to ℓ and let K and K' be their divisions. If F is ambient isotopic to F', there is a deformation chain from K to K'.*

Proof. Since F is ambient isotopic to F', there is a sequence $F = F_0, F_1, \cdots, F_s = F'$ such that F_i $(i = 1, \cdots, s)$ is obtained from F_{i-1} by an elementary move. We may assume that each F_i is in general position with respect to ℓ and obtained from F_{i-1} by an $\mathcal{E}$ operation. For each $i \in \{1, \cdots, s\}$, take divisions K_{i-1}^{**} and K_i^* of F_{i-1} and F_i such that K_i^* is obtained from K_{i-1}^{**} by an $\mathcal{E}$ operation. Put $K = K_0^*$ and $K' = K_s^{**}$. Since K_i^* and K_i^{**} are divisions of the same F_i, they are connected by a deformation chain (Lemma 26.13). Hence we have the result. ∎

26.7. An Interpretation of Markov's Theorem in Dimension Four

For a surface F in $\mathbb{R}^4$, $h(F) = 0$ denotes that F is in general position with respect to ℓ and $h(K) = 0$ for a division K of F. (It does not depend on K.)

THEOREM 26.15. *Any surface link is ambient isotopic to a surface F in $\mathbb{R}^4$ with $h(F) = 0$.*

THEOREM 26.16. *Two surface links F and F' with $h(F) = h(F') = 0$ are ambient isotopic if and only if they are Markov equivalent (in the sense of* Definition 26.9*).*

These theorems are proved in Sect. 26.8. Here we prove that they imply Alexander's and Markov's theorems in dimension four (Theorems 23.6 and 25.12).

We denote by S_r^2 the 2-sphere $\partial N_r(O; \mathbb{R}^3)$ in $\mathbb{R}^3$ with radius r $(r > 0)$ and by $M(r_1, r_2)$ the subset $\mathrm{cl}(N_{r_2}(O; \mathbb{R}^3) \backslash N_{r_1}(O; \mathbb{R}^3))$ of $\mathbb{R}^3$ for positive numbers r_1 and r_2 with $r_1 < r_2$. Identify S^2 with S_1^2. For positive integers r_1 and r_2 with $r_1 < 1 < r_2$, let $M(r_1, r_2)$ be a bi-collared neighborhood and let $c : I_2^1 \times S^2 \to M(r_1, r_2)$ be a homeomorphism with $c(0, x) = x$ for $x \in S^2$, $c(\{1\} \times S^2) = S_{r_1}^2$ and $c(\{-1\} \times S^2) = S_{r_2}^2$.

Let $P = \{x_1, \cdots, x_n\}$ be a set of points of $M(r_1, r_2)$. We say that c *respects* P if for each $i \in \{1, \cdots, n\}$, $pr_2 \circ c^{-1}(x_i) = \rho(x_i)$ where $\rho : \mathbb{R}^3 \backslash O \to S_1^2 = S^2$ is the radial projection. For any given positive number ϵ, there is a homeomorphism

c such that $pr_2 \circ c^{-1} : M(r_1, r_2) \to I_2^1 \times S^2 \to S^2$ is a PL ϵ-approximation of $\rho|_{M(r_1,r_2)}$ and c respects P (cf. [**286, 771**]). For a positive number s, we define a homeomorphism $\widetilde{c}_s : I_1^1 \times I_2^1 \times S^2 \to M(r_1, r_2) \times [-s, s](\subset \mathbb{R}^3 \times \mathbb{R}^1 = \mathbb{R}^4)$ by $\widetilde{c}_s(t, r, x) = (c(r, x), st)$.

Let F be a surface in $\mathbb{R}^4$ with $h(F) = 0$ and take positive numbers r_1, r_2 and s such that F is contained in the interior of $M(r_1, r_2) \times [-s, s]$. Let P be the set of images of vertices of a division K of F under $\pi : \mathbb{R}^4 \to \mathbb{R}^3$ and take a homeomorphism c respecting P. Since each 2-simplex of the division $\widetilde{c}_s^{-1}(K)$ is projected into S^2 homeomorphically, $\widetilde{c}_s^{-1}(F)$ is a closed surface braid in $D_1^2 \times S^2$. So F is a closed surface braid in $\mathbb{R}^4$. Hence Theorem 26.15 implies Theorem 23.6. By the tubular neighborhood theorem and by the fact that a 2-sphere in $\mathbb{R}^4$ has a unique framing up to ambient isotopy, we see that $\widetilde{c}_s^{-1}(F)$ is uniquely determined from F up to braid ambient isotopy. We call it a closed surface braid in $D_1^2 \times S^2$ corresponding to F.

LEMMA 26.17. *Let F and F' be surface links with $h(F) = h(F') = 0$ such that F' is obtained from F by an $\mathcal{R}$ operation. Then closed surface braids C and C' in $D_1^2 \times S^2$ corresponding to them are braid ambient isotopic.*

Proof. Let K and K' be divisions of F and F' such that K' is obtained from K by an $\mathcal{R}$ operation along a 3-simplex V. Let r_1, r_2 and s be positive numbers such that $F \cup F'$ is contained in the interior of $M(r_1, r_2) \times [-s, s]$ and take a homeomorphism $c : I_2^1 \times S^2 \to M(r_1, r_2)$ respecting the set of images of vertices of K and K' under π. Using $\widetilde{c}_s^{-1}(V)$, we see that $\widetilde{c}_s^{-1}(F)$ is braid ambient isotopic to $\widetilde{c}_s^{-1}(F')$ in $D_1^2 \times S^2$. ∎

LEMMA 26.18. *Let F and F' be surface links with $h(F) = h(F') = 0$ such that F' is obtained from F by a $\mathcal{W}$ (or $\mathcal{U}$) operation along V_1 and V_2. Let C and C' be closed surface braids in $D_1^2 \times S^2$ corresponding to them. Then C' is obtained from C by braid ambient isotopies and a stabilization.*

Proof. Let r_0 be a small positive number such that $N_{r_0}(O; \mathbb{R}^3)$ is disjoint from $\pi(F \cup F')$ and $N_{r_0}(O; \mathbb{R}^3) \times \{s_0\} \subset \mathbb{R}^3 \times \mathbb{R}^1 = \mathbb{R}^4$ is contained in the interior of $V_1 * V_2$ for some $s_0 \in \mathbb{R}^1$. Let $V = N_{r_0}(O; \mathbb{R}^3) \times \{s_0\}$ and $q = (O, s_0) \in \mathbb{R}^3 \times \mathbb{R}^1 = \mathbb{R}^4$, which is the intersection of V and ℓ. Take an interior point p of $F \cap V_1 * V_2 = F \cap V_1$ and put $\beta = \mathrm{cl}(\overline{pq} \backslash V)$, which is a straight segment connecting F and ∂V in $V_1 * V_2$. Take positive numbers r_1, r_2 and s such that $r_1 < r_0 < 1 < r_2$ and $F \cup F'$ is contained in the interior of $M(r_1, r_2) \times [-s, s]$. Let K and K' be divisions of F and F' such that K' is obtained from K by the $\mathcal{W}$ (or $\mathcal{U}$) operation, and let P be the set of images of vertices $K \cup K'$ under π. There is a homeomorphism $c : I_2^1 \times S^2 \to M(r_1, r_2)$ respecting P such that, if we put $C = \widetilde{c}_s^{-1}(F)$, $C' = \widetilde{c}_s^{-1}(F')$, $C_0 = \widetilde{c}_s^{-1}(\partial V)$ and $\alpha = \widetilde{c}_s^{-1}(\beta)$, then C is a closed surface braid in $D_1^2 \times S^2$, $C_0 = \{z'\} \times S^2$ for some $z' = (x', y') \in D_1^2 = I_1^1 \times I_2^1$ with $pr_1(C) \cap I_1^1 \times [y', 1] = \emptyset$ and α is a simple arc in $D_1^2 \times \{p\}$ for some $p \in S^2$ as in Lemma 25.8. By the lemma, there is a nice 1-handle h with core α attaching to $C \cup C_0$ in $D_1^2 \times S^2$ such that the surgery result $C(\alpha)$ is obtained from C by a stabilization. It is not so hard to see that $\widetilde{c}_s(C(\alpha))$, which is obtained from $F \cup \partial V$ by surgery along a 1-handle $\widetilde{c}_s(h) \subset V_1 * V_2$, is transformed into F' by $\mathcal{R}$ operations in $V_1 * V_2$. By Lemma 26.17, $C(\alpha)$ is braid ambient isotopic to C'. ∎

Theorem 25.14 follows from Lemmas 26.17 and 26.18 and Theorem 26.16. Theorem 25.12 is equivalent to Theorem 25.14. Therefore we will prove Theorem 26.16.

26.8. Proofs of Theorems 26.15 and 26.16

LEMMA 26.19 (Sawtooth Lemma). *For a surface link F which is in general position with respect to ℓ and for a 2-simplex A in F, there is a sawtooth on A avoiding F such that the surgery result is in general position with respect to ℓ.*

LEMMA 26.20 (Height Reduction Lemma I). *Let K, K' be a deformation chain of divisions of length one. If $h(K) = h(K') \neq 0$, then there is a deformation chain $K, K_1, \cdots, K_s, K'$ of length $s + 1$ for some $s \geq 1$ such that $h(K_i) < h(K)$ ($i = 1, \cdots, s$).*

LEMMA 26.21 (Height Reduction Lemma II). *Let K, K', K'' be a deformation chain of divisions. If $h(K') > h(K)$ and $h(K') > h(K'')$, then there is a deformation chain $K, K_1, \cdots, K_s, K''$ of length $s + 1$ for some $s \geq 1$ such that $h(K_i) < h(K')$ ($i = 1, \cdots, s$).*

We shall prove Lemmas 26.19, 26.20 and 26.21 in the rest of this chapter.

Proof of Theorem 26.15. Let F_0 be a surface link. Modify it by elementary moves so that it is in general position with respect to ℓ. If $h(K_0) \neq 0$ for a division K of F_0, let A_1 be a negative 2-simplex of K_0 and take a sawtooth on A_1 avoiding F_0 as in 26.19. The result K_1 by an $\mathcal{S}$ operation along the sawtooth has height $h(K_1) = h(K) - 1$. If $h(K_1) \neq 0$, consider a sawtooth on a negative 2-simplex $A_2 \in K_1$ avoiding $F_1 = |K_1|$ and let K_2 be the result by an $\mathcal{S}$ operation. Repeating this procedure, we have a deformation chain of divisions $K_0, K_1, \cdots, K_s$ with $h(K_s) = 0$ and a desired surface $F_s = |K_s|$. ∎

Proof of Theorem 26.16. Since every operation in a deformation chain is a composition of elementary moves, the if part is trivial. We prove the only if part. Suppose that F is ambient isotopic to F'. By Lemma 26.14, they have divisions K and K' connected by a deformation chain. Let h be the maximal height of the deformation chain. If $h = 0$, then K and K' are Markov equivalent. If $h \neq 0$, using Lemmas 26.20 and 26.21 we have a new deformation chain from K to K' whose maximal height is less than h. By induction on h, we have the result. ∎

26.9. Notation

For a 2-plane E in $\mathbb{R}^4$ in general position with respect to ℓ, $\pi^{-1}(\pi(E))$ is a 3-plane in $\mathbb{R}^4$, which is parallel to ℓ. We denote by W_E the component of $\mathbb{R}^4 \setminus \pi^{-1}(\pi(E))$ containing ℓ and by $H^3_{E,z}$ the 3-plane in $\mathbb{R}^4$ containing $E \cup \{z\}$ for $z \in W_E$. Let $f_E : W_E \to \ell$ be a continuous (non-PL) map defined by assigning $z \in W_E$ the intersection point of $H^3_{E,z}$ and ℓ. For a 2-simplex A in $\mathbb{R}^4$ which is in general position with respect to ℓ, let $\langle A \rangle$, $H^3_{A,z}$ and $f_A : W_A \to \ell$ stand for the 2-plane in $\mathbb{R}^4$ containing A, $H^3_{\langle A \rangle, z}$ and $f_{\langle A \rangle} : W_{\langle A \rangle} \to \ell$. For $t \in \mathbb{R}^1$ and $J \subset \mathbb{R}^1$, we abbreviate a point $(0, 0, 0, t) \in \ell$ as $t \in \ell$ and a subset $\{(0, 0, 0, s) \mid s \in J\}$ as $J \subset \ell$. Let A be a 2-simplex in $\mathbb{R}^4$ in general position with respect to ℓ, x a point of A, y a point of ℓ and ϵ a positive number. $M_\epsilon(x, A, y)$ means a join $N_\epsilon(x; A) * N_\epsilon(y; H^3_{A,y})$, which is a 3-dimensional compact convex polyhedron contained in $H^3_{A,y}$. For a subset X of $\mathbb{R}^4$, $\Phi_\epsilon(x, A, X)$ is $\{y \in \ell \mid X \cap M_\epsilon(x, A, y) = X \cap N_\epsilon(x; A)\}$.

LEMMA 26.22. (1) *A point y of ℓ is in $\Phi_\epsilon(x, A, X)$ if and only if $X \cap M_\epsilon(x, A, y) \subset A$.*

(2) If $0 \leq \epsilon' \leq \epsilon$, then $\Phi_{\epsilon'}(x, A, X) \supset \Phi_\epsilon(x, A, X)$.

(3) If $x \in A' \subset A$, then $\Phi_\epsilon(x, A', X) \supset \Phi_\epsilon(x, A, X)$.

(4) If $X' \subset X$, then $\Phi_\epsilon(x, A, X') \supset \Phi_\epsilon(x, A, X)$.

Proof. Note that $M_\epsilon(x, A, y) \cap A = N_\epsilon(x; A)$ and we have (1). (2)–(4) are trivial by the definition. ∎

26.10. Existence of a Sawtooth, I

Let F be a surface link in general position with respect to ℓ, A a 2-simplex in F, x a point of A and H a 2-plane in $\mathbb{R}^4$ containing $\ell \cup \{x\}$. Since F is in general position with respect to ℓ, $H \cap F$ consists of some points, say $p_1 = x, p_2, \cdots, p_s$ ($s \geq 1$). Let ℓ_x be $\pi^{-1}(\pi(x))$, which is a straight line through x and parallel to ℓ. The number of points of $\{p_1 = x, p_2, \cdots, p_s\}$ lying on ℓ_x is one (if $\pi(x)$ is a regular or branch point of $\pi|_F$), two (if $\pi(x)$ is a double point) or three (if $\pi(x)$ is a triple point). For a positive number ϵ, let E_ϵ be the image of $W_A \cap (\cup_{i=2}^s N_\epsilon(p_i; H))$ under $f_A : W_A \to \ell$. Taking ϵ sufficiently small, we may assume that E_ϵ is the union of some (or no) closed intervals in ℓ and J^ϵ, where J^ϵ is a subset of ℓ which is empty, $(-\infty, -b]$, $[c, \infty)$ or $(-\infty, -b] \cup [c, \infty)$ for some sufficiently large numbers b, c. As we take ϵ smaller, the compact part of E_ϵ becomes smaller and b, c larger. Since F is closed, for a given positive number ϵ_0, there is a positive number δ_0 such that for any δ with $0 < \delta \leq \delta_0$, $H \cap N_\delta(F; \mathbb{R}^4)$ is contained in $\cup_{i=1}^s N_{\epsilon_0}(p_i; H)$.

LEMMA 26.23. *In the above situation, for any point $y \in \ell \setminus E_{\epsilon_0}$ and a positive number ϵ such that $\epsilon \leq \epsilon_0$ and $\epsilon \leq \delta_0$, the following statements hold:* (1) $\overline{xy} \cap N_\epsilon(F; \mathbb{R}^4)$ *is in* $N_{\epsilon_0}(x; H)$, (2) $F \cap N_\epsilon(\overline{xy}; \mathbb{R}^4)$ *is in* $N_{2\epsilon_0}(x; \mathbb{R}^4)$ *and* (3) $F \cap M_\epsilon(x, A, y)$ *is in* $N_{2\epsilon_0}(x; \mathbb{R}^4)$.

Proof. (1) By the hypothesis of δ_0, $\overline{xy} \cap N_\epsilon(F; \mathbb{R}^4) = \overline{xy} \cap (H \cap N_\epsilon(F; \mathbb{R}^4)) \subset \overline{xy} \cap (\cup_{i=1}^s N_{\epsilon_0}(p_i; H))$. Since $y \in \ell \setminus E_{\epsilon_0}$, $\overline{xy} \cap (\cup_{i=2}^s N_{\epsilon_0}(p_i; H)) = \emptyset$. Hence $\overline{xy} \cap N_\epsilon(F; \mathbb{R}^4) \subset N_{\epsilon_0}(x; H)$. (2) Let $z \in F \cap N_\epsilon(\overline{xy}; \mathbb{R}^4)$. Take a point z' of $\overline{xy}$ with $d(z, z') \leq \epsilon$. Note that z' is in $\overline{xy} \cap N_\epsilon(F; \mathbb{R}^4)$ and hence, by (1), in $N_{\epsilon_0}(x; H)$. So $d(x, z') \leq \epsilon_0$ and $d(x, z) \leq d(x, z') + d(z, z') \leq \epsilon_0 + \epsilon \leq 2\epsilon_0$. (3) Since $M_\epsilon(x, A, y) \subset N_\epsilon(\overline{xy}; \mathbb{R}^4)$, it follows from (2). ∎

LEMMA 26.24. *Let F be a surface link in general position with respect to ℓ, let A be a 2-simplex in F, and let x be a point of A. Suppose that $\pi(x)$ is not a branch point of $\pi|_F$. Then there is a positive number ϵ such that $\Phi_\epsilon(x, A, F)$ contains a non-empty open subset of ℓ.*

REMARK 26.25. *If ϵ is as in Lemma 26.24, then so is any positive ϵ' with $\epsilon' \leq \epsilon$ (cf. Lemma 26.22).*

Proof. For convenience of notation, we abbreviate here $M_\epsilon(x, A, y)$ as $M_\epsilon(y)$. First, consider the case where x is an interior point of A. There is a positive number ϵ_0 such that $F \cap N_{2\epsilon_0}(x; \mathbb{R}^4) = N_{2\epsilon_0}(x; A)$. Making ϵ_0 smaller if necessary, we may assume that $\ell \setminus E_{\epsilon_0}$ is non-empty. Let δ_0 be as in Lemma 26.23. For any $y \in \ell \setminus E_{\epsilon_0}$ and a positive number ϵ with $\epsilon \leq \epsilon_0$ and $\epsilon \leq \delta_0$, $F \cap M_\epsilon(y)$ is in $N_{2\epsilon_0}(x; \mathbb{R}^4)$. So it is in $N_{2\epsilon_0}(x; A)$. Since $M_\epsilon(y) \cap N_{2\epsilon_0}(x; A) = N_\epsilon(x; A)$, $F \cap M_\epsilon(y) = N_\epsilon(x; A)$. Therefore $\ell \setminus E_{\epsilon_0}$, which is an open subset of ℓ, is in $\Phi_\epsilon(x, A, F)$. Secondly, consider the case where x is a boundary point of A. Take a division K of F containing A and let $\mathbb{B}$ be the set of 2-simplices of $K \setminus \{A\}$ containing x. For each 2-simplex B of $\mathbb{B}$, we define a compact subset E_B of ℓ as follows:

(Case 1) Suppose that B is adjacent to A; namely $A \cap B$ is an arc. If $B \cap W_A$ is non-empty, let $E_B = \{f_A(p)\}$ for a point $p \in B \cap W_A$. It does not depend on p. [Since $A \cap B$ is an arc, there is a 3-plane in $\mathbb{R}^4$ containing $A \cup B$. If $B \cap W_A$ is non-empty, then $B \not\subset \langle A \rangle$ and such a 3-plane is unique. $\{f_A(p)\}$ is the intersection of ℓ and this 3-plane.] In the case where $B \cap W_A$ is empty, let $E_B = \emptyset$.

Claim 1. *For any $y \in \ell \setminus E_B$ and a positive number ϵ, $B \cap M_\epsilon(y) \subset A$.*

Proof. In the case where $B \cap W_A$ is non-empty, let H^3 be the 3-plane containing $A \cup B$. Since $y \neq f_A(p)$, $H^3_{A,y} \cap H^3 = \langle A \rangle$. Noting that $B \subset H^3$ and $M_\epsilon(y) \subset H^3_{A,y}$, we have $B \cap M_\epsilon(y) \subset \langle A \rangle$. Since $M_\epsilon(y) \cap \langle A \rangle \subset A$, we have $B \cap M_\epsilon(y) \subset A$. If $B \cap W_A$ is empty and since $M_\epsilon(y)$ is in $\pi^{-1}(\pi(\langle A \rangle)) \cup W_A$, we have $B \cap M_\epsilon(y) \subset \pi^{-1}(\pi(\langle A \rangle))$. On the other hand $M_\epsilon(y) \cap \pi^{-1}(\pi(\langle A \rangle)) \subset A$. Hence $B \cap M_\epsilon(y) \subset A$. ∎

We denote by $M_\epsilon(\ell)$ the union $\cup_{y \in \ell} M_\epsilon(y)$ and by $\overline{M_\epsilon(\ell)}$ the set $\pi^{-1}(\pi(M_\epsilon(\ell)))$, which is the closure of $M_\epsilon(\ell)$ in $\mathbb{R}^4$. Note that $\overline{M_\epsilon(\ell)} \cap \langle A \rangle = N_\epsilon(x; A)$ and $\overline{M_\epsilon(\ell)} \cap \pi^{-1}(\pi(\langle A \rangle)) = \pi^{-1}(\pi(N_\epsilon(x; A)))$. Let ϵ_0 be sufficiently small such that K has no vertices in $N_{2\epsilon_0}(x; \mathbb{R}^4)$ except x and $F \cap N_{2\epsilon_0}(x; \mathbb{R}^4)$ has a cone structure from x.

(Case 2) Suppose that B is not adjacent to A; namely $A \cap B = \{x\}$. For each positive number ϵ with $\epsilon \leq 2\epsilon_0$, let D_B^ϵ be the intersection $B \cap \overline{M_\epsilon(\ell)}$, which is a k-dimensional compact convex polyhedron (for some $k \in \{0, 1, 2\}$) and x is in ∂D_B^ϵ. We regard it as a cone $x * L_B^\epsilon$ for a subset $L_B^\epsilon \subset \partial D_B^\epsilon$ homeomorphic to an interval, a point or an empty set.

Claim 2. *L_B^ϵ is contained in W_A.*

Proof. Since $\overline{M_\epsilon(\ell)}$ is in $\pi^{-1}(\pi(\langle A \rangle)) \cup W_A$, it is sufficient to show $L_B^\epsilon \cap \pi^{-1}(\pi(\langle A \rangle)) = \emptyset$. Suppose $z \in L_B^\epsilon \cap \pi^{-1}(\pi(\langle A \rangle))$; then z is in $B \cap \overline{M_\epsilon(\ell)} \cap \pi^{-1}(\pi(\langle A \rangle)) = B \cap \pi^{-1}(\pi(N_\epsilon(x; A)))$. Hence $\pi(z) \in \pi(A) \cap \pi(B)$ and the image of $\overline{xz} \subset x * L_B^\epsilon = D_B^\epsilon \subset F$ under $\pi|_F$ is in $\pi(A) \cap \pi(B)$. Since $A \cap B = \{x\}$, $\pi(x)$ must be a branch point, which contradicts the hypothesis. ∎

By Claim 2, $f_A(L_B^\epsilon)$ is defined and compact. By the cone structure of $D_B^{2\epsilon_0} = x * L_B^{2\epsilon_0}$, we see that for any positive ϵ with $\epsilon \leq 2\epsilon_0$, $f_A(L_B^\epsilon) = f_A(L_B^{2\epsilon_0})$. Put $E_B = f_A(L_B^\epsilon) = f_A(L_B^{2\epsilon_0})$.

Claim 3. *For $y \in \ell \setminus E_B$ and a positive number ϵ with $\epsilon \leq 2\epsilon_0$, $B \cap M_\epsilon(y) \subset A$.*

Proof. Let $z(\neq x) \in B \cap M_\epsilon(y)$. Since $B \cap M_\epsilon(y) \subset B \cap \overline{M_\epsilon(\ell)} = D_B^\epsilon = x * L_B^\epsilon$, $z \in \overline{xz'}$ for some $z' \in L_B^\epsilon$. If $z \in W_A$, then $z' \in W_A$ and $f_A(z) = f_A(z') \subset f_A(L_B^\epsilon) = E_B$. On the other hand $z \in M_\epsilon(y) \subset H^3_{A,y}$. So $f_A(z) = y$, which is a contradiction. Hence $z \in \pi^{-1}(\pi(\langle A \rangle))$. Since $M_\epsilon(y) \cap \pi^{-1}(\pi(\langle A \rangle)) = N_\epsilon(x; A) \subset A$, we have $z \in A$. ∎

In this way we define E_B for $B \in \mathbb{B}$. Let δ_0 and E_{ϵ_0} be as in Lemma 26.23 and ϵ a positive number with $\epsilon \leq \epsilon_0$ and $\epsilon \leq \delta_0$. Then $\ell \setminus (E_{\epsilon_0} \cup (\cup_{B \in \mathbb{B}} E_B))$ is contained in $\Phi_\epsilon(x, A, F)$. [Let y be a point of $\ell \setminus (E_{\epsilon_0} \cup (\cup_{B \in \mathbb{B}} E_B))$. By Lemma 26.23, $F \cap M_\epsilon(y)$ is in $N_{2\epsilon_0}(x; \mathbb{R}^4)$. Since $F \cap N_{2\epsilon_0}(x; \mathbb{R}^4)$ has no vertices of K except x, it is contained in the union of all 2-simplices of K containing x. For such a 2-simplex B except A, by Claims 1 and 3, $B \cap M_\epsilon(y) \subset A$. Therefore $F \cap M_\epsilon(y) \subset A$. By Lemma 26.22 (1),

y is in $\Phi_\epsilon(x, A, F)$.] Since E_{ϵ_0} becomes smaller and the compact set $\cup_{B \in \mathbb{B}} E_B$ does not change when we deform ϵ_0 smaller, we may assume that $\ell \setminus (E_{\epsilon_0} \cup (\cup_{B \in \mathbb{B}} E_B))$ is non-empty, which is desired. This completes the proof of Lemma 26.24. ∎

For any small positive number ϵ, $N_\epsilon(x; A)$ is a compact convex polyhedron and regarded as a cone $x * L_A^\epsilon$, where L_A^ϵ is a compact 1-dimensional polyhedron in $\partial N_\epsilon(x; A)$.

LEMMA 26.26. *Let F be a surface link in general position with respect to ℓ, A a 2-simplex in F and x a point of A such that $\pi(x)$ is a branch point of $\pi|_F$. There is a positive number ϵ and a division of L_A^ϵ, say $L_A^1, \cdots, L_A^m$ for some m, such that for each $i \in \{1, \cdots, m\}$, the cone $C_i = x * L_A^i$ is a 2-simplex and $\Phi_\epsilon(x, C_i, F)$ contains a non-empty open subset of ℓ.*

Proof. Let K be a division of F containing A and ϵ_0 a positive number such that $F \cap N_{2\epsilon_0}(x; \mathbb{R}^4)$ has no vertices of K except x. Since $\pi(x)$ is a branch point, if necessary making ϵ_0 smaller, we may assume that E_{ϵ_0} is compact. Fix a positive number ϵ with $\epsilon \le \epsilon_0$ and $\epsilon \le \delta_0$, where δ_0 is as in 26.23. We may assume that only double lines through $\pi(x)$ appear on $\pi(N_\epsilon(x; A))$. Let X be the restriction to $N_\epsilon(x; A)$ of the closure of the preimage of double lines under $\pi|_F$, which is a cone $x * \{p_1, \cdots, p_n\}$, where $p_1, \cdots, p_n$ are interior points of L_A^ϵ. (If $n = 0$, then $X = \{x\}$.) Divide L_A^ϵ into intervals $L_A^1, \cdots, L_A^m$ (for some m) such that each point of $\{p_1, \cdots, p_n\}$ is contained in the interior of some L_A^i ($i = 1, \cdots, m$) and each L_A^i contains at most one point. Let K^* be a subdivision of K such that $K^*|_A$ is a division of A containing $\{C_i \mid i \in \{1, \cdots, m\}\}$ and $K^*|_{\text{cl}(F \setminus A)} = K|_{\text{cl}(F \setminus A)}$. Let $\mathbb{B}_i^*$ ($i = 1, \cdots, m$) be the set of 2-simplices of $K^* \setminus \{C_i\}$ containing x. It is the union of $\mathbb{B}$ and $\{C_1, \cdots, C_{i-1}, C_{i+1}, \cdots, C_m\}$, where $\mathbb{B}$ is the set of 2-simplices of $K \setminus \{A\}$ containing x.

(Case 1) Suppose that L_A^i contains no points of $\{p_1, \cdots, p_n\}$; i.e. there is no double point of $\pi|_F$ contained in $\pi(C_i)$. For each 2-simplex B of $\mathbb{B}_i^*$, we define E_B as in 26.24. It is compact, and for any $y \in \ell \setminus E_B$, $B \cap M_\epsilon(x, C_i, y) \subset C_i$. (We can not apply directly the argument of Claim 2 in 26.24, where we used the hypothesis that $\pi(x)$ is not a branch point. But Claim 2 holds in the current case. Let z be a point as in the proof. $\pi(\overline{xz})$ is a double line on $\pi(C_i)$, which contradicts the current hypothesis.) Since E_{ϵ_0} and E_B ($B \in \mathbb{B}_i^*$) are compact, $\ell \setminus (E_{\epsilon_0} \cup (\cup_{B \in \mathbb{B}_i^*} E_B))$ is a non-empty open subset of ℓ. We show that it is contained in $\Phi_\epsilon(x, C_i, F)$. Let y be a point of $\ell \setminus (E_{\epsilon_0} \cup (\cup_{B \in \mathbb{B}_i^*} E_B))$. Since $y \notin E_{\epsilon_0}$, $F \cap M_\epsilon(x, A, y) \subset N_{2\epsilon_0}(x; \mathbb{R}^4)$ (Lemma 26.23). Hence $F \cap M_\epsilon(x, C_i, y) \subset N_{2\epsilon_0}(x; \mathbb{R}^4)$. Let B be a 2-simplex of $K^* \setminus \{C_i\}$ which meets $N_{2\epsilon_0}(x; \mathbb{R}^4)$. B is a 2-simplex of $\mathbb{B}_i^*$ or a 2-simplex of K^* contained in A. In the former case, as stated above, $B \cap M_\epsilon(x, C_i, y) \subset C_i$. In the latter case, it also holds because $M_\epsilon(x, C_i, y) \cap A = N_\epsilon(x; C_i) \subset C_i$. Hence $y \in \Phi_\epsilon(x, C_i, F)$.

(Case 2) Suppose that there is a point, say p, of $\{p_1, \cdots, p_n\}$ in the interior of L_A^i. Let q be the point of F with $\pi(\overline{xp}) = \pi(\overline{xq})$ and $p \ne q$. Let B_i be the 2-simplex of K containing $\overline{xq}$, and $D_{B_i}^\epsilon$ the intersection $B_i \cap \overline{M_\epsilon(\ell)}$, which is a cone $x * L_{B_i}^\epsilon$ for a subset $L_{B_i}^\epsilon \subset \partial D_{B_i}^\epsilon$. (Note that Claim 2 does not hold for this B_i, since $L_{B_i}^\epsilon$ intersects $\pi^{-1}(\pi(\langle A \rangle))$ on q.) $L_{B_i}^\epsilon$ is homeomorphic to an arc, whose one end point is q and the other is not in $\pi^{-1}(\pi(\langle A \rangle))$. [Otherwise B_i is in $\pi^{-1}(\pi(\langle A \rangle))$, which contradicts that $\pi(B)$ intersects $\pi(A)$ transversely.] So $L_{B_i}^\epsilon \cap W_A$ is homeomorphic to a half open interval and the open end is q. Let E_{B_i} be $f_A(L_{B_i}^\epsilon \cap W_A)$, which is

$[a, \infty)$ or $(-\infty, -a]$ for some $a \in R^1$. By the same argument as in Claim 3, we see that for $y \in \ell \setminus E_{B_i}$, $B_i \cap M_\epsilon(x, A, y) \subset A$. (In particular $B_i \cap M_\epsilon(x, A, y) = \{x\}$, since $B_i \cap A = \{x\}$.) Therefore $B_i \cap M_\epsilon(x, C_i, y) \subset C_i$ for $y \in \ell \setminus E_{B_i}$. For each 2-simplex B of $\mathbb{B}_i^* \setminus \{B_i\}$, we define a compact set E_B as in 26.24 such that $B \cap M_\epsilon(x, C_i, y) \subset C_i$ for $y \in \ell \setminus E_B$. Since E_{ϵ_0} and E_B for $B \in \mathbb{B}_i^* \setminus \{B_i\}$ are compact and $E_{B_i} = [a, \infty)$ or $(-\infty, -a]$, we see that $\ell \setminus (E_{\epsilon_0} \cup (\cup_{B \in \mathbb{B}_i^*} E_B))$ is a non-empty open subset of ℓ, which is contained in $\Phi_\epsilon(x, C_i, F)$. ∎

For a 2-simplex (or a convex 2-cell) A in $\mathbb{R}^4$ which is in general position with respect to ℓ, a point y of ℓ and a positive number ϵ, let $M_\epsilon(A, y)$ denote the join $A * N_\epsilon(y; H^3_{A,y})$ and let $\Phi_\epsilon(A, X)$ denote $\{y \in \ell \mid X \cap M_\epsilon(A, y) = A\}$. Note that $M_\epsilon(x, A, y) = M_\epsilon(N_\epsilon(x; A), y)$ and $\Phi_\epsilon(x, A, X) = \Phi_\epsilon(N_\epsilon(x; A), X)$. If $A' \subset A$, then $\Phi_\epsilon(A', X) \supset \Phi_\epsilon(A, X)$.

26.11. Proof of the Sawtooth Lemma

Proof of Lemma 26.19. For each $x \in A$, let $\epsilon(x)$ be a positive number satisfying Lemmas 26.24 and 26.26, and let U_x be an open $\epsilon(x)$-neighborhood of x in A. Take a division $\mathbb{A} = \{A_1, \cdots, A_s\}$ of A satisfying the following: (1) Each A_i is contained in some U_x; and (2) if A_i is contained in U_x for $x \in A$ such that $\pi(x)$ is a branch point of $\pi|_F$, then it is contained in $C_j = x * L_A^j$ for some j for $N_{\epsilon(x)}(x; A)$ as in Lemma 26.26. [Such a division exists. Let δ be a Lebesgue number for the covering $\{U_x | x \in A\}$ of A. For each $x \in A$ such that $\pi(x)$ is a branch point, take a subdivision of $\{C_1, \cdots, C_m\}$ of $N_{\epsilon(x)}(x; A)$ such that the diameter of each 2-simplex of the subdivision is smaller than δ. Then divide the remainder.] Let $A_1 \subset U_{x_1}$ for some $x_1 \in A$. For convenience, we abbreviate $\epsilon(x_1)$ as ϵ. Since $A_1 \subset U_{x_1} \subset N_\epsilon(x_1; A)$, $\Phi_\epsilon(A_1, F) \supset \Phi_\epsilon(N_\epsilon(x_1; A), F) = \Phi_\epsilon(x_1, A, F)$, which contains a non-empty open subset of ℓ (Lemmas 26.24 and 26.26). Take $y_1 \in \Phi_\epsilon(A_1, F)$, $z_1 \in N_\epsilon(y_1; H^3_{A_1, y_1})$ and consider a cone $V_1 = A_1 * \{z_1\} \subset A_1 * N_\epsilon(y_1; H^3_{A_1, y_1}) = M_\epsilon(A_1, y_1)$. Since $F \cap M_\epsilon(A_1, y_1) = A_1$, $F \cap V_1 = A_1$. Since the image of $N_\epsilon(y_1; H^3_{A_1, y_1})$ under π is a 3-ball whose interior contains the origin of $\mathbb{R}^3$, we can choose z_1 such that V_1 is of type γ. Modifying z_1 slightly if necessary, we may assume that the surface obtained from F by an $\mathcal{E}$ operation along V_1 is in general position with respect to ℓ. Assume that we have already constructed $V_1, \cdots, V_{i-1}$. Let x_i be a point such that A_i is contained in U_{x_i} and let ϵ be $\epsilon(x_i)$. Then $\Phi_\epsilon(A_i, F) \supset \Phi_\epsilon(x_i, A, F)$, which contains a non-empty open subset of ℓ. Take a point $y_i \in \Phi_\epsilon(A_i, F) \setminus \{y_1, \cdots, y_{i-1}\}$ and $z_i \in N_\epsilon(y_i; H^3_{A_i, y_i})$ such that $V_i = A_i * \{z_i\}$ is a type γ 3-simplex with $F \cap V_i = A_i$ and the surface obtained from F by $\mathcal{E}$ operations along $V_1, \cdots, V_i$ is in general position with respect to ℓ. (Note that $\mathcal{E}$ operations along $V_1, \cdots, V_i$ are simultaneously applicable to F, because $V_i \cap V_j = A_i \cap A_j$ for $i \neq j$, which we see below.) Since $H^3_{A_i, y_i} \cap H^3_{A_j, y_j} = \langle A \rangle$ for $i \neq j$, $V_i \cap V_j \subset \langle A \rangle$. Since $V_i \cap \langle A \rangle = A_i$ for each i, $V_i \cap V_j = A_i \cap A_j$ for $i \neq j$. ∎

26.12. Mesh Division

Throughout the rest of this chapter, every surface link is assumed to be in general position with respect to ℓ. For $a, b, c, d \in \mathbb{R}^4$ in general position, V_{abcd} means an (unoriented) 3-simplex determined by them.

Let $A = A_{abc}$ be a 2-simplex and d, e, f middle points of $\overline{ab}, \overline{bc}, \overline{ca}$. The 1-*mesh division* of A is $\{A_{adf}, A_{bed}, A_{cfe}, A_{def}\}$. For $m \geq 2$, the m-*mesh division* of A is obtained from the $(m-1)$-mesh division of A by taking the 1-mesh division for each 2-simplex. For an m-mesh division $\mathbb{A}$ of A and a vertex p of $\mathbb{A}$, $\mathrm{st}(p;\mathbb{A})$ means the set of 2-simplices of $\mathbb{A}$ containing p and $\mathrm{lk}(p;\mathbb{A})$ the set of edges of $\mathrm{st}(p;\mathbb{A})$ disjoint from p. $|\mathrm{lk}(p;\mathbb{A})|$ is an arc (if p is a boundary point of A) or a circle (if p is an interior point). For a division $\mathbb{L}_p = \{L_p^1, \cdots, L_p^s\}$ ($s \geq 1$) of $|\mathrm{lk}(p;\mathbb{A})|$, we denote by $p * \mathbb{L}_p$ the division $\{p * L_p^1, \cdots, p * L_p^s\}$ of $|\mathrm{st}(p;\mathbb{A})|$, which is a good subdivision of $\mathrm{st}(p;\mathbb{A})$. A division $\mathbb{B}$ of A is an *almost m-mesh division* of A if there are some (or no) boundary points of A, say $p_1, \cdots, p_r$, satisfying the following:

(i) $p_1, \cdots, p_r$ are vertices of the m-mesh division $\mathbb{A}$ of A such that $|\mathrm{st}(p_i;\mathbb{A})| \cap |\mathrm{st}(p_j;\mathbb{A})| = \emptyset$ for $i \neq j$.

(ii) If $|\mathrm{st}(p_i;\mathbb{A})|$ contains a vertex of A, then p_i is the vertex.

(iii) The restriction of $\mathbb{B}$ to $\mathrm{cl}(A \setminus \cup_{i=1}^r |\mathrm{st}(p_i;\mathbb{A})|)$ coincides with that of $\mathbb{A}$.

(iv) The restriction of $\mathbb{B}$ to $|\mathrm{st}(p_i;\mathbb{A})|$ for each $i \in \{1, \cdots, r\}$ is a division in the form $p_i * \mathbb{L}_{p_i}$ for a division $\mathbb{L}_{p_i}$ of $|\mathrm{lk}(p_i;\mathbb{A})|$.

A *slightly deformed almost m-mesh division* of A is a division of A which can be obtained from an almost m-mesh division of A by deforming vertices slightly.

26.13. Existence of a Sawtooth, II

LEMMA 26.27. *Let F be a surface link, A a 2-simplex in F and ξ a straight line on A which is not parallel to any edge of A. There is a sawtooth on A avoiding F with a basis $\mathbb{A}$ such that $\mathbb{A}$ is a slightly deformed almost m-mesh division of A for some m and each edge of $\mathbb{A}$ is not parallel to ξ.*

Proof. In the argument of Lemma 26.26, we may take a division $\{L_A^i\}_i$ of L_A^ϵ such that each edge through x of $\{C_i = x * L_A^i\}_i$ is not parallel to ξ. There is a slightly deformed almost m-mesh division $\mathbb{A}$ of A for some m which satisfies the condition in the proof of Lemma 26.19 and that each edge is not parallel to ξ. By the argument in the proof, we have the desired sawtooth. ∎

LEMMA 26.28. *Let F be a surface link and A a 2-simplex in F. Suppose that a 3-simplex $V = V_{abcd}$ in $\mathbb{R}^4$ satisfies that (1) a 2-face of V is contained in $\langle A \rangle$ or (2) an edge $\overline{ab}$ of V is contained in $\langle A \rangle$ and the complementary edge $\overline{cd}$ is disjoint from $\pi^{-1}(\pi(\langle A \rangle))$. There is a sawtooth on A avoiding F and V.*

Proof. In case (1), let H^3 be the 3-plane in $\mathbb{R}^4$ determined by V and put $E_V = H^3 \cap \ell$, which is empty or consists of a point. In case (2), let E_V be $f_A(\overline{cd})$ if $\overline{cd}$ is in W_A, or empty otherwise. Add the compact set E_V to the exceptional set $E_{\epsilon_0} \cup (\cup E_B)$ in proofs of Lemmas 26.24 and 26.26, and construct a sawtooth as before. $V_i \cap V$ is contained in $\langle A \rangle$ and hence it is $A_i \cap V$. ∎

LEMMA 26.29. *Let K be a division of a surface link F and let A be a 2-simplex of K. Suppose that an $\mathcal{E}$ operation along a 3-simplex V or a $\mathcal{U}^{\pm 1}$ (or $\mathcal{W}^{\pm 1}$) operation along V_1 and V_2 is applicable to K. If A is a 2-face of neither V, V_1 nor V_2, then there is a sawtooth on A avoiding F and S, where $S = V$ or $V_1 * V_2$.*

Proof. It is sufficient to prove that Lemmas 26.24 and 26.26 hold when we replace $\Phi_\epsilon(x, A, F)$ by $\Phi_\epsilon(x, A, F \cup S)$ and $\Phi_\epsilon(x, C_i, F)$ by $\Phi_\epsilon(x, C_i, F \cup S)$ in their statements. Let $x \in A$ and let H be the 2-plane in $\mathbb{R}^4$ containing $\ell \cup \{x\}$. $S \cap H$ is a compact convex polyhedron (possibly empty). First we consider the case where $x \notin S \cap H$. For a positive number ϵ_0 with $x \cap N_{\epsilon_0}(S \cap H; H) = \emptyset$, we denote by E_S the subset $f_A(W_A \cap N_{\epsilon_0}(S \cap H; H))$ of ℓ which is empty, a compact set, $(-\infty, d]$

or $[d, \infty)$ for some $d \in \mathbb{R}^1 = \ell$. Add E_S to the exceptional set $E_{\epsilon_0} \cup (\cup E_B)$ in proofs of Lemmas 26.24 and 26.26 and we have the result. Now suppose $x \in S \cap H$. Since $S \cap A$ is empty, a common vertex or a common edge of A and S, x is a boundary point of A. Let $\overline{M_\epsilon(\ell)}$ be as in Lemma 26.24. We denote by D_S^ϵ the intersection $S \cap \overline{M_\epsilon(\ell)}$, which is a k-dimensional compact convex polyhedron (for some $k \in \{0, 1, 2, 3, 4\}$) and regarded as a cone $x * L_S^\epsilon$ for a subset $L_S^\epsilon \subset \partial D_S^\epsilon$, that is a $(k-1)$-dimensional compact polyhedron. Let E_S denote the image $f_A(W_A \cap L_S^\epsilon)$. We assert that it is a compact set, $(-\infty, d]$ or $[d, \infty)$ for some $d \in \ell$. Suppose $E_S = \ell$. Then there exist points $q_1, q_2 \in L_S^\epsilon$ and $p_1, p_2 \in N_\epsilon(x; A)$ such that q_i ($i = 1, 2$) is in $\pi^{-1}(\pi(N_\epsilon(x; A)))$, the fourth coordinate of q_1 (resp. q_2) is smaller than (resp. greater than) that of p_1 (resp. p_2) and $\pi(p_i) = \pi(q_i)$. The arc $\overline{q_1 q_2}$ intersects $N_\epsilon(x; A)$ on a single point, say p. Since $\overline{q_1 q_2}$ is an arc contained in S and contains a point p of A as an interior point, it must be contained in an edge of A. This contradicts that $q_i \notin A$. Therefore $E_S \neq \ell$. Add E_S to the exceptional set $E_{\epsilon_0} \cup (\cup E_B)$ in proofs of Lemmas 26.24 and 26.26 and we have the result. ∎

26.14. Replacement of a Sawtooth

LEMMA 26.30. *Let K and K' be divisions of surface links. If K' is obtained from K by a $\mathcal{T}$ operation, then they are Markov equivalent.*

Proof. There is a division K'' of a surface link and 3-simplices V, V' in $\mathbb{R}^4$ such that K'' is obtained from K by an $\mathcal{E}_3^8$ operation along V, K' is obtained from K'' by an $\mathcal{E}_1^8$ operation along V' and $|K''| \cap V = |K''| \cap V'$, which is a negative 2-simplex A of K''. Let a, b, c, d, e be points of $\mathbb{R}^4$ such that $V = V_{abcd}$, $V' = V_{bcde}$ and $A = A_{bcd}$. The former $\mathcal{E}$ operation replaces $\{A_{abc}, A_{acd}, A_{adb}\}$ by A_{bcd} and the latter replaces A_{bcd} by $\{A_{bce}, A_{cde}, A_{dbe}\}$. Let $p \in \mathrm{int}V$ be the intersection point of V and ℓ. For $x, y \in V$ such that x, y, p span a 2-plane in $\mathbb{R}^4$, we denote the 2-plane by H_{xy}. If $z \in V$ is not in H_{xy}, then we denote by $X_{xy,z}$ the component of $V \setminus H_{xy}$ containing z.

Claim 1. *There exists a point, f, of $\mathrm{int}V$ which is in $X_{ab,c}$, $X_{ac,d}$, $X_{ad,c}$, $X_{bc,d}$, $X_{bd,c}$ and $X_{cd,a}$.*

Proof. Let $\rho : V \to A_{adb}$ be the projection along $\overrightarrow{cp}$. The interior of the intersection of images of $X_{ac,d}, X_{bc,d}, X_{cd,a}$ under ρ, say M_1, is the interior of the triangle spanning $\rho(c), d$ and a point on $\overline{ad}$ which is opposite to b with respect to $\rho(c)$. Any point in $\rho^{-1}(M_1)$ is in $X_{ac,d}, X_{cd,a}$ and $X_{bc,d}$. Take f in $\rho^{-1}(M_1)$ sufficiently near c. Then it is in $X_{ab,c}, X_{ad,c}$ and $X_{bd,c}$. ∎

(Step 1) Let f be such a point and put $K_1 = K$ and $V_1 = V_{acdf}$. Then $|K_1| \cap V_1 = A_{acd}$ and $\tau(A_{acd}) = +1$. Let K_2 be a division which is obtained from K_1 by an $\mathcal{E}$ operation $\mathcal{F}_1$ along V_1. $\mathcal{F}_1$ replaces A_{acd} by $\{A_{acf}, A_{cdf}, A_{daf}\}$. Since $f \in X_{ac,d}$, we have $\tau(A_{acf}) = \tau(A_{acd}) = +1$. Since $f \in X_{cd,a} \cap X_{ad,c}$, we have $\tau(A_{cdf}) = \tau(A_{cda}) = +1$, $\tau(A_{daf}) = \tau(A_{dac}) = +1$. Hence $\mathcal{F}_1$ is an $\mathcal{R}$ operation.

Claim 2. *There exists a point, g, of $\mathrm{int}V$ which is in $X_{bc,a}$, $X_{bd,a}$, $X_{bf,c}$, $X_{cd,b}$, $X_{cf,b}$, $X_{df,b}$ and satisfies $\tau(A_{dge}) = \tau(A_{gce}) = +1$.*

Proof. The interior of the intersection of images of $X_{bc,a}, X_{cd,b}, X_{cf,b}$ under ρ, say M_2, is the interior of the triangle spanning $\rho(c), b$ and a point on $\overline{ab}$ which is opposite to f with respect to $\rho(c)$. Any point in $\rho^{-1}(M_2)$ is in $X_{bc,a}, X_{cd,b}$ and $X_{cf,b}$. Since f is near c, $A_{abc} \cap \rho^{-1}(M_2)$ is in $X_{bf,c}$ and $X_{df,b}$. Hence, taking g in $\rho^{-1}(M_2)$ near A_{abc}, we may assume $g \in X_{bf,c} \cap X_{df,b}$. In addition, taking g near $\overline{ab}$, we may assume $g \in X_{bd,a}$. Furthermore taking g close to b, we have $\tau(A_{dge}) = \tau(A_{dbe}) = +1$ and $\tau(A_{gce}) = \tau(A_{bce}) = +1$. $\blacksquare$

(Step 2) Let g be such a point and put $V_2 = V_{cdfg}$. Then $|K_2| \cap V_2 = A_{cdf}$ and $\tau(A_{cdf}) = +1$. Let K_2' be the division obtained from K_2 by an $\mathcal{E}$ operation $\mathcal{F}_2$ along V_2. $\mathcal{F}_2$ replaces A_{cdf} by $\{A_{cdg}, A_{dfg}, A_{fcg}\}$. Noting that $g \in X_{cd,b} \cap X_{df,b} \cap X_{cf,b}$ and $f \in X_{bd,c} \cap X_{bc,d}$, we see $\tau(A_{cdg}) = \tau(A_{cdb}) = -1$, $\tau(A_{dfg}) = \tau(A_{dfb}) = \tau(A_{dcb}) = +1$ and $\tau(A_{fcg}) = \tau(A_{fcb}) = \tau(A_{dcb}) = +1$. Hence $\mathcal{F}_2$ is $\mathcal{E}_1^5$. Let $V_2' = V_{cdge}$. Then $|K_2'| \cap V_2' = A_{cdg}$ (note that $V_1 \cap V_2 \subset V$) and $\tau(A_{cdg}) = -1$. Let K_3 be a division which is obtained from K_2' by an $\mathcal{E}$ operation $\mathcal{F}_2'$ along V_2'. $\mathcal{F}_2'$ replaces A_{cdg} by $\{A_{cde}, A_{dge}, A_{gce}\}$. The latter three 2-simplices are positive, so $\mathcal{F}_2'$ is $\mathcal{E}_1^8$. Since $V_2 * V_2'$, which is the convex polyhedron spanning $\{c, d, e, f, g\}$, intersects $|K_2|$ on A_{cdf}, the composition of $\mathcal{F}_2$ and $\mathcal{F}_2'$ is a $\mathcal{W}$ operation.

Claim 3. *There exists a point, h, of* int$V$ *which is in* $X_{bc,a}$, $X_{bf,d}$ *and* $X_{cf,a}$.

Proof. The interior of the intersection of images of $X_{bc,a}$ and $X_{cf,a}$ under ρ, say M_3, is the interior of the triangle spanning $\rho(c), a$ and a point on $\overline{ad}$ which is opposite to f with respect to $\rho(c)$. Take g in $\rho^{-1}(M_3)$ near $\rho(c)$ such that $h \in X_{bf,d}$. $\blacksquare$

(Step 3) Let h be such a point. Let h' be a point of $\mathbb{R}^4$ such that it is sufficiently near h and it is not contained in 3-planes containing V and V'. Let $V_3 = V_{abcf}$. Then $|K_3| \cap V_3 = A_{abc} \cup A_{acf}$ and $\tau(A_{abc}) = \tau(A_{acf}) = +1$. Let K_3' be the division obtained from K_3 by an $\mathcal{E}$ operation $\mathcal{F}_3$ along V_3. $\mathcal{F}_3$ replaces $\{A_{abc}, A_{acf}\}$ by $\{A_{abf}, A_{bcf}\}$. Since $f \in X_{ab,c} \cap X_{bc,d}$, $\tau(A_{abf}) = \tau(A_{abc}) = +1$ and $\tau(A_{bcf}) = \tau(A_{bcd}) = -1$. Hence $\mathcal{F}_3$ is $\mathcal{E}_2^1$. Let $V_3' = V_{bcfh'}$. $|K_3'| \cap V_3' = A_{bcf}$ and $\tau(A_{bcf}) = -1$. Let K_4 be a division which is obtained from K_3' by an $\mathcal{E}$ operation $\mathcal{F}_3'$ along V_3'. $\mathcal{F}_3'$ replaces A_{bcf} by $\{A_{bch'}, A_{cfh'}, A_{fbh'}\}$. Noting that $h \in X_{bc,a} \cap X_{cf,a} \cap X_{bf,d}$, $f \in X_{ac,d} \cap X_{bd,c}$ and h' is near h, we see that $\tau(A_{abch'}) = \tau(A_{bch}) = \tau(A_{bca}) = +1$, $\tau(A_{acfh'}) = \tau(A_{cfh}) = \tau(A_{cfa}) = +1$, and $\tau(A_{afbh'}) = \tau(A_{fbh}) = \tau(A_{fbd}) = +1$. So $\mathcal{F}_3'$ is $\mathcal{E}_1^8$. Since $V_3 * V_3'$ is the 4-simplex spanning $\{a, b, c, f, h'\}$, it intersects $|K_3|$ on $A_{abc} \cup A_{acf}$ and the composition of $\mathcal{F}_3$ and $\mathcal{F}_3'$ is a $\mathcal{U}$ operation.

(Step 4) Let $V_4 = V_{abdf}$. $|K_4| \cap V_4 = A_{abf} \cup A_{adb} \cup A_{daf}$ and these three 2-simplices are positive. Let K_4' be a division which is obtained from K_4 by an $\mathcal{E}$ operation $\mathcal{F}_4$ along V_4. $\mathcal{F}_4$ replaces $\{A_{abf}, A_{adb}, A_{daf}\}$ by A_{bfd}. Since $f \in X_{bd,c}$, $\tau(A_{bfd}) = \tau(A_{bcd}) = -1$. Hence $\mathcal{F}_4$ is $\mathcal{E}_3^8$. Let $V_4' = V_{bdfg}$. Then $|K_4'| \cap V_4' = A_{bfd} \cup A_{dfg}$, $\tau(A_{bfd}) = -1$ and $\tau(A_{dfg}) = +1$. Let K_5 be a division which is obtained from K_4' by an $\mathcal{E}$ operation $\mathcal{F}_4'$ along V_4'. $\mathcal{F}_4'$ replaces $\{A_{bfd}, A_{dfg}\}$ by $\{A_{bfg}, A_{dbg}\}$. Since $g \in X_{bf,c} \cap X_{bd,a}$, we have $\tau(A_{bfg}) = \tau(A_{bfc}) = +1$ and $\tau(A_{dbg}) = \tau(A_{dba}) = +1$. Hence $\mathcal{F}_4'$ is $\mathcal{E}_2^{-1}$. Since $|K_4| \cap V_4 * V_4' = |K_4| \cap V_4$ and $|K_5| \cap V_4 * V_4' = |K_5| \cap V_4'$, the composition of $\mathcal{F}_4$ and $\mathcal{F}_4'$ is a $\mathcal{U}^{-1}$ operation.

(Step 5) Let $V_5 = V_{bcfh'}$. $|K_5| \cap V_5 = A_{bch'} \cup A_{cfh'} \cup A_{fbh'}$ and these three 2-simplices are positive. Let K_5' be a division which is obtained from K_5 by an $\mathcal{E}$ operation $\mathcal{F}_5$ along V_5. $\mathcal{F}_5$ replaces $\{A_{bch'}, A_{cfh'}, A_{fbh'}\}$ by A_{bcf}. Since $\tau(A_{bcf}) = -1$, $\mathcal{F}_5$ is $\mathcal{E}_3^8$. Let $V_5' = V_{bcfg}$. Then $|K_5'| \cap V_5' = A_{bcf} \cup A_{bfg} \cup A_{fcg}$, $\tau(A_{bcf}) = -1$ and $\tau(A_{bfg}) = \tau(A_{fcg}) = +1$. Let K_6 be a division which is obtained from K_5' by an $\mathcal{E}$ operation $\mathcal{F}_5'$ along V_5'. $\mathcal{F}_5'$ replaces $\{A_{bcf}, A_{bfg}, A_{fcg}\}$ by A_{bcg}. Since $g \in X_{bc,a}$, we have $\tau(A_{bcg}) = \tau(A_{bca}) = +1$ and $\mathcal{F}_5'$ is $\mathcal{E}_3^5$. Since $|K_5| \cap V_5 * V_5' = |K_5| \cap V_5$ and $|K_6| \cap V_5 * V_5' = |K_6| \cap V_5'$, the composition of $\mathcal{F}_5$ and $\mathcal{F}_5'$ is a type $\mathcal{W}^{-1}$ operation.

(Step 6) Let $V_6 = V_{bceg}$ in case $\tau(A_{bge}) = -1$ (resp. $V_6 = V_{bdeg}$ in case $\tau(A_{bge}) = +1$). Then $|K_6| \cap V_6 = A_{gce} \cup A_{bcg}$ (resp. $A_{dge} \cup A_{dbg}$). Let K_7 be a division which is obtained from K_6 by an $\mathcal{E}$ operation $\mathcal{F}_6$ along V_6. Since $\mathcal{F}_6$ replaces $\{A_{gce}, A_{bcg}\}$ by $\{A_{bce}, A_{gbe}\}$ (resp. $\{A_{dge}, A_{dbg}\}$ by $\{A_{bge}, A_{dbe}\}$), it is an $\mathcal{R}$ operation.

(Step 7) Let $V_7 = V_{bdeg}$ (resp. V_{bceg}). Then $|K_7| \cap V_7 = A_{gbe} \cup A_{dge} \cup A_{dbg}$ (resp. $A_{bge} \cup A_{gce} \cup A_{bcg}$). Let K_8 be a division which is obtained from K_7 by an $\mathcal{E}$ operation $\mathcal{F}_7$ along V_7. $\mathcal{F}_7$ replaces $\{A_{gbe}, A_{dge}, A_{dbg}\}$ by A_{dbe} (resp. $\{A_{bge}, A_{gce}, A_{bcg}\}$ by A_{bce}), which is an $\mathcal{R}$ operation. Since $K_8 = K'$, we have the result. ∎

LEMMA 26.31. *Let K_0 be a division of a surface link F, let A be a negative 2-simplex of K_0, and let K be a division obtained from K_0 by an $\mathcal{S}$ operation along a sawtooth $\mathbb{V}$ on A with a basis $\mathbb{A}$. For any good subdivision $\mathbb{A}'$ of $\mathbb{A}$, there is a sawtooth on A avoiding F with basis $\mathbb{A}'$ such that the division obtained from K_0 by an $\mathcal{S}$ operation along it is Markov equivalent to K.*

Proof. It is sufficient to consider the case where $\mathbb{A}'$ is obtained from $\mathbb{A}$ by a single elementary dividing operation. Let $\mathbb{A}' = (\mathbb{A} \setminus \{A_{abc}\}) \cup \{A_{abd}, A_{adc}\}$. Let V be the 3-simplex of $\mathbb{V}$ on A_{abc} and e the vertex of V with $V = A_{abc} * \{e\}$. Since V is a 3-simplex of type γ, $V \cap \ell = \{p\}$ for some $p \in \text{int} V$. Moving e slightly if necessary, by Lemma 26.30, we may assume $p \notin \overline{ad} * \{e\}$. p is an interior point of $V_1 = A_{abd} * \{e\}$ or $V_2 = A_{adc} * \{e\}$, say V_1. Then V_1 is a 3-simplex of type γ and V_2 is of type α. Let K_1 be the division obtained from K by an $\mathcal{A}$ operation replacing A_{bce} by $\{A_{bde}, A_{dce}\}$. Let K_1^* be the division obtained from K_1 by an $\mathcal{E}_2^1$ operation along V_2. Let f be a point of $\mathbb{R}^4$ such that $V_2' = A_{adc} * \{f\}$ is a 3-simplex of type γ, $|K_1| \cap V_2 * V_2' = |K_1| \cap V_2$ and $\mathbb{V}' = \mathbb{V} \setminus \{V\} \cup \{V_1, V_2'\}$ is a sawtooth on A avoiding F. [Such a point exists. Let f' be a point of R^3 such that f' is in $\pi(V)$ and the origin of $\mathbb{R}^3$ is an interior point of $\pi(A_{adc}) * \{f'\}$. Take f in the preimage $\pi^{-1}(f')$ so that it is sufficiently close to, but not in, the 3-plane in $\mathbb{R}^4$ containing V.] Let K_2 be the division obtained from K_1 by a $\mathcal{U}$ operation along V_2 and V_2', which is Markov equivalent to K_1 and hence to K. On the other hand, K_2 is obtained from K_0 by an $\mathcal{S}$ operation along $\mathbb{V}'$. ∎

LEMMA 26.32. *Let K_0 be a division of a surface link and let A be a negative 2-simplex of K_0. Let K_i $(i = 1, 2)$ be a division obtained from K_0 by an $\mathcal{S}$ operation along a sawtooth $\mathbb{V}_i$ on A. K_1 is Markov equivalent to K_2.*

Proof. Let $\mathbb{A}_i$ $(i = 1, 2)$ be the basis of $\mathbb{V}_i$. Take a common good subdivision $\mathbb{A}_3$ of $\mathbb{A}_1$ and $\mathbb{A}_2$ (Lemma 26.1). By Lemma 26.31, there is a sawtooth $\mathbb{V}_i'$ $(i = 1, 2)$ on A avoiding $|K_0|$ with basis $\mathbb{A}_3$ such that the division K_i' obtained from K_0 by an $\mathcal{S}$ operation along $\mathbb{V}_i'$ is Markov equivalent to K_i. Since K_2' is obtained from K_1' by a finite number of $\mathcal{T}$ operations, by Lemma 26.30, they are Markov equivalent. ∎

26.15. Height Reduction, I

Let H be a 2-plane in $\mathbb{R}^3$ missing the origin O, and let H_0 be the 2-plane containing O which is parallel to H. A *stereographic projection* associated with H is a continuous (non-PL) map $\mu = \mu_H : \mathbb{R}^3 \setminus H_0 \to H$ such that for $y \in \mathbb{R}^3 \setminus H_0$, $\mu(y)$ is the intersection of H and the line determined by $\overline{Oy}$. The *positive/negative domain* of μ is the component of $\mathbb{R}^3 \setminus H_0$ that does/does not contain H. Let

A_{abc} be a 2-simplex in $\mathbb{R}^4$ in general position with respect to ℓ. If $\pi(a), \pi(b)$ are contained in the positive domain of μ and $\pi(c)$ is in the positive (or negative, resp.) domain, then $\tau(A_{abc}) = \tau(A_{\mu\pi(a)\mu\pi(b)\mu\pi(c)})$ (resp. $-\tau(A_{\mu\pi(a)\mu\pi(b)\mu\pi(c)})$).

LEMMA 26.33. *Let K be a division of a surface link, let K' be a division obtained from K by an $\mathcal{E}_2^1$ operation, let A be the negative 2-simplex of K' obtained by the operation, and let K'' be a division obtained from K' by an $\mathcal{S}$ operation along a sawtooth $\mathbb{V}$ on A. Then K is Markov equivalent to K''.*

Proof. Suppose that an $\mathcal{E}_2^1$ operation is applied along $V = V_{abcd}$ and that we replace A_{bcd}, A_{adc} by a positive 2-simplex A_{adb} and a negative $A = A_{abc}$. Let H be the 2-plane in $\mathbb{R}^3$ determined by $\pi(A_{adb})$ and μ the stereographic projection associated with H. Images of a, b, c, d under $\mu\pi$ in H look like Fig 26.1, where we see H from O and abbreviate $\mu\pi(x)$ as x for $x = a, b, c, d$.

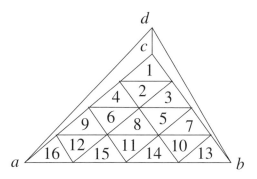

FIGURE 26.1

We assert that for a positive number ϵ, it is sufficient to prove the lemma on the assumption $d(c, d) < \epsilon$. Furthermore we may assume that the sawtooth avoids V. [Let d' be an interior point of $\overline{cd}$ with $d(c, d') < \epsilon$. Let K_1 be a division $(K \backslash \{A_{bcd}, A_{adc}\}) \cup \{A_{bcd'}, A_{bd'd}, A_{ad'c}, A_{add'}\}$, which is Markov equivalent to K (by $\mathcal{A}^{-1}$ operations). Let K_1' be a division which is obtained from K_1 by an $\mathcal{E}_2^1$ operation along $V_{abcd'}$. (Since $|K| \cap V_{abcd'} = A_{bcd'} \cup A_{ad'c}$, the operation is applicable.) Let $\mathbb{V}_1$ be a sawtooth on A avoiding $|K_1'| \cup V_{abcd}$ (Lemma 26.28(1)) and K_2 a division which is obtained from K_1' by an $\mathcal{S}$ operation along $\mathbb{V}_1$. By the hypothesis, K_2 is Markov equivalent to K_1 and hence to K. Since $\mathbb{V}_1$ avoids V_{abcd}, it avoids $V_{abdd'}$ and we can apply an $\mathcal{E}_3^1$ operation to K_2 along $V_{abdd'}$. Let K_3 be the result, which is Markov equivalent to K. It is obtained from K' by an $\mathcal{S}$ operation along $\mathbb{V}_1$. By Lemma 26.32, it is Markov equivalent to K''. So K is Markov equivalent to K''. The latter assertion follows since $\mathbb{V}_1$ avoids V_{abcd} and hence $V_{abcd'}$.] Therefore we may assume that d is arbitrarily close to c and $\mathbb{V}$ avoids V. By Lemmas 26.27 and 26.32, we may assume that $\mathbb{V}$ is a sawtooth on A avoiding $|K'|$ with a basis $\mathbb{A}$ satisfying the following: $\mathbb{A}$ is a slightly deformed almost m-mesh division of A for some m, and the image under $\mu\pi$ of every edge of $\mathbb{A}$ is not parallel to that of $\overline{cd}$. Let $\lambda : A_{abc} \to A_{abd}$ be the linear map with $\lambda(a) = a, \lambda(b) = b$ and $\lambda(c) = d$.

We first consider the case where $\mathbb{A}$ is an m-mesh division. Let $T_1 = \{B_1\}$, where B_1 is the 2-simplex of $\mathbb{A}$ containing c. For each $k \in \{2, \cdots, 2^{m+1} - 1\}$, let T_k be the set of 2-simplices B of $\mathbb{A} \backslash \cup_{i=1}^{k-1} T_i$ such that B has an edge shared with a

2-simplex of T_{k-1}. The number of 2-simplices of T_k is $(k+1)/2$ (if k is odd) or $k/2$ (if k is even), denoted by $m(k)$. Let $n(k) = \Sigma_{i=1}^{k} m(i)$. Assign indices to 2-simplices of $\mathbb{A}$ such that for each $k \in \{2, \cdots, 2^{m+1} - 1\}$, $B_{n(k-1)+1}, \cdots, B_{n(k-1)+m(k)}$ are 2-simplices of T_k ordered from $\overline{bc}$ side to $\overline{ac}$ side. (Fig 26.1 is for $m = 2$.) We construct a sequence $K = K_0, K_1, \cdots, K_{4^m} = K''$ of divisions of surface links which are Markov equivalent. Suppose that K_{i-1} has already been constructed. K_i is defined as follows: Let B_i be a 2-simplex of T_k for some k. We have four cases:

(I) k is odd and $k < 2^{m+1} - 1$.
(II) k is $2^{m+1} - 1$.
(III) k is even and $k < 2^{m+1} - 2$.
(IV) k is $2^{m+1} - 2$.

Let x, y, z be vertices of B_i such that $B_i = A_{xyz}$ and $\overline{yz}$ is either an edge of a 2-simplex of T_{k+1} (in Case I), an edge contained in $\overline{ab}$ (in Case II) or an edge of a 2-simplex of T_{k-1} (in Cases III, IV). Let x', y', z' denote $\lambda(x), \lambda(y), \lambda(z)$ respectively. Images of x, y, z, x', y', z' under $\mu\pi$ are as in Fig 26.2.

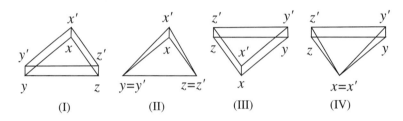

(I)   (II)   (III)   (IV)

FIGURE 26.2

Let V_i be the 3-simplex of $\mathbb{V}$ on B_i and P_i the set of 2-simplices of K'' which are 2-faces of V_i. We define K_i by

$$\begin{cases} (K_{i-1} \setminus (R_{xyy'x'} \cup R_{zxx'z'})) \cup \{A_{x'z'y'}\} \cup P_i \cup R_{zyy'z'} & \text{for Cases I, II,} \\ (K_{i-1} \setminus R_{yzz'y'}) \cup \{A_{x'z'y'}\} \cup P_i \cup R_{yxx'y'} \cup R_{xzz'x'} & \text{for Cases III, IV,} \end{cases}$$

where $R_{pqq'p'}$ is $\{A_{pqq'}, A_{pq'p'}\}$ or $\{A_{pqp'}, A_{p'qq'}\}$ for $p, q \in \{x, y, z\}$. (We regard $R_{pqq'p'} = \{A_{pqq'}\}$ if $p = p'$ and $R_{pqq'p'} = \{A_{pqp'}\}$ if $q = q'$.) There is an ambiguity in the definition of K_i due to division of $R_{pqq'p'}$. Since 2-simplices in $R_{pqq'p'}$ are positive, this ambiguity is not important in our argument. (We may regard two divisions of a surface that differ by positive 2-simplices as the same, since they are connected by $\mathcal{A}^{\pm 1}$ operations (Lemma 26.13).) We prove that K_{i-1} and K_i are Markov equivalent. Suppose $B_i \in T_k$ and consider each case I–IV.

Case I. Apply an $\mathcal{E}$ operation to K_{i-1} along $V_{xx'yz}$, which replaces $\{A_{zxx'}, A_{xyx'}\}$ by $\{A_{xyz}, A_{x'zy}\}$. Since $\tau(A_{zxx'}) = \tau(A_{xyx'}) = \tau(A_{x'zy}) = 1$ and $\tau(A_{xyz}) = -1$, the operation is $\mathcal{E}_2^1$. Apply an $\mathcal{E}$ operation along V_i, which replaces A_{xyz} by P_i. Since x' is sufficiently close to x, we may assume $|K_{i-1}| \cap V_{xx'yz} * V_i = |K_{i-1}| \cap V_{xx'yz}$. The composition of these $\mathcal{E}$ operations is a $\mathcal{U}$ operation. Let $K^{(1)}$ be the result and $K^{(2)}$ the division obtained from $K^{(1)}$ by an $\mathcal{E}_2^3$ operation along $V_{x'yzz'}$, which replaces $\{A_{x'zy}, A_{zx'z'}\}$ by $\{A_{zyz'}, A_{yx'z'}\}$. (By the assumption that $\mathbb{V}$ avoids V, this operation is applicable, i.e. $|K^{(1)}| \cap V_{x'yzz'} = A_{x'zy} \cup A_{zx'z'}$.) Apply an $\mathcal{E}_2^3$ operation along $V_{x'yy'z'}$ to $K^{(2)}$, replacing $\{A_{yx'z'}, A_{x'yy'}\}$ by $\{A_{x'z'y'}, A_{yy'z'}\}$. The result is K_i.

Case II. Apply the above argument until we have $K^{(1)} = K_i$.

Case III. An $\mathcal{E}$ operation along $V_{xyy'z}$ to K_{i-1} replaces $A_{yzy'}$ by $\{A_{xyz}, A_{xy'y}, A_{xzy'}\}$. Since $\tau(A_{yzy'}) = \tau(A_{xy'y}) = \tau(A_{xzy'}) = 1$ and $\tau(A_{xyz}) = -1$, the operation is $\mathcal{E}_1^5$. Apply an $\mathcal{E}_1^8$ operation along V_i, replacing A_{xyz} by P_i. Since y' is sufficiently close to y, the composition of these two operations is a $\mathcal{W}$ operation. Let $K^{(1)}$ be the result. Apply an $\mathcal{E}_2^3$ operation along $V_{xy'zz'}$ to $K^{(1)}$ and let $K^{(2)}$ be the result. Finally applying an $\mathcal{E}_1^1$ operation along $V_{xx'y'z'}$ to $K^{(2)}$, we obtain K_i.

Case IV. Do the above argument until we have $K^{(2)} = K_i$.

This completes the case where $\mathbb{A}$ is an m-mesh division.

Consider the case where $\mathbb{A}$ is an almost m-mesh division. Let $\mathbb{A}_0$ be the m-mesh division of A and $B_1, \cdots, B_{4^m}$ 2-simplices of $\mathbb{A}_0$ ordered as before. Let $p_1, \cdots, p_r$ be vertices of $\mathbb{A}_0$ such that the restriction of $\mathbb{A}$ to $|\mathrm{st}(p_s; \mathbb{A}_0)|$ $(s = 1, \cdots, r)$ is a division in the form $\{p_s\} * \mathbb{L}_{p_s}$. We construct a sequence $K = K_0, K_1, \cdots, K_{4^m} = K''$ of divisions of surface links which are Markov equivalent. If B_i is not contained in $|\mathrm{st}(p_s; \mathbb{A}_0)|$ for any s, define K_i as before. If B_i is contained in $|\mathrm{st}(p_s; \mathbb{A}_0)|$ for some s, define K_i as follows: Let $B_i^1, \cdots, B_i^{j(i)}$ be 2-simplices of $\{p_s\} * \mathbb{L}_{p_s}$ and V_i^j $(j = 1, \cdots, j(i))$ the 3-simplex of $\mathbb{V}$ on B_i^j. Denote by P_i^j the set of 2-simplices of K'' which are 2-faces of V_i^j. Put $P_i = \cup_{j=1}^{j(i)} P_i^j$ and define K_i as before. In order to prove that K_{i-1} is Markov equivalent to K_i, we construct a sequence $K_{i-1} = K^0, K^1, \cdots, K^{j(i)} = K_i$ of divisions of surface links such that K^{j-1} $(j = 1, \cdots, j(i))$ is Markov equivalent to K^j.

Consider the case where Case I occurs for B_i and p_s is x. Since the image under $\mu\pi$ of each edge of $\mathbb{A}$ is not parallel to that of $\overline{cd}$ (and hence that of $\overline{xx'}$), there is a unique 2-simplex in $\{p_s\} * \mathbb{L}_{p_s}$ that is divided into two 2-simplices by the line on H determined by $\mu\pi(\overline{xx'})$. Re-index B_i^j $(j = 1, \cdots, j(i))$ as follows:

(1) The above 2-simplex is $B_i^{j(i)}$.
(2) $B_i^1, \cdots, B_i^r$ are on the z side of $B_i^{j(i)}$ and arranged from the $\overline{xz}$ side. (If there are no 2-simplices on the z side, then $r = 0$.)
(3) $B_i^{r+1}, \cdots, B_i^{j(i)-1}$ are on the y side of $B_i^{j(i)}$ and arranged from the $\overline{xy}$ side.

For example, see Fig 26.3. Let u, v, w be vertices of B_i^j such that $B_i^j = A_{uvw}$ and $u = x$. Let u', v', w' denote $\lambda(u), \lambda(v), \lambda(w)$ respectively. Images of u, v, w, u', v', w' under $\mu\pi$ are either (A) if $1 \leq j \leq r$, (B) if $r + 1 \leq j \leq j(i) - 1$ or (C) if $j = j(i)$ in Fig 26.4. In this notation, K^j is defined by

$$\begin{cases} (K^{j-1} \setminus R_{wuu'w'}) \cup \{A_{u'w'v'}\} \cup P_i^j \cup R_{wvv'w'} \cup R_{vuu'v'} & \text{for Case A,} \\ (K^{j-1} \setminus R_{uvv'u'}) \cup \{A_{u'w'v'}\} \cup P_i^j \cup R_{uww'u'} \cup R_{wvv'w'} & \text{for Case B,} \\ (K^{j-1} \setminus (R_{uvv'u'} \cup R_{wuu'w'})) \cup \{A_{u'w'v'}\} \cup P_i^j \cup R_{wvv'w'} & \text{for Case C.} \end{cases}$$

For Cases A and B, we see that K^{j-1} and K^j are Markov equivalent by a similar argument to Case III. For Case C, we see the result by an argument as in Case I.

Consider the case where Case I occurs for B_i and p_s is y. Re-index B_i^j $(j = 1, \cdots, j(i))$ such that $B_i^1, \cdots, B_i^{j(i)}$ are arranged from the $\overline{xy}$ side in this order; see Fig 26.5. Let u, v, w be vertices of B_i^j with $B_i^j = A_{uvw}$ and $v = y$. Let u', v', w' denote $\lambda(u), \lambda(v), \lambda(w)$. Images of them under $\mu\pi$ are (C) in Fig 26.4. K^j is defined as in Case C and, by the same reasoning, it is Markov equivalent to K_{j-1}. For the case where Case I occurs for B_i and p_s is z, apply a similar argument to the case above. By an analogous argument, we obtain the result in Cases II, III and

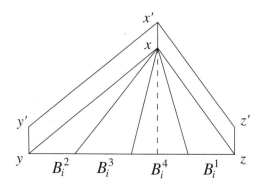

FIGURE 26.3

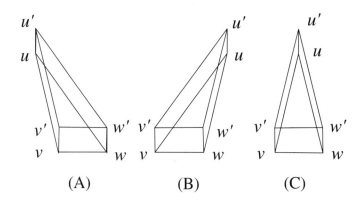

(A)   (B)   (C)

FIGURE 26.4

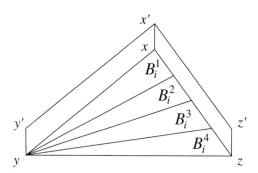

FIGURE 26.5

IV. (Details are left to the reader.) This argument is valid directly for a slightly deformed almost m-mesh division. ∎

LEMMA 26.34. *Let K be a division of a surface link, K' a division obtained from K by an $\mathcal{E}_2^1$ operation, A the negative 2-simplex of K' obtained by the operation and K'' a division obtained from K' by an $\mathcal{E}_2^{-1}$ operation deleting A. Then K is Markov equivalent to K''.*

Proof. Let $\widetilde{K}'$ be a division obtained from K' by an $\mathcal{S}$ operation along a sawtooth on A. By Lemma 26.33, K and K'' are Markov equivalent via $\widetilde{K}'$. ∎

26.16. Height Reduction, II

In this section we prove the following lemma.

LEMMA 26.35. *Let K be a division of a surface link, K' a division obtained from K by an $\mathcal{E}_1^3$ operation along a 3-simplex V and A, A' the negative 2-simplices of K, K' which are 2-faces of V. Let $\widetilde{K}$ (resp. $\widetilde{K}'$) be a division obtained from K (resp. K') by an $\mathcal{S}$ operation along a sawtooth $\mathbb{V}$ (resp. $\mathbb{V}'$) on A (resp. A'). Then $\widetilde{K}$ and $\widetilde{K}'$ are Markov equivalent.*

Let a, b, c, d be vertices of V such that the $\mathcal{E}_1^3$ operation replaces a negative 2-simplex $A = A_{abc}$ by a negative 2-simplex $A' = A_{abd}$ and positive 2-simplices A_{bcd}, A_{adc}. Let H be the 2-plane in $\mathbb{R}^3$ containing $\pi(A')$ and μ the stereographic projection associated with it. For $e, e' \in \overline{cd}$, let $\lambda_e^{e'} : A_{abe} \to A_{abe'}$ be the linear map with $\lambda_e^{e'}(a) = a$, $\lambda_e^{e'}(b) = b$ and $\lambda_e^{e'}(e) = e'$. We abbreviate λ_c^d as λ. For $e \in \overline{cd}$, let F_e denote a surface link which is obtained from $F = |K|$ by an $\mathcal{E}$ operation along V_{abce}.

LEMMA 26.36. *There is a sequence of points $c = e(0), e(1), \cdots, e(r) = d$ appearing on $\overline{cd}$ in this order and sawteeth $\mathbb{V}_i$ $(i = 0, \cdots, r-1)$ on $A_{abe(i)}$ avoiding $F_{e(i)}$ with a basis $\mathbb{A}_{abe(i)}$ satisfying the following:*

(1) *For each $i \in \{0, \cdots, r-1\}$, $\mathbb{A}_{abe(i)}$ is a slightly deformed almost m_i-mesh division of $A_{abe(i)}$ for some m_i.*

(2) *The image under $\mu\pi$ of each edge of $\mathbb{A}_{abe(i)}$ $(i = 0, \cdots, r-1)$ is not parallel to that of $\overline{cd}$.*

(3) *For each 2-simplex B of $\mathbb{A}_{abe(i)}$ $(i = 0, \cdots, r-1)$, let g_B be the point such that $B * \{g_B\}$ is a 3-simplex of $\mathbb{V}_i$. Then $\lambda_{e(i)}^{e(i+1)}(\mathbb{V}_i) = \{\lambda_{e(i)}^{e(i+1)}(B) * \{g_B\} \mid B \in \mathbb{A}_{abe(i)}\}$ is a sawtooth on $A_{abe(i+1)}$ avoiding $F_{e(i+1)}$ with basis $\lambda_{e(i)}^{e(i+1)}(\mathbb{A}_{abe(i)}) = \{\lambda_{e(i)}^{e(i+1)}(B) \mid B \in \mathbb{A}_{abe(i)}\}$.*

(4) *$\mathbb{V}_i$ $(i = 0, \cdots, r-1)$ avoids the 3-simplex $V_{abe(i)e(i+1)}$.*

Proof. By Lemma 26.27, for each $e \in \overline{cd}$, there is a sawtooth $\mathbb{V}_{abe}$ on A_{abe} avoiding F_e with a basis $\mathbb{A}_{abe}$ satisfying conditions (1) and (2). Furthermore there is a small positive number $\epsilon = \epsilon(e)$ such that if $e' \in N_\epsilon(e; \overline{cd})$, then $\{\lambda_e^{e'}(B) * \{g_B\} \mid B \in \mathbb{A}_{abe}\}$ is a sawtooth on $A_{abe'}$ avoiding $F_{e'}$ with basis $\lambda_e^{e'}(\mathbb{A}_{abe})$. Take a sequence of points $c = e(0), e(1), \cdots, e(r) = d$ such that, for each $i \in \{0, \cdots, r-1\}$, $e(i)$ and $e(i+1)$ are contained in $N_{\epsilon(e)}(e; \overline{cd})$ for some e. Put $\mathbb{A}_{abe(i)} = \lambda_e^{e(i)}(\mathbb{A}_{abe})$ and $\mathbb{V}_i = \lambda_e^{e(i)}(\mathbb{V}_{abe})$. By the argument of Lemma 26.28(1), we may assume (4). ∎

LEMMA 26.37. *Suppose that there is a sawtooth $\mathbb{V}$ on A avoiding $|K|$ with a basis $\mathbb{A}$ satisfying the following:*

(1) *$\mathbb{A}$ is a slightly deformed almost m-mesh division of A for some m.*

(2) *The image under $\mu\pi$ of each edge of $\mathbb{A}$ is not parallel to that of $\overline{cd}$.*

(3) *For each 2-simplex B of $\mathbb{A}$, let g_B be the point such that $B * \{g_B\}$ is a 3-simplex of $\mathbb{V}$. Then $\lambda(\mathbb{V}) = \{\lambda(B) * \{g_B\} \mid B \in \mathbb{A}\}$ is a sawtooth on A' avoiding $|K'|$ with basis $\lambda(\mathbb{A}) = \{\lambda(B) \mid B \in \mathbb{A}\}$.*

(4) *$\mathbb{V}$ avoids $V = V_{abcd}$.*

Let $\widetilde{K}$ (resp. $\widetilde{K}'$) be the division obtained from K (resp. K') by an $\mathcal{S}$ operation along $\mathbb{V}$ (resp. $\lambda(\mathbb{V})$). Then $\widetilde{K}$ and $\widetilde{K}'$ are Markov equivalent.

Proof. We construct a sequence $\widetilde{K} = \widetilde{K}_0, \cdots, \widetilde{K}_{4^m} = \widetilde{K}'$ of divisions of surface links which are Markov equivalent. Consider the case where $\mathbb{A}$ is an m-mesh division. Assign indices to 2-simplices of $\mathbb{A}$ in the opposite order to that in Lemma 26.33, see Fig 26.6 where $m = 2$. Let $T_1 = \{B_{4^m}\}$, where B_{4^m} is the 2-simplex of $\mathbb{A}$ containing c and define T_k for $k \in \{2, \cdots, 2^{m+1} - 1\}$ inductively as before. Cases I–IV and points x, y, z, x', y', z' are as in Lemma 26.33. Images of x, y, z, x', y', z' under $\mu\pi$ are as in Fig 26.2.

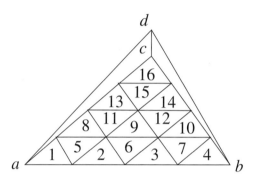

FIGURE 26.6

Let V_i be the 3-simplex of $\mathbb{V}$ on $B_i = A_{xyz}$, and let V_i' be the 3-simplex of $\lambda(\mathbb{V})$ on $\lambda(B_i) = A_{x'y'z'}$. Then $V_i = A_{xyz} * \{g\}$ and $V_i' = A_{x'y'z'} * \{g\}$ for some $g \in \mathbb{R}^4$. Put $P_i = \{A_{xyg}, A_{yzg}, A_{zxg}\}$, $P_i' = \{A_{x'y'g}, A_{y'z'g}, A_{z'x'g}\}$ and define $\widetilde{K}_i$ by

$$\begin{cases} (\widetilde{K}_{i-1} \setminus (P_i \cup R_{zyy'z'})) \cup P_i' \cup R_{xyy'x'} \cup R_{zxx'z'} & \text{for Cases I, II,} \\ (\widetilde{K}_{i-1} \setminus (P_i \cup R_{yxx'y'} \cup R_{xzz'x'})) \cup P_i' \cup R_{yzz'y'} & \text{for Cases III, IV,} \end{cases}$$

where $R_{pqq'p'}$ is $\{A_{pqq'}, A_{pq'p'}\}$ or $\{A_{pqp'}, A_{p'qq'}\}$ for $p, q \in \{x, y, z\}$ as before. Since $V_i = A_{xyz} * \{g\}, V_i' = A_{x'y'z'} * \{g\}$ are 3-simplices of type γ and since $\pi(A_{xyz}), \pi(A_{x'y'z'})$ are in the positive domain of μ, we see that $\pi(g)$ is in the negative domain of μ and $\mu\pi(g)$ is an interior point of $\mu\pi(A_{xyz})$ and of $\mu\pi(A_{x'y'z'})$. Modifying g slightly if necessary, by Lemma 26.27, we may assume that $\mu\pi(g)$ is not on the line in H determined by $\mu\pi(\overline{xx'})$. Consider Cases I–IV as before.

Case I. Apply an $\mathcal{E}$ operation along $V_{yzz'g}$ to $\widetilde{K}_{i-1}$, which replaces $\{A_{yzg}, A_{zyz'}\}$ by $\{A_{yz'g}, A_{z'zg}\}$. Since $\tau(A_{yzg}) = \tau(A_{zyz'}) = \tau(A_{yz'g}) = 1$ and $\tau(A_{z'zg}) = -1$, the operation is $\mathcal{E}_2^1$. Apply an $\mathcal{E}$ operation along $V_{xzz'g}$ to the result, which replaces $\{A_{zxg}, A_{z'zg}\}$ by $\{A_{z'xg}, A_{zxz'}\}$. Since $\tau(A_{zxg}) = \tau(A_{z'zg}) = \tau(A_{z'xg}) = 1$ and $\tau(A_{z'zg}) = -1$, the operation along $V_{xzz'g}$ is $\mathcal{E}_2^{-1}$. Let $\widetilde{K}_{i-1}^{(1)}$ be the result. By Lemma 26.34, $\widetilde{K}_{i-1}$ is Markov equivalent to $\widetilde{K}_{i-1}^{(1)}$. Apply an $\mathcal{E}$ operation along $V_{yy'z'g}$ to $\widetilde{K}_{i-1}^{(1)}$ followed by an $\mathcal{E}$ operation along $V_{xyy'z}$, and let $\widetilde{K}_{i-1}^{(2)}$ be the result. The former replaces $\{A_{yz'g}, A_{yy'z'}\}$ by $\{A_{y'z'g}, A_{yy'g}\}$, which is $\mathcal{E}_2^1$ with $\tau(A_{yy'z}) = -1$. The latter replaces $\{A_{xyg}, A_{yy'g}\}$ by $\{A_{xy'g}, A_{xyy'}\}$, which is $\mathcal{E}_2^{-1}$. By Lemma 26.34, $\widetilde{K}_{i-1}^{(1)}$ is Markov equivalent to $\widetilde{K}_{i-1}^{(2)}$. In case $\mu\pi(g)$ is on the same side as $\mu\pi(z)$ with respect to the line determined by $\mu\pi(\overline{xx'})$, $\tau(A_{x'xg}) = 1$. Apply

$\mathcal{E}$ operations along $V_{xx'z'g}$ to $\widetilde{K}_{i-1}^{(2)}$ and along $V_{xx'yg}$. The former replaces $A_{z'xg}$ by $\{A_{z'x'g}, A_{z'xx'}, A_{x'xg}\}$ and the latter $\{A_{xy'g}, A_{x'xg}\}$ by $\{A_{x'y'g}, A_{xy'x'}\}$, both of which are $\mathcal{R}$ operations. The result is $\widetilde{K}_i$. In case $\mu\pi(g)$ is on the opposite side to $\mu\pi(z)$, $\tau(A_{xx'g}) = 1$. Apply $\mathcal{R}$ operations along $V_{xx'yg}$ to $\widetilde{K}_{i-1}^{(2)}$ and along $V_{xx'z'g}$ to obtain $\widetilde{K}_i$. (From the assumption (4), all $\mathcal{E}$ operations above are applicable.)

Case II. Starting the above argument from $\widetilde{K}_{i-1}^{(2)}$, we have the result.

Case III. In case $\mu\pi(g)$ is on the same side as $\mu\pi(y)$ with respect to the line determined by $\mu\pi(\overline{xx'})$, $\tau(A_{x'xg}) = 1$. Apply $\mathcal{E}$ operations along $V_{xx'zg}$ to $\widetilde{K}_{i-1}$ and then along $V_{xx'yg}$. The former replaces $\{A_{zxg}, A_{xzx'}\}$ by $\{A_{zx'g}, A_{x'xg}\}$ and the latter $\{A_{xyg}, A_{xx'y}, A_{x'xg}\}$ by $A_{x'yg}$, both of which are $\mathcal{R}$ operations. Let $\widetilde{K}_{i-1}^{(1)}$ be the result. In case $\mu\pi(g)$ is on the side opposite to $\mu\pi(y)$, apply $\mathcal{R}$ operations along $V_{xx'yg}$ to $\widetilde{K}_{i-1}$ and along $V_{xx'zg}$ to obtain $\widetilde{K}_{i-1}^{(1)}$. Apply $\mathcal{E}$ operations along $V_{x'yy'g}$ to $\widetilde{K}_{i-1}^{(1)}$ and along $V_{yy'zg}$. The former replaces $\{A_{x'yg}, A_{x'y'y}\}$ by $\{A_{x'y'g}, A_{y'yg}\}$, which is $\mathcal{E}_2^1$ with $\tau(A_{y'yg}) = -1$ and the latter $\{A_{yzg}, A_{y'yg}\}$ by $\{A_{y'zg}, A_{y'yz}\}$, which is $\mathcal{E}_2^{-1}$. By Lemma 26.34, the result $\widetilde{K}_{i-1}^{(2)}$ is Markov equivalent to $\widetilde{K}_{i-1}^{(1)}$. Apply $\mathcal{E}$ operations along $V_{x'zz'g}$ to $\widetilde{K}_{i-1}^{(2)}$ and along $V_{y'zz'g}$. By Lemma 26.34 again, the result, which is $\widetilde{K}_i$, is Markov equivalent to $\widetilde{K}_{i-1}^{(2)}$. So $\widetilde{K}_{i-1}$ is Markov equivalent to $\widetilde{K}_i$.

Case IV. Starting the above argument from $\widetilde{K}_{i-1}^{(1)}$, we have the result.

This completes the case where $\mathbb{A}$ is an m-mesh division. By a similar argument to Lemma 26.33, we obtain the result in the case where $\mathbb{A}$ is a (slightly deformed) almost m-mesh division. ∎

Proof of Lemma 26.35. Let $c = e(0), e(1), \cdots, e(r) = d$ be a sequence of points on $\overline{cd}$ as in Lemma 26.36 and, for $i \in \{0, \cdots, r-1\}$, let K_{i+1} be the division obtained from K_i by an $\mathcal{E}_1^3$ operation along $V_{abe(i)e(i+1)}$, where $K_0 = K$. For $i \in \{0, \cdots, r-1\}$, let $\mathbb{V}_i$ be a sawtooth on $A_{abe(i)}$ avoiding $F_{e(i)}$ as in Lemma 26.36, $K_{e(i)}^*$ the division obtained from K_i by an $\mathcal{S}$ operation along $\mathbb{V}_i$ and $K_{e(i+1)}^{**}$ the division obtained from K_{i+1} by an $\mathcal{S}$ operation along $\lambda_{e(i)}^{e(i+1)}(\mathbb{V}_i) = \{\lambda_{e(i)}^{e(i+1)}(B) * \{g_B\} \mid B \in \mathbb{A}_{abe(i)}\}$. By Lemma 26.37, $K_{e(i)}^*$ is Markov equivalent to $K_{e(i+1)}^{**}$. Let $K_{e(0)}^{**} = \widetilde{K}$ and $K_{e(r)}^*$ the division obtained from K_r by an $\mathcal{S}$ operation along $\mathbb{V}'$. By Lemma 26.32, $K_{e(i)}^{**}$ and $K_{e(i)}^*$ ($i = 0, \cdots, r$) are Markov equivalent. So $\widetilde{K}$ is Markov equivalent to $K_{e(r)}^*$. Since $\widetilde{K}'$ differs from $K_{e(r)}^*$ by positive 2-simplices on $A_{bcd} \cup A_{adc}$, they are Markov equivalent. ∎

26.17. Height Reduction, III

In this section we prove the following lemma.

LEMMA 26.38. *Let K be a division of a surface link, let K' be a division obtained from K by an $\mathcal{E}_2^4$ operation along a 3-simplex V, and let A, A' be the negative 2-simplices of K, K' which are 2-faces of V. Let $\widetilde{K}$ (resp. $\widetilde{K}'$) be a division obtained from K (resp. K') by an $\mathcal{S}$ operation along a sawtooth $\mathbb{V}$ (resp. $\mathbb{V}'$) on A (resp. A'). Then $\widetilde{K}$ is Markov equivalent to $\widetilde{K}'$.*

Let a, b, c, d be vertices of V such that the $\mathcal{E}_2^4$ operation replaces $\{A_{abc}, A_{acd}\}$ by $\{A_{abd}, A_{bcd}\}$ where $A = A_{abc}$, $A' = A_{abd}$. In the notation of Sect. 26.4, $J_1 \cup J_4 =$

$J_2 \cup J_3$ holds and $J_1 \cap J_2 \cap J_3 \cap J_4$ consists of a single ray. Take a 2-plane H in $\mathbb{R}^3$ intersecting the ray perpendicularly. The image $\pi(V)$ is contained in the positive domain of the stereographic projection μ associated with H. Let f be a unique point of $\overline{bc}$ such that $\pi(f)$ is on the ray. Images of a, b, c, d, f under $\mu\pi$ are as in Fig 26.7.

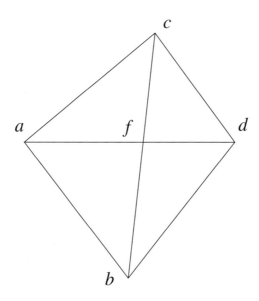

FIGURE 26.7

It is sufficient to prove the lemma under the assumption that c is sufficiently close to f. [Let c' be an interior point of $\overline{cf}$ which is close to f. Let $\mathbb{W}$ be a sawtooth on A_{abc} avoiding $|K|$ with a basis which is a subdivision of $\{A_{abc'}, A_{ac'c}\}$. Let $\mathbb{W}_{abc'}$ and $\mathbb{W}_{ac'c}$ be subsets of $\mathbb{W}$ which are on $A_{abc'}$ and $A_{ac'c}$, respectively. We may assume that $\mathbb{W}_{abc'}$ avoids $V_{acc'd}$ (Lemma 26.28). Let K_1 be the division obtained from K by an $\mathcal{S}$ operation along $\mathbb{W}$, which is Markov equivalent to $\widetilde{K}$ (Lemma 26.32). Apply to K_1 an $\mathcal{S}^{-1}$ operation along $\mathbb{W}_{ac'c}$ followed by an $\mathcal{E}_2^{-1}$ operation along $V_{acc'd}$, and let K_2 be the result. It is Markov equivalent to K_1 (Lemma 26.33). Let K_3 be the division obtained from K_2 by an $\mathcal{S}^{-1}$ operation along $\mathbb{W}_{abc'}$ followed by an $\mathcal{E}_2^4$ operation along $V_{abc'd}$. Let K_4 be the division obtained from K_3 by an $\mathcal{S}$ operation along $\mathbb{V}'$. By the assumption, K_2 is Markov equivalent to K_4. $\widetilde{K}'$ is obtained from K_4 by an $\mathcal{A}^{-1}$ operation replacing $\{A_{bc'd}, A_{c'cd}\}$ by $\{A_{bcd}\}$. Hence $\widetilde{K}$ is Markov equivalent to $\widetilde{K}'$.] Therefore we may assume that c is sufficiently close to f. For points e, e' of $\overline{cd}$, let $\lambda_e^{e'} : A_{abe} \rightarrow A_{abe'}$ be the linear map with $\lambda_e^{e'}(a) = a, \lambda_e^{e'}(b) = b$ and $\lambda_e^{e'}(e) = e'$. λ means λ_c^d. For $e \in \overline{cd}$, denote by F_e the surface link obtained from $F = |K|$ by an $\mathcal{E}$ operation along V_{abce}.

LEMMA 26.39. *In the above situation, the same statement as in* Lemma 26.36 *holds.*

Proof. By the same argument as in Lemma 26.36, we have the result. ∎

LEMMA 26.40. *In the above situation, the same statement as in* Lemma 26.37 *holds.*

Proof. We construct a sequence $\widetilde{K} = \widetilde{K}_0, \cdots, \widetilde{K}_{4^m} = \widetilde{K}'$ of divisions of surface links which are Markov equivalent. Consider the case where $\mathbb{A}$ is an m-mesh division. We assign indices to 2-simplices of $\mathbb{A}$ as in Fig 26.8 where $m = 2$. Let $T_1 = \{B_{4^m}\}$ and define T_k for $k \in \{2, \cdots, 2^{m+1} - 1\}$ inductively as before. Cases I, II, III, IV and points x, y, z, x', y', z' are as in Lemma 26.33. Since c is close to f, we may assume that $\angle \mu\pi(d)\mu\pi(c)\mu\pi(b) > \angle \mu\pi(a)\mu\pi(b)\mu\pi(c)$. Since $\mu\pi(\overline{xx'}) \parallel \mu\pi(\overline{yy'}) \parallel \mu\pi(\overline{zz'}) \parallel \mu\pi(\overline{cd})$ and $\mu\pi(\overline{yz}) \parallel \mu\pi(\overline{y'z'}) \parallel \mu\pi(\overline{ab})$, we have $\angle \mu\pi(y)\mu\pi(z)\mu\pi(z') < \pi$ radians in Case I. Similarly we have $\angle \mu\pi(z)\mu\pi(y)\mu\pi(y') < \pi$ radians in Case III. So images of x, y, z, x', y', z' under $\mu\pi$ are as in Fig 26.9.

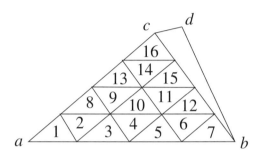

FIGURE 26.8

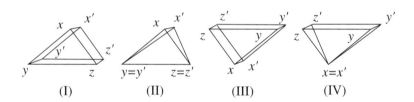

FIGURE 26.9

Let V_i be the 3-simplex of $\mathbb{V}$ on $B_i = A_{xyz}$ and V_i' the 3-simplex of $\lambda(\mathbb{V})$ on $\lambda(B_i) = A_{x'y'z'}$ for some $g = g_i \in \mathbb{R}^4$. Then $V_i = A_{xyz} * \{g\}$ and $V_i' = A_{x'y'z'} * \{g\}$. We denote by P_i the set $\{A_{xyg}, A_{yzg}, A_{zxg}\}$ and by P_i' the set $\{A_{x'y'g}, A_{y'z'g}, A_{z'x'g}\}$. We define $\widetilde{K}_i$ by

$$\begin{cases} (\widetilde{K}_{i-1} \setminus (P_i \cup R_{yxx'y'} \cup R_{zyy'z'})) \cup P_i' \cup R_{zxx'z'} & \text{for Cases I, II,} \\ (\widetilde{K}_{i-1} \setminus (P_i \cup R_{xzz'x'})) \cup P_i' \cup R_{xyy'x'} \cup R_{yzz'y'} & \text{for Cases III, IV.} \end{cases}$$

We show that $\widetilde{K}_{i-1}$ and $\widetilde{K}_i$ are Markov equivalent. Since $V_i = A_{xyz} * \{g\}, V_i' = A_{x'y'z'} * \{g\}$ are 3-simplices of type γ and $\pi(A_{xyz}), \pi(A_{x'y'z'})$ are in the positive domain of μ, we see that $\pi(g)$ is in the negative domain of μ and $\mu\pi(g)$ is an interior point of $\mu\pi(A_{xyz})$ and of $\mu\pi(A_{x'y'z'})$. Modifying g slightly if necessary, using Lemma 26.32, we may assume that $\mu\pi(g)$ is not on the line in H determined by $\mu\pi(\overline{yy'})$ (for Cases I, III, IV).

Case I. If $\mu\pi(g)$ is on the same side as $\mu\pi(x)$ with respect to the line determined by $\mu\pi(\overline{yy'})$, then $\tau(A_{yy'g}) = 1$. Apply an $\mathcal{R}$ operation along $V_{yy'zg}$ to $\widetilde{K}_{i-1}$, which replaces $\{A_{yzg}, A_{yy'z}\}$ by $\{A_{y'zg}, A_{yy'g}\}$. Then apply an $\mathcal{R}$ operation along $V_{xyy'g}$, which replaces $\{A_{xyg}, A_{xy'y}, A_{yy'g}\}$ by $A_{xy'g}$, and let $\widetilde{K}_{i-1}^{(1)}$ be the result, which is Markov equivalent to $\widetilde{K}_{i-1}$. If $\mu\pi(g)$ is on the same side as $\mu\pi(z)$, then $\tau(A_{y'yg}) = 1$. Apply $\mathcal{E}$ operations along $V_{xyy'g}$ to $\widetilde{K}_{i-1}$ and along $V_{yy'zg}$. The result is also $\widetilde{K}_{i-1}^{(1)}$ above. Apply an $\mathcal{E}$ operation along $V_{y'zz'g}$ to $\widetilde{K}_{i-1}^{(1)}$, which replaces $\{A_{y'zg}, A_{y'z'z}\}$ by $\{A_{y'z'g}, A_{z'zg}\}$. Since $\tau(A_{y'zg}) = \tau(A_{y'z'z}) = \tau(A_{y'z'g}) = 1$ and $\tau(A_{z'zg}) = -1$, the $\mathcal{E}$ operation along $V_{y'zz'g}$ is $\mathcal{E}_2^1$. Apply an $\mathcal{E}_2^{-1}$ operation along $V_{xzz'g}$ which replaces $\{A_{z'zg}, A_{zxg}\}$ by $\{A_{z'xg}, A_{xz'z}\}$. By Lemma 26.34, the result $\widetilde{K}_{i-1}^{(2)}$ is Markov equivalent to $\widetilde{K}_{i-1}^{(1)}$. Apply $\mathcal{E}$ operations along $V_{xx'y'g}$ to $\widetilde{K}_{i-1}^{(2)}$ and along $V_{xx'z'g}$. By Lemma 26.34 again, the result, which coincides with $\widetilde{K}_i$, is Markov equivalent to $\widetilde{K}_{i-1}^{(2)}$.

Case II. Starting the above argument from $\widetilde{K}_{i-1}^{(2)}$, we have the result.

Case III. Apply $\mathcal{E}$ operations along $V_{xx'zg}$ to $\widetilde{K}_{i-1}$ and along $V_{xx'yg}$. The former replaces $\{A_{zxg}, A_{xzx'}\}$ by $\{A_{zx'g}, A_{x'xg}\}$ which is $\mathcal{E}_2^1$ with $\tau(A_{x'xg}) = -1$, and the latter $\{A_{xyg}, A_{x'xg}\}$ by $\{A_{x'yg}, A_{x'xy}\}$ which is $\mathcal{E}_2^{-1}$. Let $\widetilde{K}_{i-1}^{(1)}$ be the result, which is Markov equivalent to $\widetilde{K}_{i-1}$ (Lemma 26.34). Apply $\mathcal{E}$ operations along $V_{x'zz'g}$ to $\widetilde{K}_{i-1}^{(1)}$ and along $V_{yzz'g}$. The former replaces $\{A_{zx'g}, A_{x'zz'}\}$ by $\{A_{z'x'g}, A_{zz'g}\}$ which is $\mathcal{E}_2^1$ with $\tau(A_{zz'g}) = -1$, and the latter replaces $\{A_{zz'g}, A_{yzg}\}$ by $\{A_{yz'g}, A_{yzz'}\}$ which is $\mathcal{E}_2^{-1}$. The result $\widetilde{K}_{i-1}^{(2)}$ is Markov equivalent to $\widetilde{K}_{i-1}^{(1)}$ (Lemma 26.34). Finally apply $\mathcal{E}$ operations along $V_{x'yy'g}$ (or $V_{yy'z'g}$, resp.) to $\widetilde{K}_{i-1}^{(2)}$ and along $V_{yy'z'g}$ (resp. $V_{x'yy'g}$) if $\mu\pi(g)$ is on the same side as $\mu\pi(x)$ (resp. $\mu\pi(z)$) with respect to the line determined by $\mu\pi(\overline{yy'})$. These operations are $\mathcal{R}$ operations, and the result is $\widetilde{K}_i$.

Case IV. Starting the above argument from $\widetilde{K}_{i-1}^{(1)}$, we have the result.

This completes the case where $\mathbb{A}$ is an m-mesh division. For the case where $\mathbb{A}$ is a (slightly deformed) almost m-mesh division, by a similar argument to Lemma 26.33, arranging B_i^j $(j = 1, \cdots, j(i))$ in an appropriate order and with an appropriate definition of $\widetilde{K}_i^j$, we have the result. Details are left to the reader. ∎

Proof of Lemma 26.38. Using Lemmas 26.39 and 26.40, by a similar argument to Lemma 26.35, we have the result. ∎

26.18. Height Reduction, IV

LEMMA 26.41. *Let K be a division of a surface link, let K' be a division obtained from K by an $\mathcal{E}_1^5$ operation, let A be the negative 2-simplex of K' obtained by the operation, and let K'' be a division obtained from K' by an $\mathcal{S}$ operation along a sawtooth $\mathbb{V}$ on A. Then K is Markov equivalent to K''.*

Let the $\mathcal{E}_1^5$ operation be applied along $V = V_{abcd}$ and replace A_{abd} by $\{A_{acd}, A_{bdc}, A_{abc}\}$ where $A = A_{abc}$. In the notation of Sect. 26.4, $J_1 \cap J_2 \cap J_3 \cap J_4$ consists of a single ray. Take a 2-plane H in $\mathbb{R}^3$ intersecting the ray perpendicularly. $\pi(V)$ is contained in the positive domain of the stereographic projection μ associated with H. Let f be a unique point of $\overline{cd}$ such that $\pi(f)$ is on the ray. Images of a, b, c, d, f under $\mu\pi$ are as in Fig 26.10.

Proof. It is sufficient to prove the lemma on the assumption that for any given positive numbers ϵ and ϵ', $d(c, f) < \epsilon$ and $d(d, f) < \epsilon'$. [Take interior points c' of $\overline{cf}$ and d' of $\overline{df}$ with $d(c', f) < \epsilon$, $d(d', f) < \epsilon'$. Let K_1 be the division

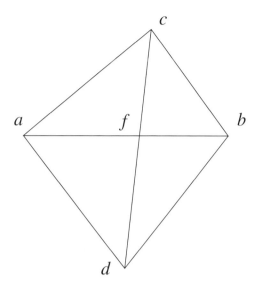

FIGURE 26.10

obtained from K by an $\mathcal{R}$ operation along $V_{abdd'}$, K_1' the division obtained from K_1 by an $\mathcal{E}_1^5$ operation along $V_{abc'd'}$ and K_1'' an division obtained from K_1' by an $\mathcal{S}$ operation along a sawtooth on $A_{abc'}$. By the assumption, K_1 is Markov equivalent to K_1''. Let K_2 be the division obtained from K_1' by an $\mathcal{E}_1^3$ operation along $V_{abcc'}$ and $\widetilde{K}_2$ the division obtained from K_2 by an $\mathcal{S}$ operation along $\mathbb{V}$. By Lemma 26.35, K_1'' is Markov equivalent to $\widetilde{K}_2$. By Lemma 26.13, $\widetilde{K}_2$ is Markov equivalent to K''.] Therefore, we may assume that c, d are sufficiently close to f and $\angle \mu\pi(c)\mu\pi(a)\mu\pi(b) < \angle \mu\pi(d)\mu\pi(b)\mu\pi(a)$. By Lemmas 26.28 and 26.32, we may assume that $\mathbb{V}$ is a sawtooth on A with a basis $\mathbb{A}$ satisfying the following: $\mathbb{A}$ is a slightly deformed almost m-mesh division of A for some m, the image under $\mu\pi$ of each edge of $\mathbb{A}$ is not parallel to that of $\overline{cd}$ and $\mathbb{V}$ avoids V.

First we consider the case where $\mathbb{A}$ is an m-mesh division. Assign indices to 2-simplices of $\mathbb{A}$ in the same way as in Lemma 26.37 and let subsets T_k ($k = 1, \cdots, 2^{m+1} - 1$), Cases I–IV and points x, y, z, x', y', z' be similar. Let $\lambda : A_{abc} \to A_{abd}$ be the linear map with $\lambda(a) = a, \lambda(b) = b$ and $\lambda(c) = d$. Then $\lambda(\mathbb{A}) = \{\lambda(B_i) \,|\, B_i \in \mathbb{A}\}$ is a division of A_{abd}. Images of $\mathbb{A}$, $\lambda(\mathbb{A})$ (for $m = 2$) and of x, y, z, x', y', z' under $\mu\pi$ are as in Figs. 26.11 and 26.12.

Let K_0 be a division $(K \setminus \{A_{abd}\}) \cup \lambda(\mathbb{A})$, which is Markov equivalent to K (Lemma 26.13). For each $i \in \{1, \cdots, 4^m\}$, we define K_i by

$$\begin{cases} (K_{i-1} \setminus (\{A_{x'y'z'}\} \cup R_{yzz'y'})) \cup P_i \cup R_{yxx'y'} \cup R_{xzz'x'} & \text{for Cases I, II,} \\ (K_{i-1} \setminus (\{A_{x'y'z'}\} \cup R_{xyy'x'} \cup R_{zxx'z'})) \cup P_i \cup R_{zyy'z'} & \text{for Cases III, IV.} \end{cases}$$

Since K_{4^m} is Markov equivalent to K'' (Lemma 26.13), it suffices to prove that, for each $i \in \{1, \cdots, 4^m\}$, K_{i-1} is Markov equivalent to K_i. Consider each case I–IV.

Case I. Apply $\mathcal{R}$ operations along $V_{x'y'zz'}$ to K_{i-1} and along $V_{x'yy'z}$. The former replaces $\{A_{x'y'z'}, A_{y'zz'}\}$ by $\{A_{x'zz'}, A_{x'y'z}\}$, and the latter $\{A_{y'yz}, A_{x'y'z}\}$ by $\{A_{x'y'y}, A_{x'yz}\}$. Let $K_{i-1}^{(1)}$ be the result, which is Markov equivalent to K_{i-1}. Apply $\mathcal{E}$ operations along $V_{xx'yz}$ to K_{i-1} and along V_i, replacing $A_{x'yz}$

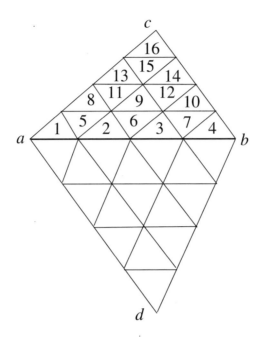

FIGURE 26.11

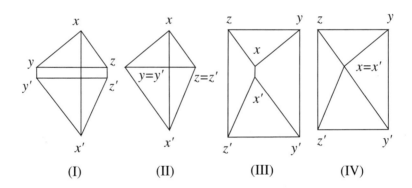

FIGURE 26.12

by $\{A_{x'yx}, A_{xzx'}, A_{xyz}\}$ and A_{xyz} by P_i. The composition is a $\mathcal{W}$ operation. (Since c is sufficiently close to d and since $\mathbb{V}$ avoids V, we may assume $|K_{i-1}^{(1)}| \cap V_{xx'yz} * V_i = |K_{i-1}^{(1)}| \cap V_{xx'yz}$.) The result is K_i.

Case II. Start the above argument from $K_{i-1}^{(1)}$.

Case III. Apply an $\mathcal{E}_1^1$ operation along $V_{xx'y'z'}$ to K_{i-1}, which replaces $\{A_{x'y'z'}, A_{xy'x'}, A_{xx'z'}\}$ by $A_{xy'z'}$. Let $K_{i-1}^{(1)}$ be the result. Since $\mu\pi(\overline{yz}) \parallel \mu\pi(\overline{y'z'}) \parallel \mu\pi(\overline{ab}), \mu\pi(\overline{xy}) \parallel \mu\pi(\overline{ac}), \mu\pi(\overline{x'z'}) \parallel \mu\pi(\overline{bd})$ and the assumption $\angle\mu\pi(c)\mu\pi(a)\mu\pi(b) < \angle\mu\pi(d)\mu\pi(b)\mu\pi(a)$, we have that $\mu\pi(\overline{xy'})$ and $\mu\pi(\overline{yz'})$ intersect each other; i.e., $V_{xyy'z'}$ is a 3-simplex of type β. Apply to $K_{i-1}^{(1)}$ an $\mathcal{E}_2^3$ operation along $V_{xyy'z'}$ replacing $\{A_{xy'z'}, A_{xyy'}\}$ by $\{A_{xyz'}, A_{yy'z'}\}$. Let $K_{i-1}^{(2)}$ be the result.

Apply $\mathcal{E}$ operations along $V_{xyzz'}$ to $K_{i-1}^{(2)}$ and along V_i. The former is $\mathcal{E}_2^1$ replacing $\{A_{xz'z}, A_{xyz'}\}$ by $\{A_{yz'z}, A_{xyz}\}$, and the latter is $\mathcal{E}_1^8$ replacing A_{xyz} by P_i. The composition of them is a $\mathcal{U}$ operation. (Note that $|K_{i-1}^{(1)}| \cap V_{xyzz'} * V_i = |K_{i-1}^{(1)}| \cap V_{xyzz'}$.) The result is K_i.

Case IV. Start the above argument from $K_{i-1}^{(1)}$.

For the case where $\mathbb{A}$ is a (slightly deformed) almost m-mesh division, by a similar argument to Lemma 26.33, arranging B_i^j ($j = 1, \cdots, j(i)$) in an appropriate order and with an appropriate definition of K_i^j, we have the result. ∎

26.19. Height Reduction, V

LEMMA 26.42. *Let K be a division of a surface link, let K' be a division obtained from K by an operation $\mathcal{F}$ with $h(K) < h(K')$, let A be a negative 2-simplex of K' obtained by $\mathcal{F}$, and let K'' be a division obtained from K' by an $\mathcal{S}$ operation along a sawtooth on A. Then there is a deformation chain from K to K'' whose maximal height is smaller than $h(K')$.*

Proof. If $\mathcal{F}$ is an $\mathcal{S}^{-1}$ or $\mathcal{E}_3^8$ operation, by Lemma 26.32, K and K'' are connected by a deformation chain with a constant height which is smaller than $h(K')$. The case where $\mathcal{F}$ is $\mathcal{E}_1^5$ (or $\mathcal{E}_2^1$) is Lemma 26.41 (Lemma 26.33). Suppose that $\mathcal{F}$ is a $\mathcal{B}$ operation, $K' = (K \setminus \{A_0\}) \cup \{A_1, A_2\}$ and $A = A_2$. There is a sawtooth $\mathbb{V}_0$ on A_0 avoiding $|K|$ with a basis which is a subdivision of $\{A_1, A_2\}$ (Lemmas 26.1, 26.19, 26.31). Let $\mathbb{V}_1$ and $\mathbb{V}_2$ be subsets of $\mathbb{V}_0$ which are sawteeth on A_1 and A_2, respectively. Let K_1 be the division obtained from K by an $\mathcal{S}$ operation along $\mathbb{V}_0$. Then $h(K_1) = h(K) - 1$. Applying an $\mathcal{S}^{-1}$ operation along $\mathbb{V}_1$ to K_1, we have a division K_2 with $h(K_2) = h(K)$. Since K_2 is obtained from K' by an $\mathcal{S}$ operation along $\mathbb{V}_2$, by Lemma 26.32, it is Markov equivalent to K''. It remains to consider whether $\mathcal{F}$ is an $\mathcal{E}_1^2$, $\mathcal{E}_1^4$, $\mathcal{E}_1^6$, $\mathcal{E}_1^7$, $\mathcal{E}_2^{-2}$ or $\mathcal{E}_2^6$ operation. Let V be the 3-simplex along which $\mathcal{F}$ is applied.

Suppose that $\mathcal{F}$ is $\mathcal{E}_1^2$. Let a, b, c, d be vertices of V such that $\mathcal{F}$ replaces A_{abc} by $\{A_{abd}, A_{bcd}, A_{cad}\}$ where $\tau(A_{abc}) = \tau(A_{abd}) = 1$ and $\tau(A_{cad}) = \tau(A_{bcd}) = -1$. A is either A_{bcd} or A_{cad}, say A_{bcd}. Let e be an interior point of $\overline{ab}$ which is close to b so that V_{bcde} is a 3-simplex of type β and V_{acde} is of type α. Let K_1 be the division obtained from K by an $\mathcal{A}$ operation replacing A_{abc} by $\{A_{cae}, A_{bce}\}$, and let K_2 be the division obtained from K_1 by an $\mathcal{E}_1^5$ operation along V_{bcde} which replaces A_{bce} by $\{A, A_{ced}, A_{ebd}\}$. Take a sawtooth $\mathbb{V}$ on A avoiding $|K_2| \cup V$ (and hence V_{acde}); cf. Lemma 26.28(1). Let K_3 be the division obtained from K_2 by an $\mathcal{S}$ operation along $\mathbb{V}$. We can apply an $\mathcal{E}_2^1$ operation along V_{acde} to K_3 and let K_4 be the result. Let K_5 be a division $(K_4 \setminus \{A_{ebd}, A_{eda}\}) \cup \{A_{abd}\}$ which is obtained from K_4 by an $\mathcal{A}^{-1}$ operation. On the other hand, K_5 is also obtained from K' by an $\mathcal{S}$ operation along $\mathbb{V}$. By Lemma 26.32, K_5 is Markov equivalent to K'' and there is a deformation chain $K_5, \cdots, K''$ with constant height. Since $h(K) = h(K_1) = h(K_3) < h(K')$ and $h(K_2) = h(K_4) = h(K_5) = h(K'') = h(K) + 1 < h(K')$, we have the result.

Suppose that $\mathcal{F}$ is $\mathcal{E}_1^4$. Let a, b, c, d be vertices of V such that $\mathcal{F}$ replaces A_{abc} by $\{A_{abd}, A_{bcd}, A_{cad}\}$ and $A = A_{abd}$. Let H be the 2-plane in $\mathbb{R}^3$ determined by $\pi(A_{abc})$ and μ the stereographic projection associated with H. Let e be an interior point of A_{abc} such that $\mu\pi(e)$ is in the interior of $\mu\pi(A_{abd})$ and on the same side as $\mu\pi(b)$ with respect to the line on H determined by $\mu\pi(\overline{cd})$; see Fig 26.13(1) where we omit the symbol $\mu\pi$. Let $\mathbb{W}$ be a sawtooth on A_{abc} avoiding $|K|$ with a basis which is a subdivision of $\{A_{abe}, A_{bce}, A_{cae}\}$, and let $\mathbb{W}_{abe}, \mathbb{W}_{bce}, \mathbb{W}_{cae}$ be subsets of $\mathbb{W}$ on $A_{abe}, A_{bce}, A_{cae}$ respectively. Let K_1, K_2, K_3 and K^* be the division obtained from K by an $\mathcal{S}$ operation along $\mathbb{W}$, the division obtained from K_1 by an $\mathcal{S}^{-1}$ operation along $\mathbb{W}_{cae}$, the division obtained from K_2 by an $\mathcal{S}^{-1}$ operation along $\mathbb{W}_{bce}$ and the division obtained from K_3 by an $\mathcal{S}^{-1}$ operation along $\mathbb{W}_{abe}$. Then $h(K_1) = h(K) - 1, h(K_2) = h(K), h(K_3) = h(K) + 1$ and $h(K^*) = h(K) + 2$. K^* is $(K \setminus \{A_{abc}\}) \cup \{A_{abe}, A_{bce}, A_{cae}\}$. Let K^{**} be the division obtained from K^* by an $\mathcal{E}_1^3$ operation along V_{abde}. Take a sawtooth $\mathbb{V}$ on A avoiding $|K^{**}| \cup V$ and let K_4 be the division obtained from K^{**} by an $\mathcal{S}$ operation along it. By Lemma 26.35, there is a deformation chain $K_3, \cdots, K_4$ with constant height. Apply an $\mathcal{E}_2^4$ operation along V_{bcde} to K_4, which replaces $\{A_{bce}, A_{bed}\}$ by $\{A_{bcd}, A_{ced}\}$, and let K_5 be the result with $h(K_5) = h(K_4)$. Apply an $\mathcal{E}_3^3$ operation along V_{acde} to K_5 and let K_6 be the result. Since K_6 is also obtained from K' by an $\mathcal{S}$ operation along $\mathbb{V}$, by Lemma 26.32, there is a deformation chain $K_6, \cdots, K''$ with constant height. Hence we have a desired deformation chain $K, K_1, K_2, K_3, \cdots, K_4, K_5, K_6, \cdots, K''$.

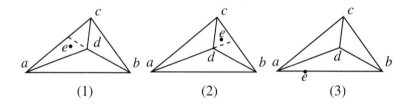

$$\text{(1)} \qquad \text{(2)} \qquad \text{(3)}$$

FIGURE 26.13

Suppose that $\mathcal{F}$ is $\mathcal{E}_1^6$. Let a, b, c, d be vertices of V such that $\mathcal{F}$ replaces A_{abc} by $\{A_{abd}, A_{bcd}, A_{cad}\}$ where $\tau(A_{abc}) = \tau(A_{abd}) = \tau(A_{cad}) = -1$ and $\tau(A_{bcd}) = 1$. A is A_{abd} or A_{cad}, say A_{abd}. Let e be an interior point of $\overline{bc}$ which is close to b so that V_{abde} is a 3-simplex of type α and V_{acde} is of type β. Let $\mathbb{W}$ be a sawtooth on A_{abc} avoiding $|K|$ with a basis which is a subdivision of $\{A_{abe}, A_{cae}\}$, and let $\mathbb{W}_{abe}, \mathbb{W}_{cae}$ be subsets of $\mathbb{W}$ on A_{abe}, A_{cae} respectively. Let K_1 be the division obtained from K by an $\mathcal{S}$ operation along $\mathbb{W}$, K_2 the division obtained from K_1 by an $\mathcal{S}^{-1}$ operation along $\mathbb{W}_{cae}$ and K^* the division obtained from K_2 by an $\mathcal{S}^{-1}$ operation along $\mathbb{W}_{abe}$, which is $(K \setminus \{A_{abc}\}) \cup \{A_{abe}, A_{cae}\}$. Let K^{**} be the division obtained from K^* by an $\mathcal{E}_1^3$ operation along V_{abde}. Take a sawtooth $\mathbb{V}$ on A avoiding $|K^{**}| \cup V$ (and hence V_{acde}) and let K_3 be the division obtained from K^{**} by an $\mathcal{S}$ operation along it. By Lemma 26.35, there is a deformation chain $K_2, \cdots, K_3$ with constant height. Let K_4 be the division obtained from K_3 by an $\mathcal{E}_2^4$ operation along V_{acde} and K_5 the division obtained from K_4 by an $\mathcal{A}^{-1}$ operation replacing $\{A_{bed}, A_{dec}\}$ by A_{bcd}. K_5 is also obtained from K' by a type $\mathcal{S}$ operation along $\mathbb{V}$. By Lemma 26.32, there is a deformation chain $K_5, \cdots, K''$ with constant height. We have the desired deformation chain $K, K_1, K_2, \cdots, K_3, K_4, K_5, \cdots, K''$.

Suppose that $\mathcal{F}$ is $\mathcal{E}_1^7$. Let a, b, c, d be vertices of V such that $\mathcal{F}$ replaces A_{abc} by $\{A_{abd}, A_{bcd}, A_{cad}\}$ and $A = A_{abd}$. Let H be the 2-plane in $\mathbb{R}^3$ determined by $\pi(A_{abc})$ and μ the stereographic projection associated with H. Since V is a 3-simplex of type γ, $\pi(d)$ is in the negative domain of μ and $\mu\pi(d)$ is an interior point of $\mu\pi(A_{abd})$. Let e be an interior point of A_{abc} such that $\mu\pi(e)$ is in the interior of $\mu\pi(A_{abd})$ and on the same side as $\mu\pi(a)$ with respect to the line on H

determined by $\mu\pi(\overline{cd})$; see Fig 26.13(2). Let $K_1 = (K \setminus \{A_{abc}\}) \cup \{A_{abe}, A_{bce}, A_{cae}\}$, which is Markov equivalent to K (Lemma 26.13). Let K_2 be the division obtained from K_1 by an $\mathcal{E}$ operation along V_{abde} replacing A_{abe} by $\{A_{abd}, A_{bed}, A_{ead}\}$. Since $\pi(a), \pi(b), \pi(c)$ are in the positive domain of μ and $\pi(d)$ is in the negative, we see from Fig 26.13(2) that $\tau(A_{bed}) = \tau(A_{ead}) = 1$. Hence the $\mathcal{E}$ operation is $\mathcal{E}_1^5$. Take a sawtooth $\mathbb{V}$ on A avoiding $|K_2| \cup V$ and let K_3 be the division obtained from K_2 by an $\mathcal{S}$ operation along it. Apply an $\mathcal{E}$ operation along V_{acde} to K_3 (say K_4 is the result) and along V_{bcde} to K_4. Let K_5 be the result. The former is $\mathcal{E}_2^1$ and the latter is $\mathcal{E}_3^8$. Since K_5 is also obtained from K' by an $\mathcal{S}$ operation along $\mathbb{V}$, by Lemma 26.32 there is a deformation chain $K_5, \cdots, K''$ with constant height.

Suppose that $\mathcal{F}$ is $\mathcal{E}_2^{-2}$. Let a, b, c, d be vertices of V such that $\mathcal{F}$ replaces $\{A_{abc}, A_{cbd}\}$ by $\{A_{abd}, A_{cad}\}$ with $\tau(A_{abc}) = -1$, $\tau(A_{cbd}) = 1$. Assume $A = A_{abd}$. (The case $A = A_{cad}$ is similarly shown.) Let H be the 2-plane in $\mathbb{R}^3$ determined by $\pi(A_{abc})$ and μ the stereographic projection associated with H. Take an interior point e of $\overline{bc}$ such that $\mu\pi(e)$ is on the same side as $\mu\pi(b)$ with respect to the line on H determined by $\mu\pi(\overline{ad})$; see Fig 26.13(3). Let K_1 be the division obtained from K by an $\mathcal{A}$ operation replacing A_{cbd} by $\{A_{bde}, A_{dce}\}$. Let $\mathbb{W}$ be a sawtooth on A_{abc} avoiding $|K_1|$ with a basis which is a subdivision of $\{A_{abe}, A_{cae}\}$, and let $\mathbb{W}_{abe}, \mathbb{W}_{cae}$ be subsets of $\mathbb{W}$ on A_{abe}, A_{cae} respectively. Apply an $\mathcal{S}$ operation along $\mathbb{W}$ to K_1 and let K_2 be the result. Apply an $\mathcal{S}^{-1}$ operation along $\mathbb{W}_{cae}$ to K_2 and let K_3 be the result. Let K^* be the division obtained from K_3 by an $\mathcal{S}^{-1}$ operation along $\mathbb{W}_{abe}$, which is $(K \setminus \{A_{abc}\}) \cup \{A_{abe}, A_{cae}\}$, and K^{**} the division obtained from K^* by an $\mathcal{E}_2^4$ operation along V_{abde} replacing $\{A_{bde}, A_{abe}\}$ by $\{A_{ade}, A_{abd}\}$. Take a sawtooth $\mathbb{V}$ on A avoiding $|K^{**}| \cup V$ and let K_4 be the surgery result. By Lemma 26.38, there is a deformation chain $K_3, \cdots, K_4$ with constant height. Apply an $\mathcal{E}_3^3$ operation along V_{acde} to K_4 and let K_5 be the result with $h(K_5) = h(K_4)$. On the other hand K_5 is also obtained from K' by an $\mathcal{S}$ operation along $\mathbb{V}$. By Lemma 26.32, there is a deformation chain $K_5, \cdots, K''$ with constant height.

Suppose that $\mathcal{F}$ is $\mathcal{E}_2^6$. Let a, b, c, d be vertices of V such that $\mathcal{F}$ replaces $\{A_{abc}, A_{acd}\}$ by $\{A_{abd}, A_{bcd}\}$. Assume $A = A_{abd}$. Let e be an interior point of $\overline{ac}$ which is close to a such that V_{abde} is a 3-simplex of type α and V_{bcde} is of type γ. Let K_1 be the division obtained from K by an $\mathcal{A}$ operation replacing A_{abc} by $\{A_{abe}, A_{bce}\}$, let K_2 be the division obtained from K_1 by an $\mathcal{A}$ operation replacing A_{acd} by $\{A_{cde}, A_{dae}\}$, and let K_3 be the division obtained from K_2 by an $\mathcal{E}_2^1$ operation along V_{abde}. Take a sawtooth $\mathbb{V}$ on A avoiding $|K_3| \cup V$ and let K_4 be the surgery result. Apply an $\mathcal{E}_3^8$ operation along V_{bcde} to K_4 and let K_5 be the result. Since K_5 is also obtained from K' by an $\mathcal{S}$ operation along $\mathbb{V}$, by Lemma 26.32, there is a deformation chain $K_5, \cdots, K''$ with constant height. ∎

26.20. Proof of the Height Reduction Lemma II

Proof of Lemma 26.21. Let K, K', K'' be a deformation chain with $h(K) < h(K')$ and $h(K') > h(K'')$. Let $\mathcal{F}_1$ and $\mathcal{F}_2$ be operations which transform K into K' and K' into K'' respectively. (1) Suppose that there is a negative 2-simplex A of K' which is obtained by $\mathcal{F}_1$ and deleted by $\mathcal{F}_2$. If $\mathcal{F}_1$ is $\mathcal{S}^{-1}$ or $\mathcal{F}_2$ is $\mathcal{S}$, then by Lemma 26.42 we have the result. If neither $\mathcal{F}_1^{-1}$ nor $\mathcal{F}_2$ is $\mathcal{S}$, then consider a division $\widetilde{K}'$ obtained from K' by an $\mathcal{S}$ operation along a sawtooth on A. Deformation chains $K, K', \widetilde{K}'$ and $\widetilde{K}', K', K''$ are in the above situation, and hence we have deformation chains $K, \cdots, \widetilde{K}'$ and $\widetilde{K}', \cdots, K''$ whose maximal heights are smaller than $h(K')$. (2) Suppose that there is no negative 2-simplex of K' that is obtained by $\mathcal{F}_1$ and deleted by $\mathcal{F}_2$. Apply $\mathcal{F}_2$ to K and then apply $\mathcal{F}_1$. (If necessary, combine appropriate $\mathcal{S}^{\pm 1}$ operations and use Lemma 26.42.) ∎

We note that the deformation chain between K and K'' constructed above has no $\mathcal{E}_2^5$ operations.

26.21. Proof of the Height Reduction Lemma I

Proof of Lemma 26.20. Let K, K' be a deformation chain with $h(K) = h(K') \neq 0$. Let $\mathcal{F}$ be the operation transforming K onto K'. It is an $\mathcal{A}^{\pm 1}$, $\mathcal{E}$, $\mathcal{R}$, $\mathcal{U}^{\pm 1}$ or $\mathcal{W}^{\pm 1}$ operation.

Case I. The case where $\mathcal{F}$ is not $\mathcal{E}_2^5$. If $\mathcal{F}$ is $\mathcal{A}^{\pm 1}$, $\mathcal{R}$, $\mathcal{U}^{\pm 1}$ or $\mathcal{W}^{\pm 1}$, let S be empty (if $\mathcal{F}$ is $\mathcal{A}^{\pm 1}$), a 3-simplex V (if $\mathcal{F}$ is $\mathcal{R}$ and applied along V), or $V_1 * V_2$ (if $\mathcal{F}$ is $\mathcal{U}^{\pm 1}$ or $\mathcal{W}^{\pm 1}$ and applied along V_1 and V_2). Since $h(K) \neq 0$, there is a negative 2-simplex A of K which is not contained in S. S and A satisfy the condition of Lemma 26.29, and there is a sawtooth $\mathbb{V}$ on A avoiding $|K| \cup S$. Let K_1 be the division obtained from K by an $\mathcal{S}$ operation along $\mathbb{V}$ and K_2 the division obtained from K_1 by applying $\mathcal{F}$. Since K' is obtained from K_2 by an $\mathcal{S}^{-1}$ operation along $\mathbb{V}$, we have the desired sequence K, K_1, K_2, K'. If $\mathcal{F}$ is $\mathcal{E}_1^3$ or $\mathcal{E}_3^3$, then the result follows from Lemmas 26.19 and 26.35. If $\mathcal{F}$ is $\mathcal{E}_2^4$, it follows from Lemmas 26.19 and 26.38. Note that all deformation chains constructed above have no $\mathcal{E}_2^5$ operations.

Case II. The case where $\mathcal{F}$ is $\mathcal{E}_2^5$. Let $\mathcal{F}$ be applied along $V = V_{abcd}$ and replace $\{A_{abc}, A_{bad}\}$ by $\{A_{bcd}, A_{cad}\}$. Let e be an interior point of $\overline{cd}$ which is close to d such that V_{abce} is a 3-simplex of type β and V_{abde} is type α. Let $K_1^{(1)}$ be the division obtained from $K = K_0^{(1)}$ by an $\mathcal{E}_1^6$ operation along V_{abce} and $K_2^{(1)}$ the division obtained from $K_1^{(1)}$ by an $\mathcal{E}_2^{-2}$ operation along V_{abde}. Then $K_2^{(1)} = (K \setminus \{A_{abc}, A_{bad}\}) \cup \{A_{cae}, A_{ade}, A_{bce}, A_{dbe}\}$ and $h(K_2^{(1)}) = h(K) + 2$. Let $\mathbb{V}$ be a sawtooth on A_{cad} avoiding $|K'|$ with a basis which is a subdivision of $\{A_{cae}, A_{ade}\}$ and let $\mathbb{V}_{cae}, \mathbb{V}_{ade}$ be subsets of $\mathbb{V}$ which are on A_{cae}, A_{ade} respectively. Let $K_3^{(1)}$ be the division obtained from $K_2^{(1)}$ by an $\mathcal{S}$ operation along $\mathbb{V}_{cae}$, $K_4^{(1)}$ the division obtained from $K_3^{(1)}$ by an $\mathcal{S}$ operation along $\mathbb{V}_{ade}$ and $K_5^{(1)}$ the division obtained from $K_4^{(1)}$ by an $\mathcal{S}^{-1}$ operation along $\mathbb{V}$. $|K_5^{(1)}| = |K'|$. Let $\mathbb{W}$ be a sawtooth on A_{bcd} avoiding $|K'|$ with a basis which is a subdivision of $\{A_{bce}, A_{dbe}\}$ and $\mathbb{W}_{bce}, \mathbb{W}_{dbe}$ subsets of $\mathbb{W}$ on A_{bce}, A_{dbe} respectively. Let $K_6^{(1)}$ be the division obtained from $K_5^{(1)}$ by an $\mathcal{S}$ operation along $\mathbb{W}_{bce}$, $K_7^{(1)}$ the division obtained from $K_6^{(1)}$ by an $\mathcal{S}$ operation along $\mathbb{W}_{dbe}$ and $K_8^{(1)}$ the division obtained from $K_7^{(1)}$ by an $\mathcal{S}^{-1}$ operation along $\mathbb{W}$. Then $K_8^{(1)} = K'$. So we have a deformation chain $K = K_0^{(1)}, \cdots, K_8^{(1)} = K'$ with maximal height $h(K) + 2$ at $K_2^{(1)}$, which has no $\mathcal{E}_2^5$ operations. Apply Lemma 26.21 to $K_1^{(1)}, K_2^{(1)}, K_3^{(1)}$, and we have a new deformation chain $K = K_0^{(2)}, \cdots, K_{s(2)}^{(2)} = K'$ (for some $s(2)$) whose maximal height is $h(K) + 1$ without introducing $\mathcal{E}_2^5$ operations. Since we have already proved Lemma 26.20 except the case where $\mathcal{F}$ is $\mathcal{E}_2^5$ (Case I) and since all deformation chains constructed in Case I and in Lemma 26.21 have no $\mathcal{E}_2^5$ operations, we have the desired deformation chain, without introducing $\mathcal{E}_2^5$ operations. ∎

Part 5

Surface Braids and Surface Links

CHAPTER 27

Knot Groups

27.1. Classical Knot Groups

Let b be an m-braid and let $\mathrm{Artin}(b) : F_m \to F_m$ be Artin's automorphism of the free group $F_m = \langle x_1, \ldots, x_m \rangle$ associated with the braid b.

This is interpreted as follows: Assume that b is a geometric m-braid in $D^2 \times I$ and the free group F_m is $\pi_1(D^2 \setminus Q_m, q_0)$ for a point $q_0 \in \partial D^2$ as before (Sect. 2.6). Take copies of the generators $x_1, \ldots, x_m$ in $\pi_1(D^2[0] \setminus Q_m[0], q_0[0])$ and call these *old generators*, where $Q_m[0]$ stands for a copy of Q_m in $pr_2^{-1}(0) = D^2 \times \{0\}$, etc. Similarly, take copies of the generators $x_1, \ldots, x_m$ in $\pi_1(D^2[1] \setminus Q_m[1], q_0[1])$ and call these *new generators*. Precisely speaking, consider two families of generators in the fundamental group $\pi_1(D^2 \times I \setminus b, q_0 \times [0,1])$ as in Fig. 27.1. Each old generator x_j is written as a word $\mathrm{Artin}(b)(x_j)$ in the new generators.

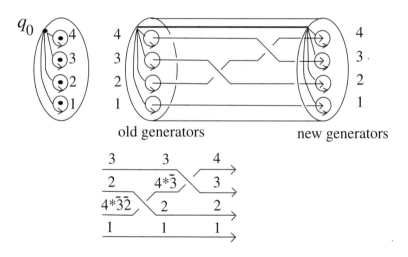

old generators   new generators

FIGURE 27.1. Artin's automorphism

We can use the above interpretation to calculate Artin's automorphism. Draw a diagram of a braid b and put labels $x_1, \ldots, x_m$ to the arcs at the terminal of b; see Fig. 27.1, where labels are abbreviated to $1, \ldots, m$. By the rule of upper presentation (Fig. 3.12), label all the arcs of the diagram. Then the labels on the arcs at the initial b give the images of $x_1, \ldots, x_m$ by $\mathrm{Artin}(b)$. For example, see Fig. 27.1. The symbol $*$ is used here for a conjugation: $4 * \bar{3}\,\bar{2}$ means $x_2 x_3 x_4 x_3^{-3} x_1^{-2}$, etc. So $\text{Artin}(\sigma_2\sigma_3)$ maps

$$\begin{aligned}
x_1 &\mapsto x_1, \\
x_2 &\mapsto x_2 x_3 x_4 x_3^{-3} x_1^{-2}, \\
x_3 &\mapsto x_2, \\
x_4 &\mapsto x_3.
\end{aligned}$$

Let L be a closed m-braid in $D^2 \times S^1$ associated with b. Then $\pi_1(D^2 \times S^1 \setminus L, q_0[1])$ is given by

$$\langle x_1, \ldots, x_m, t \mid t^{-1} x_j t = \text{Artin}(b)(x_j) \text{ for } j = 1, \ldots, m \rangle.$$

Put $D^2 \times S^1$ in $\mathbb{R}^3$ in a standard way; i.e., identify it with a tubular neighborhood of a trivial knot in $\mathbb{R}^3$ such that t corresponds to a preferred longitude which bounds a disk in $\mathbb{R}^3$. Then we have

$$\pi_1(\mathbb{R}^3 \setminus L) = \langle x_1, \ldots, x_m \mid x_j = \text{Artin}(b)(x_j) \text{ for } j = 1, \ldots, m \rangle.$$

Since any classical link in $\mathbb{R}^3$ is presented as a closed braid, we have the following.

PROPOSITION 27.1. *A group G is a link group $\pi_1(\mathbb{R}^3 \setminus L)$ of a link L with c components if and only if*

(1) *$G/[G, G]$ is a free abelian group of rank c,*

(2) *there is an m-braid b for some m such that G has a presentation*

$$\langle x_1, \ldots, x_m \mid x_j = \text{Artin}(b)(x_j) \text{ for } j = 1, \ldots, m \rangle.$$

The first condition reflects the statement that L has c components. Thus it may be replaced by a condition on b that the number of orbits of the permutation associated with b is c.

27.2. Knot Groups of Surface Braids

Let S be a simple surface braid of degree m. Recall that a braid system of S is written as

$$(w_1^{-1} \sigma_{k_1}^{\epsilon_1} w_1, w_2^{-1} \sigma_{k_2}^{\epsilon_2} w_2, \ldots, w_n^{-1} \sigma_{k_n}^{\epsilon_n} w_n),$$

where n is the number of branch points, $w_1, \ldots, w_n$ are m-braids and $\epsilon_1, \ldots, \epsilon_n \in \{+1, -1\}$.

PROPOSITION 27.2. *The fundamental group $\pi_1(D_1^2 \times D_2^2 \setminus S)$ has a group presentation with generators $x_1, \ldots, x_m$ and relations*

$$\text{Artin}(w_i)(x_{k_i}) = \text{Artin}(w_i)(x_{k_i+1}).$$

Proof. Recall the situation of Sect. 18.10. We may assume that S is a surface braid $S(\Gamma)$ associated with a chart Γ. Let $\mathcal{A} = (a_1, \ldots, a_n)$ be a Hurwitz arc system starting from the branch point set $\Sigma(S) = \{y_1, \ldots, y_n\}$ toward a point $y_0 \in \partial D_2^2$. Let $N(y_i)$ be a regular neighborhood of y_i and let a_i^* be an arc obtained from a_i by removing $a_i \cap \text{int} N(y_i)$. We denote by y_i^* the starting point of a_i^*. Then

$$w_\Gamma(c_i) = w_\Gamma(a_i^*)^{-1} w_\Gamma(\partial N(y_i)) w_\Gamma(a_i^*),$$

where w_Γ is the intersection braid word and $c_1 \ldots, c_n$ are loops in $(D_2^2 \setminus \Sigma(S), y_0)$ associated with $\mathcal{A}$ as in Sect. 2.6 which represent the Hurwitz generator system $\eta_1, \ldots, \eta_n$ of $\pi_1(D_2^2 \setminus \Sigma(S), y_0)$. In this situation,

$$w_i = w_\Gamma(a_i^*) \quad \text{and} \quad \sigma_{k_i}^{\epsilon_i} = w_\Gamma(\partial N(y_i)).$$

The restriction of S over a_i^* is a geometric m-braid whose word expression is w_i.

We denote by $D_1^2[y], Q_m[y]$ and $q_0[y]$ the copies of D_1^2, Q_m and q_0 in $pr_2^{-1}(y) = D_1^2 \times \{y\} \subset D_1^2 \times D_2^2$ for $y \in D_2^2$.

Let $x_1^{(i)}, \ldots, x_m^{(i)}$ be the generators of $\pi_1(D_1^2[y_i^*] \setminus Q_m[y_i^*], q_0[y_i^*])$ corresponding to $x_1, \ldots, x_n$ under a natural identification $D_1^2[y_i^*] = D_1^2$. The restriction of S to $N(y_i)$ is a simple braided surface whose braid monodromy is $\sigma_{k_i}^{\epsilon_i}$, and hence the group $\pi_1(D_1^2 \times N(y_i) \setminus S, q_0[y_i^*])$ is generated by $x_1^{(i)}, \ldots, x_m^{(i)}$ with a single relation $x_{k_i}^{(i)} = x_{k_i+1}^{(i)}$.

Let $x_1, \ldots, x_m$ denote also the generators of $\pi_1(D_1^2[y_0] \setminus Q_m[y_0], q_0[y_0])$ corresponding to $x_1, \ldots, x_n$. When we connect the base point $q_0[y_i^*]$ of $\pi_1(D_1^2[y_i^*] \setminus Q_m[y_i^*], q_0[y_i^*])$ to the base point $q_0[y_0]$ by an arc $\{q_0\} \times a_i^*$, each generator $x_j^{(i)}$ is mapped to $\text{Artin}(w_i)(x_j)$. Thus $\pi_1(D_1^2 \times \cup_{i=1}^n(N(y_i) \cup a_i^*) \setminus S, q_0[y_0])$ has a desired presentation. Since $D_1^2 \times \cup_{i=1}^n(N(y_i) \cup a_i^*) \setminus S$ is a deformation retract of $D_1^2 \times D_2^2 \setminus S$, we have the result. ∎

EXAMPLE 27.3. Let S be the surface braid of degree 4 described by a chart as in Fig. 18.14. Let $\mathcal{A} = (a_1, a_2, a_3, a_4)$ be the Hurwitz arc system in the figure. Then

$$
\begin{aligned}
w_1 &= \sigma_1 & \sigma_{k_1}^{\epsilon_1} &= \sigma_1 \\
w_2 &= 1 & \sigma_{k_1}^{\epsilon_2} &= \sigma_3^{-1} \\
w_3 &= \sigma_1\sigma_2 & \sigma_{k_1}^{\epsilon_3} &= \sigma_2^{-1} \\
w_4 &= 1 & \sigma_{k_1}^{\epsilon_4} &= \sigma_3
\end{aligned}
$$

Thus $\pi_1(D_1^2 \times D_2^2 \setminus S)$ has a presentation with generators $x_1, \ldots, x_4$ and relations

$$
\begin{aligned}
\text{Artin}(\sigma_1)(x_1) &= \text{Artin}(\sigma_1)(x_2) \\
\text{Artin}(1)(x_3) &= \text{Artin}(1)(x_4) \\
\text{Artin}(\sigma_1\sigma_2)(x_2) &= \text{Artin}(\sigma_1\sigma_2)(x_3) \\
\text{Artin}(1)(x_3) &= \text{Artin}(1)(x_4).
\end{aligned}
$$

Fig. 27.2 shows Artin's automorphisms for σ_1 and $\sigma_1\sigma_2$.

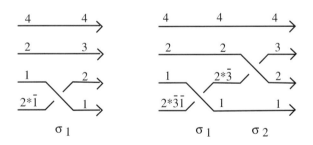

FIGURE 27.2. Artin's automorphisms for σ_1 and $\sigma_1\sigma_2$

The relations are

$$
\begin{aligned}
x_1 x_2 x_1^{-1} &= x_1 \\
x_3 &= x_4 \\
x_1 &= x_2 \\
x_3 &= x_4.
\end{aligned}
$$

Therefore the group is a rank-2 free group

$$\langle x_1, x_3 \rangle.$$

EXAMPLE 27.4. Let S be the surface braid of degree 4 described by a chart as in Fig. 27.3. For a Hurwitz arc system as in the figure, we have a braid system $(w_1^{-1}\sigma_{k_1}^{\epsilon_1}w_1, w_2^{-1}\sigma_{k_2}^{\epsilon_2}w_2, \ldots, w_6^{-1}\sigma_{k_6}^{\epsilon_6}w_6)$ given by

$$\begin{aligned}
w_1 &= \sigma_1\sigma_3, & \sigma_{k_1}^{\epsilon_1} &= \sigma_2, \\
w_2 &= \sigma_2^2\sigma_3, & \sigma_{k_2}^{\epsilon_2} &= \sigma_1^{-1}, \\
w_3 &= \sigma_2^2\sigma_3, & \sigma_{k_3}^{\epsilon_3} &= \sigma_3, \\
w_4 &= \sigma_2, & \sigma_{k_4}^{\epsilon_4} &= \sigma_3^{-1}, \\
w_5 &= \sigma_2, & \sigma_{k_5}^{\epsilon_5} &= \sigma_1, \\
w_6 &= \sigma_1^2, & \sigma_{k_6}^{\epsilon_6} &= \sigma_2^{-1}.
\end{aligned}$$

Thus $\pi_1(D_1^2 \times D_2^2 \setminus S)$ has a presentation with generators $x_1, \ldots, x_4$ and relations

$$\begin{aligned}
\mathrm{Artin}(\sigma_1\sigma_3)(x_2) &= \mathrm{Artin}(\sigma_1\sigma_3)(x_3) \\
\mathrm{Artin}(\sigma_2^2\sigma_3)(x_1) &= \mathrm{Artin}(\sigma_2^2\sigma_3)(x_2) \\
\mathrm{Artin}(\sigma_2^2\sigma_3)(x_3) &= \mathrm{Artin}(\sigma_2^2\sigma_3)(x_4) \\
\mathrm{Artin}(\sigma_2)(x_3) &= \mathrm{Artin}(\sigma_2)(x_4) \\
\mathrm{Artin}(\sigma_2)(x_1) &= \mathrm{Artin}(\sigma_2)(x_2) \\
\mathrm{Artin}(\sigma_1^2)(x_2) &= \mathrm{Artin}(\sigma_1^2)(x_3).
\end{aligned}$$

Fig. 27.4 shows Artin's automorphisms for $\sigma_1\sigma_3$, $\sigma_2^2\sigma_3$, σ_2 and σ_1^2. The relations are

$$\begin{aligned}
x_1 &= x_3x_4x_3^{-1} \\
x_1 &= x_2x_3x_4x_3^{-1}x_2x_3x_4^{-1}x_3^{-1}x_2^{-1} \\
x_2x_3x_4x_3^{-1}x_2^{-1} &= x_3 \\
x_2 &= x_4 \\
x_1 &= x_2x_3x_2^{-1} \\
x_1x_2x_1^{-1} &= x_3.
\end{aligned}$$

Therefore the group has a presentation

$$\langle x_1, x_2 \mid x_1x_2x_1 = x_2x_1x_2, \ x_1 = x_2^{-2}x_1x_2^2 \rangle.$$

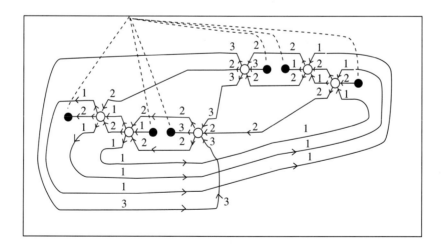

FIGURE 27.3. 2-twist-spun trefoil

Let F be the closure in $\mathbb{R}^4$ of a simple surface braid S in $D_1^2 \times D_2^2$. Since F is obtained from $S \subset D_1^2 \times D_2^2 \subset \mathbb{R}^4$ by attaching 2-disks trivially, $\pi_1(\mathbb{R}^4 \setminus F)$ is

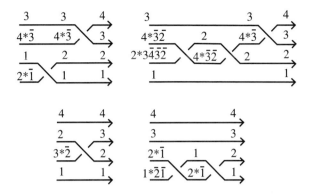

FIGURE 27.4. Artin's automorphisms for $\sigma_1\sigma_3$, $\sigma_2^2\sigma_3$, σ_2 and σ_1^2

isomorphic to $\pi_1(D_1^2 \times D_2^2 \setminus S)$. Thus the presentation given in Proposition 27.2 is a presentation of the knot group $\pi_1(\mathbb{R}^4 \setminus F)$.

27.3. Knot Groups of Surface Links

An *n-knot* K is an n-sphere embedded smoothly or PL and locally flatly in $\mathbb{R}^{n+2}$ or in S^{n+2}. The fundamental group $\pi_1(\mathbb{R}^{n+2} \setminus K)$ is called the *knot group* of an *n-knot* K or an *n-knot group*.

There is a famous algebraic characterization of n-knot groups for $n \geq 3$.

THEOREM 27.5 ([**392**]). *A group G is the knot group of an n-knot if and only if the following conditions are satisfied.*

(1) *G is finitely presented.*
(2) *$G/[G, G]$ (which is $H_1(G; \mathbb{Z})$) is an infinite cyclic group.*
(3) *$H_2(G; \mathbb{Z}) = 0$.*
(4) *There is an element x whose normal closure $\langle\langle x \rangle\rangle^G$ is G.*

These conditions are called the *Kervaire conditions*.

For a positive integer n, we denote by KG_n the set of knot groups of n-knots. By spinning construction introduced by Artin in [**16**] (Sect. 10.1), we have

$$KG_1 \subset KG_2 \subset KG_3 \subset KG_4 \subset \cdots.$$

The Kervaire theorem (Theorem 27.5) asserts that

$$KG_3 = KG_4 = \cdots.$$

On the other hand, it is known that

$$KG_1 \neq KG_2 \quad \text{and} \quad KG_2 \neq KG_3.$$

The former is easily seen. There exist a lot of 2-knot groups whose first elementary ideals are not principal. The latter is seen by considering the Farber-Levine torsion pairings [**142**, **143**, **474**].

Proposition 27.1 (with $c = 1$) gives a characterization of KG_1, and Theorem 27.5 is a characterization of KG_n ($n \geq 3$). The following theorem (with $c = 1$ and $\chi = 2$) gives a characterization of KG_2.

THEOREM 27.6. *Let c be a positive integer and χ an even number with $\chi \leq 2c$. A group G is the knot group of an oriented surface link F of c components with $\chi(F) = \chi$ if and only if*

(1) *$G/[G, G]$ is a free abelian group of rank c,*
(2) *there exist m-braids $w_1, \ldots, w_n$ and numbers $k_1, \ldots, k_n \in \{1, \ldots, m-1\}$ for some m and n with $2m - n = \chi$ such that G has a presentation*

$$\langle x_1, \ldots, x_m \mid \mathrm{Artin}(w_i)(x_{k_i}) = \mathrm{Artin}(w_i)(x_{k_i+1}) \text{ for } j = 1, \ldots, n \rangle$$

and that

$$\prod_{j=1}^{n} w_j^{-1} \sigma_{k_j}^{\epsilon_j} w_j = 1 \quad \text{in } B_m$$

for some $\epsilon_1, \ldots, \epsilon_n \in \{+1, -1\}$.

Proof. The only if part follows from Alexander's theorem in dimension four (Theorem 23.6 or Theorem 21.4) and Proposition 27.2. Conversely, if G satisfies (2), construct a surface braid whose braid system is

$$(w_1^{-1} \sigma_{k_1}^{\epsilon_1} w_1, w_2^{-1} \sigma_{k_2}^{\epsilon_2} w_2, \ldots, w_n^{-1} \sigma_{k_n}^{\epsilon_n} w_n).$$

The closure of the surface braid is a surface link whose knot group is G. Condition (1) and the equation $2m - n = \chi$ determine the Euler characteristic and the number of components. See [**316**] for the details. ∎

A finite presentation $\langle x_1, \ldots, x_m \mid r_1, \ldots, r_n \rangle$ is called a *Wirtinger presentation* if each relation r_i is in a form $x_c = x_b^{-\epsilon} x_a x_b^{\epsilon}$ for some x_a, x_b, x_c and $\epsilon \in \{+1, -1\}$. It is known that the knot group of any closed oriented n-manifold embedded in $\mathbb{R}^{n+2}$ has a Wirtinger presentation (cf. [**328, 831, 961**]).

PROPOSITION 27.7. *Let c be a positive integer. A group G is the knot group of an oriented surface link of c components if and only if*

(1) *$G/[G, G]$ is a free abelian group of rank c, and*
(2) *G has a Wirtinger presentation.*

Proof. If G is the knot group of an oriented surface link of c components, then by the Alexander duality theorem we have (1). And G has a Wirtinger presentation (Sect. 12.1). Conversely if G has a Wirtinger presentation, we can construct an oriented surface link F whose knot group is G (Exercise). Furthermore, if G satisfies (1), then the number of components of F must be c. ∎

EXERCISE 27.8. For a group G with a Wirtinger presentation, there exists an oriented (ribbon) surface link whose knot group is G. [*Hint*: Consider a trivial 2-link with m components. The knot group is a free group generated by $x_1, \ldots, x_m$. For each relation, apply a 1-handle surgery. See [**960**].]

The *deficiency* of a finite presentation is the number of the generators minus that of the relations (relators).

THEOREM 27.9 ([**960**]). *A group G is the knot group of a ribbon 2-knot if and only if the following conditions are satisfied.*

(1) *G is finitely presented.*
(2) *$G/[G, G]$ is an infinite cyclic group.*
(3) *G has a Wirtinger presentation of deficiency one.*

For a non-negative integer g, let $\mathcal{F}_g$ denote the set of knot groups of surface knots of genus g. Since surgery along a trivial 1-handle changes a surface knot of genus g into a surface knot of genus $g + 1$ without changing the knot group, we have

$$KG_2 = \mathcal{F}_0 \subset \mathcal{F}_1 \subset \mathcal{F}_2 \subset \cdots.$$

For a group $G \in \mathcal{F}_g$, the second homology group $H_2(G)$ is a subgroup A of an abelian group of rank $2g$. Conversely, for any subgroup A, there exists a group $G \in \mathcal{F}_g$ with $H_2(G) \cong A$; cf. [527]. Thus

$$\mathcal{F}_g \neq \mathcal{F}_{g+1} \quad \text{for each } g.$$

For a non-negative integer g, let $\mathcal{K}_g^2$ denote the set of knot groups G of surface knots of genus g such that $H_2(G) = 0$. Then we have

$$KG_2 = \mathcal{K}_0^2 \subset \mathcal{K}_1^2 \subset \mathcal{K}_2^2 \subset \cdots.$$

It is known that

$$\mathcal{K}_g^2 \neq \mathcal{K}_{g+1}^2 \quad \text{for each } g$$

(cf. Theorem 14.1.9 of [369]). Theorem 27.6 (with $c = 1$ and $\chi = 2 - 2g$) gives a characterization of $\mathcal{F}_g$, and moreover, together with the condition that $H_2(G) = 0$, it gives a characterization of $\mathcal{K}_g^2$.

CHAPTER 28

Unknotted Surface Braids and Surface Links

Throughout this chapter surface links are oriented.

28.1. Unknotted Surface Braids

Recall that a free edge of a chart is an edge whose endpoints are black vertices.

DEFINITION 28.1. A chart is *unknotted* if it is an empty chart or consists of some free edges. A surface braid is *unknotted* if it is described by an unknotted chart.

A trivial surface braid is described by an empty chart. Therefore it is an unknotted surface braid.

28.2. Surface Braids of Degree 2

LEMMA 28.2. *Any surface braid S of degree 2 is described by an unknotted chart as in* Fig. 28.1 (*or an empty chart*).

Proof. Let n be the number of branch points of S, which is even. If $n = 0$, then S is a trivial surface braid and described by an empty chart. If $n \neq 0$, then it is equivalent to a surface braid described by a chart as in Fig. 28.1, since equivalence classes of degree-2 surface braids are determined by the number of branch points (Proposition 17.18). ∎

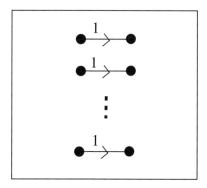

FIGURE 28.1. Standard chart description of degree 2

This lemma is also verified as follows. Let Γ be a chart description of S. Since every edge of Γ is labeled 1, there are no degree-4 and degree-6 vertices. Thus Γ is a union of some free edges and some simple loops. If γ has no free edges, then Γ is replaced with an empty chart by a CI-move (a chart move of type I). Thus S is unknotted. If Γ has a free edge, then we can remove simple loops inductively from the nearest one from the free edge as follows: (1) Move the free edge toward a loop, (2) apply a CI-move (as in Fig. 18.17(2)), and (3) get a long free edge. Shrink it to regain the free edge. This process eliminates a simple loop near a free edge. Thus Γ becomes a union of free edges.

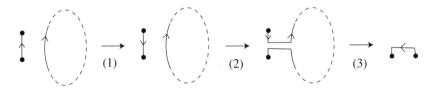

FIGURE 28.2. A free edge absorbs a simple loop

PROPOSITION 28.3. (1) *Let F be a surface link which is a simple closed surface braid of degree 2. Then F is trivial (unknotted) as a surface link.*
(2) *For a surface link F,* $\mathrm{Braid}(F) = 2$ *if and only if F is either*

    (i) *a trivial surface knot whose genus is positive, or*
    (ii) *a trivial surface link which is a pair of trivial 2-spheres.*

Proof. Let F be a closure of a surface braid S of degree 2, and let n be the number of branch points of S. If $n = 0$, S is a trivial surface braid of degree 2 and its closure is a trivial surface link which is a pair of trivial 2-spheres. If $n \neq 2$, S has a chart description as in Fig. 28.1. Then F is described by a motion picture as in Fig. 28.3. The closure is a trivial surface knot of genus $(n-2)/2$. Thus we have (1). And, since a trivial 2-knot is a unique surface link whose braid index is one, we have (2). ∎

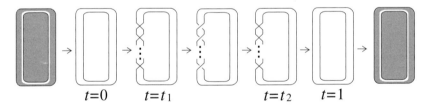

FIGURE 28.3. Surface link of braid index 2

28.3. Chart Descriptions of Unknotted Surfaces

For a while, we denote by Λ_n an unknotted chart of degree 2 with n black vertices for $n \geq 0$ (Fig. 28.1).

LEMMA 28.4. *An unknotted chart of degree $m \geq 3$ is obtained from $m-1$ unknotted charts of degree 2 by connected sum.*

Proof. Let Γ be an unknotted chart of degree m. For each $i \in \{1, \ldots, m-1\}$, let n_i be the number of the black vertices with label i. Then

$$\Gamma = \iota_0^{m-2}(\Lambda_{n_1}) \cdot \iota_1^{m-1}(\Lambda_{n_2}) \cdots \iota_{m-2}^{0}(\Lambda_{n_{m-1}}).$$

Thus Γ is a connected sum of $\Lambda_{n_1}, \ldots, \Lambda_{n_{m-1}}$. ∎

THEOREM 28.5. *A surface link described by an unknotted chart is unknotted* (*trivial*). *Conversely, any unknotted surface link can be described by an unknotted chart.*

Proof. Suppose that a surface link F is described by an unknotted chart. An unknotted chart is a connected sum of some (or no) charts of degree 2 (Lemma 28.4). By Proposition 24.6, F is a connected sum of surface links described by charts of degree 2. Since surface braids of degree 2 are unknotted, F is unknotted. Conversely let F be an unknotted surface link. It is a split sum of some trivial surface knots. Every unknotted surface knot can be described by an unknotted chart of degree 2. Thus F can be described by a chart which is a pile product of some unknotted chart of degree 2. This is an unknotted chart. ∎

This theorem is the reason that we use the term "unknotted charts". We avoided calling them "trivial charts" because they can be confused with charts for trivial surface braids. Instead we use "unknotted surface links" rather than "trivial surface links" so that this theorem is easily memorized without confusion.

28.4. The Braid Index of an Unknotted Surface

PROPOSITION 28.6. *Let F be a surface link. Let p be the number of spherical components of F, and let q be the number of components of F with positive genus. If F is an unknotted surface link, then the braid index is*

$$\mathrm{Braid}(F) = p + 2q.$$

Proof. An unknotted 2-knot is of braid index 1 and an unknotted surface knot with positive genus is of braid index 2. By Proposition 24.3, we have the result. ∎

28.5. The Braid System of an Unknotted Braid

DEFINITION 28.7. *An n-tuple of m-braids, $(b_1, b_2, \ldots, b_n)$, is an unknotted braid system if n is an even number, say $n = 2k$, and for each $j \in \{1, \ldots, k\}$, $b_{2j-1} = b_{2j}^{-1}$ and it is a standard generator of B_m.*

LEMMA 28.8. *A surface braid is unknotted if and only if its braid system is slide equivalent to an unknotted braid system.*

Proof. If a surface braid is unknotted and described by an unknotted chart, then it is easily seen that it has an unknotted braid system. Any braid system is slide equivalent to this. Conversely, if a surface braid has an unknotted braid system, then there is an unknotted chart which describes it. ∎

CHAPTER 29

Ribbon Surface Braids and Surface Links

Throughout this chapter surface links are oriented.

29.1. Ribbon Surface Braids

For convenience, we assume that a chart is in the 2-disk $D = \{(x_1, x_2) \in \mathbb{R}^2 \mid x_1, x_2 \in [-1, 1]\}$. Let ℓ be the line in D determined by $x_2 = 0$. Let $p_i : D \to [-1, 1]$ be the ith factor projection.

For a chart Γ in D, we denote by Γ^* its mirror image with respect to the line ℓ in D, and by $-\Gamma$ the chart obtained from Γ by reversing the orientations of all edges (and loops).

DEFINITION 29.1. A chart is a *ribbon chart* if it is ambient isotopic to a chart Γ with $\Gamma = -\Gamma^*$. A surface braid is *ribbon* if it is described by a ribbon chart.

THEOREM 29.2. *For a surface braid S, the following conditions are mutually equivalent.*

(1) *It has a chart description Γ such that $\Gamma = -\Gamma^*$.*
(2) *It has a chart description Γ such that every black vertex of Γ is an endpoint of a free edge.*
(3) *It has a chart description Γ such that Γ is a split union of some oval nests.*
(4) *It has a chart description Γ such that there exist no white vertices.*

Proof. (1) $\Rightarrow$ (2) Assume that $\Gamma = -\Gamma^*$. Since $\Gamma = -\Gamma^*$, there are no vertices of Γ on ℓ. Modifying Γ by an ambient isotopy of D which is symmetric with respect to ℓ, we assume that the images of all vertices of Γ under p_1 are all distinct and that every edge adjacent to a black vertex is parallel to ℓ. Consider an arc connecting a black vertex and its mirror image. This arc may intersect some edges of Γ. Apply a CI-move in a neighborhood of this arc as in Fig. 29.1 to make a free edge as in Fig. 29.1. Now every black vertex is an endpoint of a free edge.

(2) $\Rightarrow$ (3) Consider an arc in D connecting a free edge and a point of ∂D. This arc may intersect some edges of Γ transversely. Move the free edge along the arc toward ∂D. As the free edge goes across an edge of Γ, it is surrounded by a simple loop and becomes an oval nest; see Fig. 29.2. Finally, we can separate a free edge from Γ although it becomes an oval nest. Apply the same argument to all free edges. Now Γ is a union of a chart without black vertices and some oval nests. By a CI-move, we can remove the chart without black vertices and we have a chart as in (3).

(3) $\Rightarrow$ (1) It is easy to deform a chart as in (3) into a chart as in (1) by an ambient isotopy of D.

(3) $\Rightarrow$ (4) This is obvious.

(4) $\Rightarrow$ (2) Suppose that Γ has no white vertices. Every black vertex is connected to another by an arc in Γ which is a union of edges. If this arc consists of a single

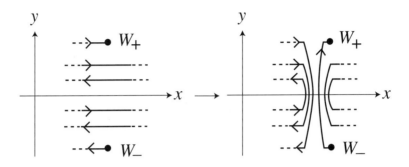

FIGURE 29.1. Making a free edge

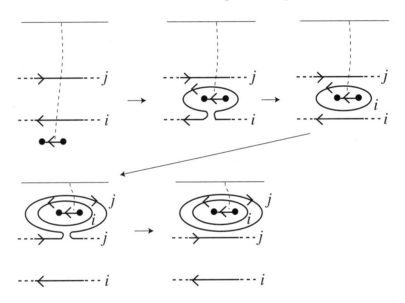

FIGURE 29.2. Pushing a free edge out

edge, then it is a free edge. Otherwise, shorten this arc by CII-moves and make it a single edge. Doing this for all pairs of black vertices, we have a chart as in (2).

This completes the proof. ∎

29.2. Chart Descriptions of Ribbon Surfaces

THEOREM 29.3. *A surface link described by a ribbon chart is ribbon. Conversely, any ribbon surface link can be described by a ribbon chart.*

Proof. Suppose that a surface link F is described by a ribbon chart Γ. It is symmetric to the hyperplane $\mathbb{R}^3[0]$ of $\mathbb{R}^4$. By Theorem 11.4, it is a ribbon surface link. Conversely, suppose that F is a ribbon surface link. By Theorem 23.7, it is ambient isotopic to a closure of a ribbon surface braid. ∎

Suppose that a surface link F is described by a ribbon chart Γ of degree m. We may assume Γ is as in (3) of Theorem 29.2. Γ is obtained from an empty chart by adding oval nests. An empty chart describes a trivial closed surface braid in $\mathbb{R}^4$. That is an unknotted 2-link with m components. Recall that adding an oval nest corresponds to a 1-handle surgery along a nice 1-handle. Thus F is obtained from a trivial 2-link of m components by k 1-handle surgeries, where k is half the number of black vertices.

29.3. The Braid System of a Ribbon Braid

DEFINITION 29.4. An n-tuple of m-braids, $(b_1, b_2, \ldots, b_n)$, is a *ribbon braid system* if n is an even number ($n = 2k$), if for each $j \in \{1, \ldots, k\}$, $b_{2j-1} = b_{2j}^{-1}$, and if each b_i (for $i \in \{1, \ldots, n\}$) is a conjugate of a standard generator of B_m.

LEMMA 29.5. *A surface braid is ribbon if and only if its braid system is slide equivalent to a ribbon braid system.*

Proof. If a surface braid is ribbon and described as a ribbon chart as in (3) of Theorem 29.2, then it is easily seen that it has a ribbon braid system. Any braid system is slide equivalent to this. Conversely, if a surface braid has a ribbon braid system, then there is a ribbon chart as in (3) of Theorem 29.2. ∎

29.4. Example

EXAMPLE 29.6. Let F be a spun trefoil. Recall that it has a chart description of degree 3 as in Fig. 29.3. We have a braid system

$$(\sigma_2, \sigma_2^{-1}, \sigma_1^{-3}\sigma_2\sigma_1^3, \sigma_1^{-3}\sigma_2^{-1}\sigma_1^3)$$

of degree 3.

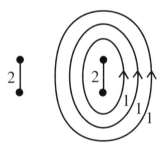

FIGURE 29.3. A chart description of a spun trefoil

EXAMPLE 29.7. Let F be a 2-knot whose motion picture is illustrated as in Fig. 29.4. This is Fox's example 10 in [**168**]. It has a chart description of degree 4 as in Fig. 29.5.

29.5. Reduced Ribbon Braid Form

Recall a normal ribbon braid form (Definition 22.1). Any ribbon surface link is ambient isotopic to a surface link F in a normal ribbon braid form which is a closed realizing surface of a hyperbolic transformation $L_+ \to L_0 \to L_+$ satisfying the condition in Definition 22.1. If the degree is m and the number of saddle bands is n, there exist m-braids $b_1, \ldots, b_n, c_1, \ldots, c_n$ such that

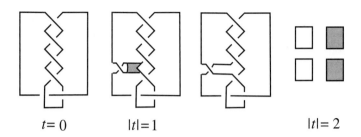

$t=0$ $\qquad$ $|t|=1$ $\qquad\qquad$ $|t|=2$

FIGURE 29.4. Fox's example 10

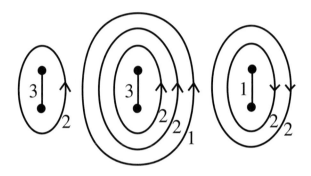

FIGURE 29.5. A chart description of Fox's example 10

(1) L_0 is the closed braid that corresponds to $b_1 b_1^{-1} \ldots b_n b_n^{-1}$,
(2) L_+ is the closed braid that corresponds to $b_1 c_1 b_1^{-1} \ldots b_n c_n b_n^{-1}$, and
(3) each c_i is a standard generator or its inverse.

We denote a surface link in such a normal ribbon braid form by

$$R[(b_1, c_1), \ldots, (b_n, c_n)]_m;$$

see Fig. 29.6. If $n = 0$, i.e., if it is a trivial closed braid, then we denote it by $R[\emptyset]_m$.

The number n is determined by the Euler characteristic $\chi(F)$ of the surface link F and the degree m so that $\chi(F)/2 = m - n$.

In chart description, $R[(b_1, c_1), \ldots, (b_n, c_n)]_m$ is described by a chart of degree m which is a union of n oval nests. The ith oval nest is a free edge whose label is the subscript of c_i surrounded by some simple loops corresponding to the m-braid b_i. Recall the process to get a braid system from a chart given in Sect. 18.10. For example, the chart of degree 3 in Fig. 29.3 describes $R[(1, \sigma_2), (\sigma_1^{-3}, \sigma_2)]_3$.

EXERCISE 29.8. The (ribbon) chart of degree 4 illustrated in Fig. 29.5 describes $R[(\sigma_2^{-1}, \sigma_3), (\sigma_1^{-1}\sigma_2^{-2}, \sigma_3), (\sigma_2^2, \sigma_1)]_4$. Draw a picture of this surface link in the motion picture method as in Fig. 29.6.

Let τ be the automorphism of B_m with $\tau(\sigma_i) = \sigma_{m-i}$ for $i \in \{1, \ldots, m-1\}$.

LEMMA 29.9. For $F = R[(b_1, c_1), \ldots, (b_n, c_n)]_m$, the following hold:

(1) $F \cong R[(b_2, c_2), \ldots, (b_n, c_n), (b_1, c_1)]_m$.
(2) $F \cong R[(bb_1, c_1), \ldots, (bb_n, c_n)]_m$ for any $b \in B_m$.
(3) $F \cong R[(b_1, c_1), \ldots, (b'_i, c'_i), \ldots, (b_n, c_n)]_m$ provided $b_i c_i b_i^{-1} = b'_i c'_i b'^{-1}_i$.

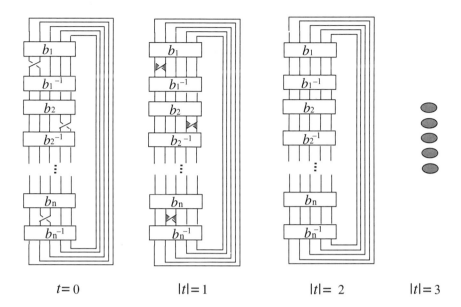

$t = 0$ $|t| = 1$ $|t| = 2$ $|t| = 3$

FIGURE 29.6. Normal ribbon braid form $R[(b_1, c_1), \ldots, (b_n, c_n)]_m$

(4) $F \cong R[(\tau(b_1), \tau(c_1)), \ldots, (\tau(b_n), \tau(c_n))]_m$.
(5) $F \cong R[(b_1, c_1), \ldots, (b_i, c_i^{-1}), \ldots, (b_n, c_n)]_m$.

Here $F \cong F'$ means that F and F' are ambient isotopic in $\mathbb{R}^4$.

Proof. (1) $\sim$ (4) are easily verified from the definition. (5) follows from the fact that two surfaces illustrated in Fig. 14.1 are ambient isotopic. ∎

REMARK 29.10. In the chart description, operations (1) and (5) of Lemma 29.9 are realized by an ambient isotopy of D. (3) is realized by chart moves. (2) is realized by a conjugation by b^{-1}, which introduces some large simple loops corresponding to b^{-1} surrounding the chart, and by CI-moves. (4) is also realized by conjugation by Garside's m-braid

$$\Delta = (\sigma_1 \sigma_2 \cdots \sigma_{m-1})(\sigma_1 \sigma_2 \cdots \sigma_{m-2})(\sigma_1 \sigma_2)\sigma_1$$

and by CI-moves.

DEFINITION 29.11. A surface link is in a *reduced ribbon braid form* of degree m if it is

$$R[(1, \sigma_1), (b_1, \sigma_1), \ldots, (b_k, \sigma_1)]_m$$

for some k and $b_1, \ldots, b_k \in B_m$. We denote by .

$$RR[b_1, \ldots, b_k]_m$$

this surface link.

The number k is determined by the Euler characteristic $\chi(F)$ of the surface link F and the degree m so that $\chi(F)/2 = m - k - 1$.

EXERCISE 29.12. If F is a surface link in a normal ribbon form $R[(b_1, c_1), \ldots, (b_n, c_n)]_m$ with $m \geq 2$ and $n > 0$, then it is transformed into a reduced ribbon braid form of degree m. [*Hint*: Use Lemma 29.9 (2), (3) and (5).]

EXAMPLE 29.13. Applying $\tau : B_3 \to B_3$, we can change the chart in Fig. 29.3 into a chart describing a reduced ribbon braid form, Fig. 29.7. This is achieved by applying $\tau : B_3 \to B_3$. Thus a spun trefoil is described as $RR[\sigma_2^{-3}]_3$.

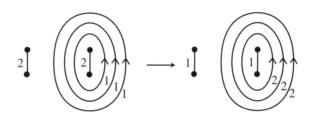

FIGURE 29.7. A spun trefoil in a reduced ribbon braid from

EXAMPLE 29.14. Fig. 29.8 shows the process to change the chart illustrated in Fig. 29.5 into the chart describing a reduced ribbon braid form. First, we apply $\tau : B_4 \to B_4$. The second step is a conjugation. The third and the fourth steps are operation (3) of Lemma 29.9; see Fig. 18.21(B). Thus the 2-knot called Fox's example 10 is described as $RR[\sigma_2\sigma_3^{-1}\sigma_2^{-2}, \sigma_2^2\sigma_3^{-1}\sigma_1\sigma_2]_4$.

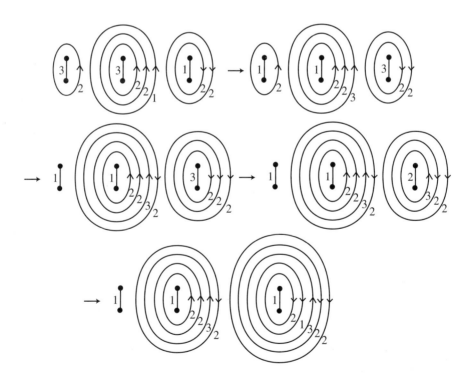

FIGURE 29.8. Fox's example 10 in a reduced ribbon braid form

29.6. Mirror Image of Reduced Ribbon Braid Form

Let μ be the automorphism of B_m with $\mu(\sigma_i) = \sigma_i^{-1}$ for $i \in \{1, \ldots, m-1\}$.

LEMMA 29.15. $RR[\mu(b_1), \ldots, \mu(b_k)]_m$ *is a mirror image of* $RR[b_1, \ldots, b_k]_m$.

Proof. A mirror image of $R[(1, \sigma_1), (b_1, \sigma_1), \ldots, (b_k, \sigma_1)]_m$ is given as

$$R[(1, \sigma_1^{-1}), (\mu(b_1), \sigma_1^{-1}), \ldots, (\mu(b_k), \sigma_1^{-1})]_m.$$

By Lemma 29.9, this is $R[(1, \sigma_1), (\mu(b_1), \sigma_1), \ldots, (\mu(b_k), \sigma_1)]_m$. ∎

CHAPTER 30

3-Braid 2-Knots

Throughout this chapter surface links are oriented.

30.1. Surface Braids of Degree 3 Are Ribbon

A 2-knot of braid index 1 is a trivial 2-knot. There are no 2-knots of braid index 2. We investigate 2-knots of braid index 3. The results in this chapter are found in [**312, 324**].

For a chart Γ, we denote by $w(\Gamma)$ the number of white vertices of Γ and by $f(\Gamma)$ the number of free edges of Γ.

LEMMA 30.1. *For any chart* Γ *of degree* 3 *with* $w(\Gamma) > 0$, *there exists a chart* Γ' *such that* Γ *and* Γ' *are chart move equivalent and*

$$w(\Gamma') - f(\Gamma') < w(\Gamma) - f(\Gamma).$$

EXERCISE 30.2. Prove this lemma. (Refer to [**312**] for the solution.)

THEOREM 30.3. *Surface braids of degree* 3 *are ribbon.*

Proof. Let S be a surface braid of degree 3 with n branch points. For any chart description Γ, $w(\Gamma) \geq 0$ and $f(\Gamma) \leq n/2$. Thus $w(\Gamma) - f(\Gamma) \geq -n/2$. Therefore there is a chart description Γ such that $w(\Gamma) - f(\Gamma)$ is the minimum among all chart descriptions of S. If $w(\Gamma) > 0$, by Lemma 30.1, S has a chart description Γ' with $w(\Gamma') - f(\Gamma') < w(\Gamma) - f(\Gamma)$. This is a contradiction. Thus $w(\Gamma) = 0$. By Theorem 29.2, S is ribbon. ∎

THEOREM 30.4. *A surface link* F *with* $\mathrm{Braid}(F) \leq 3$ *is a ribbon surface link.*

Proof. This is a consequence of Theorems 29.3 and 30.3. Recall that if $\mathrm{Braid}(F) \leq 2$, then F is unknotted. ∎

In Sect. 21.4, we saw that a 2-twist-spun trefoil has a braid presentation of degree 4. This 2-knot is not ribbon (since its first elementary ideal is not a principal ideal.) Therefore the braid index is four.

Now a question comes up:

What is the braid index of Fox's example 10?

It has a braid presentation of degree 4 (Example 29.7) and it is a ribbon 2-knot. The answer is given in Sect. 30.4.

30.2. 3-Braid 2-Knots

Let F be a surface link with $\chi(F) = 2$ and $\mathrm{Braid}(F) \leq 3$. By Theorem 30.3, it is ribbon and it is described by a normal ribbon braid form $R[(b_1, c_1), \ldots, (b_n, c_n)]_m$ with $m = 3$ and $n = 2$. By Exercise 29.12, it can be described by a reduced ribbon braid form $RR[b]_3 = R[(1, \sigma_1), (b, \sigma_1)]_3$ for some $b \in B_3$. We denote by $J(F)$ the subset of B_m consisting of all elements $b \in B_3$ such that $F \cong RR[b]_3$.

LEMMA 30.5.   (1) *If $b \in J(F)$, then $b^{-1} \in J(F)$.*
    (2) *If $b \in J(F)$ and $b\sigma_1 b^{-1} = b'\sigma_1 b'^{-1}$, then $b' \in J(F)$.*

Proof.   Recall Lemma 29.9.   (1) $R[(1, \sigma_1), (b^{-1}, \sigma_1)]_3 \cong R[(b, \sigma_1), (1, \sigma_1)]_3 \cong R[(1, \sigma_1), (b, \sigma_1)]_3$. (2) $R[(1, \sigma_1), (b, \sigma_1)]_3 \cong R[(1, \sigma_1), (b', \sigma_1)]_3$. ∎

The *length* of a 3-braid b is the minimum length of a word expression of b on $\{\sigma_1, \sigma_1^{-1}, \sigma_2, \sigma_2^{-1}\}$. A 3-braid word $s_1 \cdots s_n$ is *principal* if all $s_1, \ldots, s_n$ are either in $\{\sigma_1^{-1}, \sigma_2\}$ or in $\{\sigma_1, \sigma_2^{-1}\}$. An *oddly principal* 3-braid word is a principal one whose initial and terminal letters are σ_2 or σ_2^{-1}. We call a 3-braid b a *principal* (resp. *oddly principal*) 3-braid if it has a word expression which is principal (resp. oddly principal). We assume that the empty word is oddly principal and so is the identity element $e \in B_3$.

REMARK 30.6. For a principal 3-braid b, a word expression is principal if and only if the length of the word is the length of b (cf. [**324**]).

LEMMA 30.7. *If $b \in J(F)$ has the minimum length among $J(F)$, then it is oddly principal.*

Proof. Let α be the length of b and let $b = s_1 \cdots s_\alpha$ be a word expression, where $s_i \in \{\sigma_1, \sigma_1^{-1}, \sigma_2, \sigma_2^{-1}\}$ for $i \in \{1, \ldots, \alpha\}$. If $\alpha = 0$, then $b = e$ and by definition this is oddly principal. So we assume $\alpha > 0$. By Lemma 30.5 and the minimality of α, we see that $s_1, s_\alpha \in \{\sigma_2, \sigma_2^{-1}\}$. If $\alpha = 1$, then $b = \sigma_2$ or σ_2^{-1}, which is oddly principal. We consider the case where $\alpha \geq 2$. First, we assert that if $s_1 = \sigma_2$, then $s_1, \ldots, s_\alpha \in \{\sigma_1^{-1}, \sigma_2\}$. Suppose that there exists an integer k with $1 \leq k < \alpha$ such that $s_1, \ldots, s_k \in \{\sigma_1^{-1}, \sigma_2\}$ and $s_{k+1} \in \{\sigma_1, \sigma_2^{-1}\}$. Put $x = s_1 \cdots s_{k+1}$ and $y = s_{k+2} \cdots s_\alpha$. There are three cases:

(1) $x = \sigma_2^{a_1} \sigma_1$,
(2) $x = \sigma_2^{a_1} \sigma_1^{-a_2} \sigma_2^{a_3} \sigma_1^{-a_4} \cdots \sigma_1^{-a_{n-1}} \sigma_2^{a_n} \sigma_1$ ($n > 1$, odd),
(3) $x = \sigma_2^{a_1} \sigma_1^{-a_2} \sigma_2^{a_3} \sigma_1^{-a_4} \cdots \sigma_2^{a_{n-1}} \sigma_1^{-a_n} \sigma_2^{-1}$ ($n > 0$, even),

where $a_1, \ldots, a_n$ are positive integers. According to (1)–(3), let x' be the 3-braid expressed by

(1) $x' = \sigma_2^{-1} \sigma_1^{a_1 - 2} \sigma_2^{-1}$,
(2) $x' = \sigma_2^{-1} \sigma_1^{a_1 - 1} \sigma_2^{-a_2} \sigma_1^{a_3} \sigma_2^{-a_4} \cdots \sigma_2^{-a_{n-1}} \sigma_1^{a_n - 1} \sigma_2^{-1}$,
(3) $x' = \sigma_2^{-1} \sigma_1^{a_1 - 1} \sigma_2^{-a_2} \sigma_1^{a_3} \sigma_2^{-a_4} \cdots \sigma_1^{a_{n-1}} \sigma_2^{-a_n + 1} \sigma_1$.

Since $x^{-1} \sigma_1 x = x'^{-1} \sigma_1 x'$, by Lemma 30.5 we have $x'y \in J(F)$. The length of $x'y$ is smaller than α unless $x = \sigma_2 \sigma_1$. Hence $x = \sigma_2 \sigma_1$, $\alpha \geq 3$ and $s_3 \in \{\sigma_1, \sigma_2, \sigma_2^{-1}\}$. Put $z = s_4 \cdots s_\alpha$. If $s_3 = \sigma_1$, then $b^{-1} \sigma_1 b = (\sigma_2^{-1} z)^{-1} \sigma_1 (\sigma_2^{-1} z)$ and hence $(\sigma_2^{-1} z) \in J(F)$. It is a contradiction, for the length of $\sigma_2^{-1} z$ is smaller than α. If $s_3 = \sigma_2^{\pm 1}$, then $b^{-1} \sigma_1 b = (\sigma_2 \sigma_1 z)^{-1} \sigma_1 (\sigma_2 \sigma_1 z)$ and $\sigma_2 \sigma_1 z \in J(F)$, which yields a contradiction. Thus we have the assertion. The case where $s_1 = \sigma_2^{-1}$ is proved similarly. ∎

30.3. The Alexander Polynomial of a 3-Braid 2-Knot

Let $E = \mathbb{R}^4 \setminus F$ be the exterior of an (oriented) surface link F. The homomorphism $H_1(E; \mathbb{Z}) \to \mathbb{Z}$ sending each oriented meridian of F to $1 \in \mathbb{Z}$ determines an infinite cyclic covering $\widetilde{E} \to E$ and $H_1(\widetilde{E}; \mathbb{Z})$ is a Λ-module in a natural way, where $\Lambda = \mathbb{Z}[t, t^{-1}]$. In this chapter, the *Alexander polynomial* of F means the greatest common divisor of the elements of its 0th elementary ideal, which is unique up to multiplication by units of Λ. (If F is surface knot, this is the same as in Sect. 12.3. For a surface link, this is a reduced Alexander polynomial.)

We call the *span* of an Alexander polynomial the maximum degree minus the minimum degree. If the polynomial is zero, we assume the span to be -1 for convenience.

Let λ be an element of Λ and let $A = (a_{ij})$ be an (m, n)-matrix over Λ. We denote it by $A \in L_{m \times n}(\lambda)$ if there exists a (not necessarily strictly) increasing function $f : \{1, \ldots, n\} \to \{1, \ldots, m\}$ such that $a_{ij} = \lambda$ if $i = f(j)$ and $a_{ij} = 0$ otherwise.

LEMMA 30.8. *Let* $b = s_1 \cdots s_n$ *be an oddly principal 3-braid such that* $s_1, \ldots, s_n \in \{\sigma_1^{-1}, \sigma_2\}$ *and* $n \geq 2$. *Let* u *and* v *be numbers of* σ_1^{-1}*'s and* σ_2*'s appearing in* b. *Then for any surface link* F *with* $F \cong RR[b]_3$, $H_1(\widetilde{E}; \mathbb{Z})$ *has a square* Λ-*presentation matrix*

$$
\begin{bmatrix}
1 & & & & & -t & & & 0 \\
t & 1 & & & & 0 & & & 0 \\
 & t & \ddots & & & \vdots & & A_1 & \vdots \\
 & & \ddots & 1 & & 0 & & & 0 \\
 & & & & t & 0 & & & -t \\
\hline
 & & & & & 1 & t & & \\
 & & & & & & 1 & t & \\
 & & A_2 & & & & & \ddots & \ddots \\
 & & & & & & & & 1 & t
\end{bmatrix}
$$

of size $n(= u + v)$ *for some* $A_1 \in L_{(u+1) \times (v-2)}(-t)$ *and* $A_2 \in L_{(v-1) \times u}(-1)$.

Proof. For convenience sake, we assume that F is $R[(1, \sigma_1^{-1}), (b, \sigma_1^{-1})]_3$ rather than $RR[b]_3$. Let R_j $(j = 1, 2, 3)$ be a rectangle $\{(x, y, z) \in \mathbb{R}_+^3 \mid x \in [0, 1], y = j, z \in [0, 1]\}$ in $\mathbb{R}_+^3 = \{(x, y, z) \in \mathbb{R}^3 \mid z \geq 0\}$. Let $h_0, h_1, \ldots, h_{n+1}$ be simple bands attached to the $(x = 1)$-boundary of $R_1 \cup R_2 \cup R_3$ in this order (from the top) such that h_0 and h_{n+1} are bands corresponding to σ_1^{-1} and for each $i \in \{1, \ldots, n\}$, the band h_i corresponds to $s_i \in \{\sigma_1^{-1}, \sigma_2\}$. For example, if $b = \sigma_2 \sigma_1^{-1} \sigma_2 \sigma_2$, then the bands $h_0, \ldots, h_5$ are as in Fig. 30.1. For $\theta \in (-\pi, \pi]$, let

$$\rho_\theta : \mathbb{R}_+^3 \to \mathbb{R}^4 \quad ; \quad (x, y, z) \mapsto (x, y, z\cos\theta, z\sin\theta).$$

Put

$$M_0 = \cup_{\theta \in (-\pi, \pi]} \rho_\theta(R_1 \cup R_2 \cup R_3),$$

and

$$H_i = \begin{cases} \cup_{\theta \in (-\pi, \pi]} \rho_\theta(h_i) & \text{for } i \in \{1, \ldots, n\} \\ \cup_{\theta \in [-\epsilon, \epsilon]} \rho_\theta(h_i) & \text{for } i \in \{0, n+1\}, \end{cases}$$

where ϵ is a small positive number. Then F is ambient isotopic in $\mathbb{R}^4$ to the boundary of a 3-manifold $M = M_0 \cup H_0 \cup \cdots \cup H_{n+1}$. ($M_0$ is a union of three 3-disks. For each $i \in \{1, \ldots, n\}$, H_i is a "round 1-handle" attaching to M_0, which is a solid torus attached to M_0 with a pair of annuli homotopic to a longitude as attaching area. For $i \in \{0, n+1\}$, H_i is a 1-handle in the ordinary sense. This is a special case of a method to construct a Seifert manifold which is given in Sect. 32.2.) Let j_+ and $j_- : H_1(M; \mathbb{Z}) \to H_1(\mathbb{R}^4 \setminus M; \mathbb{Z})$ be homomorphisms obtained by sliding 1-cycles in M in the positive and negative normal directions of M respectively. By the Mayer-Vietoris theorem we have a Λ-isomorphism

$$H_1(\widetilde{E}; \mathbb{Z}) \cong H_1(\mathbb{R}^4 \setminus M; \mathbb{Z}) \otimes_{\mathbb{Z}} \Lambda / (j_+ \otimes t - j_- \otimes 1)(H_1(M; \mathbb{Z}) \otimes_{\mathbb{Z}} \Lambda).$$

Put

$$\Sigma = R_1 \cup R_2 \cup R_3 \cup h_0 \cup \cdots \cup h_{n+1}.$$

Rename bands $h_0, \ldots, h_{n+1}$ by $A_1, \ldots, A_{u+2}$, $B_1, \ldots, B_v$ as in Fig. 30.1 such that $A_1, \ldots, A_{u+2}$ (resp. $B_1, \ldots, B_v$) are attached to $R_1 \cup R_2$ (resp. $R_2 \cup R_3$). Define 1-cycles $a_1, \ldots, a_{u+1}, b_1, \ldots, b_{v-1}$ in Σ as follows: For each $i \in \{1, \ldots, u+1\}$ (resp. $j \in \{1, \ldots, v-1\}$), the 1-cycle a_i (resp. b_j) consists of cores of A_i and A_{i+1} (resp. B_j and B_{j+1}) and two straight segments in $R_1 \cup R_2$ (resp. $R_2 \cup R_3$) connecting endpoints of the cores. Assign a_i (resp. b_j) an orientation whose restriction to the core of A_i (resp. B_j) is from R_1 to R_2 (resp. R_2 to R_3); see Fig. 30.1. Then $H_1(\Sigma; \mathbb{Z})$ is a free abelian group with basis $\{a_1, \ldots, a_{u+1}, b_1, \ldots, b_{v-1}\}$, where we use the same symbols for 1-cycles and their homology classes. Let $\{\alpha_1, \ldots, \alpha_{u+2}, \beta_1, \ldots, \beta_v\}$ be a basis of $H_1(\mathbb{R}^3_+ \setminus \Sigma; \mathbb{Z})$ such that α_i ($i \in \{1, \ldots, u+2\}$) and β_j ($j \in \{1, \ldots, v\}$) are represented by small loops around A_i and B_j with $\mathrm{lk}(\alpha_i, a_i) = 1$ and $\mathrm{lk}(\beta_j, b_j) = 1$ respectively, where lk means the linking number.

Let k_+ and $k_- : H_1(\Sigma; \mathbb{Z}) \to H_1(\mathbb{R}^3_+ \setminus \Sigma; \mathbb{Z})$ be homomorphisms obtained by sliding 1-cycles in Σ in the positive and negative normal directions of Σ. By construction, the following hold:

(I-1) For each $i \in \{1, \ldots, u+1\}$, α_i is involved in $k_+(a_i)$ and never in $k_+(a_{i'})$ for $i' \neq i$. α_{u+2} does not appear in $k_+(a_i)$ for any i.

(I-2) For each $j \in \{1, \ldots, v\}$, the term on β_j appears as $-\beta_j$ in $k_+(a_i)$ for a unique $i = i(j)$. If $j_1 < j_2$, then $i(j_1) \leq i(j_2)$.

(I-3) $k_+(a_1)$ involves $-\beta_1$ and $k_+(a_{u+1})$ involves $-\beta_v$.

(II) For each $j \in \{1, \ldots, v-1\}$, $k_+(b_j) = \beta_{j+1}$.

(III) For each $i \in \{1, \ldots, u+1\}$, $k_-(a_i) = -\alpha_{i+1}$.

(IV-1) For each $j \in \{1, \ldots, v-1\}$, the term on β_j appears as $-\beta_j$ in $k_-(b_j)$ and never in $k_-(b_{j'})$ for $j' \neq j$. β_v is not involved in $k_-(b_j)$ for any j.

(IV-2) For each $i \in \{2, \ldots, u+1\}$, α_i appears in $k_-(b_j)$ for a unique $j = j(i)$. If $i_1 < i_2$, then $j(i_1) \leq j(i_2)$.

(IV-3) α_1 and α_{u+2} are not involved in $k_-(b_j)$ for any j.

The map $\rho_0 : \mathbb{R}^3_+ \to \mathbb{R}^4$ induces homomorphisms

$$\rho_{0*} : H_1(\Sigma; \mathbb{Z}) \to H_1(M; \mathbb{Z})$$

and

$$\rho_{0*} : H_1(\mathbb{R}^3_+ \setminus \Sigma; \mathbb{Z}) \to H_1(\mathbb{R}^4 \setminus M; \mathbb{Z}).$$

The images of $a_i, b_j, \alpha_i, \beta_j$ under ρ_{0*} will be denoted by the same symbols. By construction of M, $H_1(M; \mathbb{Z})$ and $H_1(\mathbb{R}^4 \setminus M; \mathbb{Z})$ are free abelian groups with basis $\{a_1, \ldots, a_{u+1}, b_1, \ldots, b_{v-1}\}$ and $\{\alpha_2, \ldots, \alpha_{u+1}, \beta_1, \ldots, \beta_v\}$. Notice that $\rho_{0*}(\alpha_1) = \rho_{0*}(\alpha_{u+2}) = 0$. From the commutative diagram

$$
\begin{array}{ccc}
H_1(\Sigma; \mathbb{Z}) & \xrightarrow{k_+, k_-} & H_1(\mathbb{R}^3_+ \setminus \Sigma; \mathbb{Z}) \\
\cong \downarrow {\scriptstyle \rho_{0*}} & & \downarrow {\scriptstyle \rho_{0*}} \\
H_1(M; \mathbb{Z}) & \xrightarrow{j_+, j_-} & H_1(\mathbb{R}^4 \setminus M; \mathbb{Z}),
\end{array}
$$

we see that $H_1(\widetilde{E}; \mathbb{Z})$ has the desired Λ-presentation matrix. $\blacksquare$

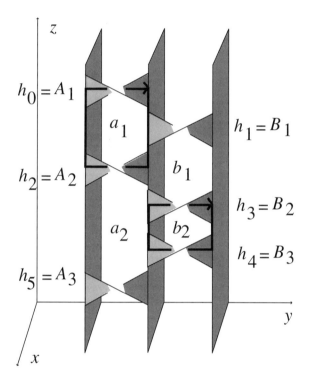

FIGURE 30.1

EXAMPLE 30.9. Let $b = \sigma_2 \sigma_1^{-1} \sigma_2 \sigma_2$ and $F = RR[b]_3$ (precisely speaking $R[(1, \sigma_1^{-1}), (b, \sigma_1^{-1})]_3$). The Σ is as in Fig. 30.1. Then

$$
\begin{array}{ll}
k_+(a_1) = \alpha_1 - \beta_1, & k_-(a_1) = -\alpha_2, \\
k_+(a_2) = \alpha_2 - \beta_2 - \beta_3, & k_-(a_2) = -\alpha_3, \\
k_+(b_1) = \beta_2, & k_-(b_1) = \alpha_2 - \beta_1, \\
k_+(b_2) = \beta_3, & k_-(b_2) = -\beta_2
\end{array}
$$

and

$$
\begin{array}{ll}
t j_+(a_1) - j_-(a_1) & = t(-\beta_1) - (-\alpha_2), \\
t j_+(a_2) - j_-(a_2) & = t(\alpha_2 - \beta_2 - \beta_3) - 0, \\
t j_+(b_1) - j_-(b_1) & = t\beta_2 - (\alpha_2 - \beta_1), \\
t j_+(b_2) - j_-(b_2) & = t\beta_3 - (-\beta_2).
\end{array}
$$

Therefore $H_1(\widetilde{E}; \mathbb{Z})$ has a presentation with generators $\alpha_2, \beta_1, \beta_2, \beta_3$ and relations

$$
\begin{aligned}
\alpha_2 - t\beta_1 &= 0, \\
t\alpha_2 - t\beta_2 - t\beta_3 &= 0, \\
-\alpha_2 + \beta_1 + t\beta_2 &= 0, \\
\beta_2 + t\beta_3 &= 0.
\end{aligned}
$$

Thus we have a presentation matrix

$$
\left[
\begin{array}{c|ccc}
1 & -t & & \\
t & & -t & -t \\
\hline
-1 & 1 & t & \\
& & 1 & t
\end{array}
\right].
$$

The determinant, which is the Alexander polynomial of F, is $t^4 - t^3 + 2t^2 - t$.

Let F be a surface link with $\chi(F) = 2$ and $\mathrm{Braid}(F) \leq 3$ and let $J(F)$ be the subset of B_m consisting of all elements $b \in B_3$ such that $F \cong RR[b]_3$.

LEMMA 30.10. *If $b \in J(F)$ is oddly principal, then the span of the Alexander polynomial of F is the length of b minus one.*

Proof. If $b = 1$, then F is an unknotted surface link which is a union of a trivial 2-sphere and a trivial torus. Its Alexander polynomial is zero and the degree is understood to be -1 here. If $b = \sigma_2$ or σ_2^{-1}, then F is a trivial 2-knot whose Alexander polynomial is 1. If the length n of b is greater than one, by Lemma 29.15, we may assume that b is as in Lemma 30.8. The Alexander polynomial of F is the determinant of a square matrix of size n as in the lemma. The span is $n - 1$. ∎

THEOREM 30.11. *For $b \in J(F)$, the following three conditions are mutually equivalent.*

(1) *b is oddly principal.*
(2) *b has the minimum length among $J(F)$.*
(3) *The length of the Alexander polynomial of F is the length of b minus one.*

Proof. It is a consequence of Lemma 30.7 and 30.10. ∎

THEOREM 30.12. *Every 3-braid 2-knot has a nontrivial Alexander polynomial.*

Proof. By Theorem 30.11, if a 2-knot F with $\mathrm{Braid}(F) \leq 3$ has trivial Alexander polynomial, then it is ambient isotopic to $RR[\sigma_2]_3$ or $RR[\sigma_2^{-1}]_3$. This is a trivial 2-knot and the braid index is one. ∎

EXERCISE 30.13. For any 3-braid 2-knot, the coefficients of the Alexander polynomial appearing at the maximum and minimum degrees are ± 1.

EXERCISE 30.14. The number of 3-braid 2-knots whose Alexander polynomials have the same span is finite.

30.4. Table of 3-Braid 2-Knots

For a surface link F with $\chi(F) = 2$ and $\text{Braid}(F) \leq 3$, let $\alpha(F)$ denote the length of $b \in J(F)$ as in Theorem 30.11, which is the span of the Alexander polynomial of F plus one. We classify such surface links according to $\alpha(F)$.

We denote by $[F]$ the equivalence class of F, and by $[F]^*$ the further equivalence class including mirror images. Let

$$H_\alpha = \{[F] \,|\, F \text{ is a surface link with } \chi(F) = 2, \text{ Braid}(F) \leq 3 \text{ and } \alpha(F) = \alpha\}$$

and

$$H_\alpha^* = \{[F]^* \,|\, F \text{ is a surface link with } \chi(F) = 2, \text{ Braid}(F) \leq 3 \text{ and } \alpha(F) = \alpha\}.$$

Both H_0 and H_0^* consist of the class of an unknotted surface link which is $S^2 \amalg T^2$. H_1 and H_1^* consist of the class of an unknotted 2-knot. So we assume that $\alpha \geq 2$ in what follows.

For each integer $\alpha \geq 2$, let G_α be the power set of $\{1, 2, \ldots, \alpha - 2\}$ and define a map

$$\varphi : G_\alpha \to B_3 \quad ; \quad g \mapsto s_1 \ldots s_{\alpha-2} \quad \text{where } s_i = \begin{cases} \sigma_1^{-1} & \text{if } i \in g \\ \sigma_2 & \text{otherwise.} \end{cases}$$

By Theorem 30.11 (and Lemma 29.15) we have a surjection

$$\Phi_\alpha : G_\alpha \to H_\alpha^*, \quad g \mapsto [RR[\sigma_2\varphi(g)\sigma_2]_3]^*.$$

For $g \in G_\alpha$, let $g^{\mathbf{co}} = \{1, \ldots, \alpha - 2\}\backslash g$ and $g^{\mathbf{op}} = \{\alpha - 1 - j \,|\, j \in g\}$.

Then $\varphi(g^{\mathbf{co}}) = \tau \circ \mu(\varphi(g))$ and $\varphi(g^{\mathbf{op}}) = \mu(\varphi(g))^{-1}$, where τ and μ are automorphisms of B_3 with $\tau(\sigma_1) = \sigma_2$, $\tau(\sigma_2) = \sigma_1$, $\mu(\sigma_1) = \sigma_1^{-1}$, and $\mu(\sigma_2) = \sigma_2^{-1}$.

LEMMA 30.15. $RR[\sigma_2\varphi(g^{\mathbf{co}})\sigma_2]_3$ and $RR[\sigma_2\varphi(g^{\mathbf{op}})\sigma_2]_3$ are ambient isotopic to the mirror image of $RR[\sigma_2\varphi(g)\sigma_2]_3$.

Proof. Note that a mirror image of $RR[\sigma_2\varphi(g)\sigma_2]_3$ is $RR[\sigma_2^{-1}\mu(\varphi(g))\sigma_2^{-1}]_3$.

$$\begin{aligned}
RR[\sigma_2^{-1}\mu(\varphi(g))\sigma_2^{-1}]_3 &= R[(1, \sigma_1^{-1}), (\sigma_2^{-1}\mu(\varphi(g))\sigma_2^{-1}, \sigma_1^{-1})]_3 \\
&\cong R[(\sigma_1\sigma_2, \sigma_1^{-1}), (\sigma_1\mu(\varphi(g))\sigma_2^{-1}, \sigma_1^{-1})]_3 \\
&\cong R[(1, \sigma_2^{-1}), (\sigma_1\mu(\varphi(g))\sigma_2^{-1}, \sigma_1^{-1})]_3 \\
&\cong R[(1, \sigma_2^{-1}), (\sigma_1\mu(\varphi(g))\sigma_1, \sigma_2^{-1})]_3 \\
&\cong R[(1, \sigma_1^{-1}), (\sigma_2\tau \circ \mu(\varphi(g))\sigma_2, \sigma_1^{-1})]_3 \\
&= RR[\sigma_2 g^{\mathbf{co}}\sigma_2]_3
\end{aligned}$$

and

$$\begin{aligned}
RR[\sigma_2^{-1}\mu(\varphi(g))\sigma_2^{-1}]_3 &= R[(1, \sigma_1^{-1}), (\sigma_2^{-1}\mu(\varphi(g))\sigma_2^{-1}, \sigma_1^{-1})]_3 \\
&\cong R[(\sigma_2^{-1}\mu(\varphi(g))\sigma_2^{-1}, \sigma_1^{-1}), (1, \sigma_1^{-1})]_3 \\
&\cong R[(1, \sigma_1^{-1}), (\sigma_2\mu(\varphi(g))^{-1}\sigma_2, \sigma_1^{-1})]_3 \\
&= RR[\sigma_2 g^{\mathbf{op}}\sigma_2]_3. \blacksquare
\end{aligned}$$

COROLLARY 30.16. *If $g = g^{\mathbf{op}}$, then $RR[\sigma_2\varphi(g)\sigma_2]_3$ is amphicheiral (i.e., ambient isotopic to its mirror image).*

Define an equivalence relation $\sim$ on G_α by $g \sim g^{\mathbf{co}} \sim g^{\mathbf{op}} \sim g^{\mathbf{coop}} = g^{\mathbf{opco}}$. We denote by $[g]^*$ the equivalence class of g and by G_α^* the quotient set of G_α. By the above lemma, the surjection $\Phi_\alpha : G_\alpha \to H_\alpha^*$ induces a surjection

$$\Phi_\alpha^* : G_\alpha^* \to H_\alpha^*.$$

CONJECTURE 30.17. *The map* $\Phi_\alpha^* : G_\alpha^* \to H_\alpha^*$ *is a bijection for any* α.

It is known that $\Phi_\alpha^* : G_\alpha^* \to H_\alpha^*$ is a bijection for $\alpha \le 10$; cf. [**324**]. For $\alpha \le 8$ this is verified by calculating Alexander polynomials. For $\alpha = 9$, there is a pair which cannot be distinguished by Alexander polynomials. For $\alpha = 10$, there are two such pairs. In [**324**], these exceptional cases were solved by using the computer program "Knot" by Kouji Kodama.

Here we extract a table for $\alpha \le 9$ from [**324**]. (For $\alpha = 10$, see [**324**].)

In the first column the index α is given. In the second column an element $g \in G_\alpha$ with $\Phi_\alpha^*([g]^*) = [F]^*$ is given. This information enables us to recover the configuration of F. For the third, we divide H_α^* into two families. The symbol S (resp. T) means that F is a 2-knot (resp. a surface link consisting of a 2-sphere and a torus). The first subscript indicates α and the second the order of $[F]^*$ in the subset. In the fourth column the coefficients of the Alexander polynomial of $[F]^*$ are given. In the last column A (resp. N) stands for F being amphicheiral (resp. non-amphicheiral).

REMARK 30.18. Two Laurent polynomials $f(t)$ and $g(t)$ are *equivalent in the weak sense* if $f(t) = \pm t^k g(t)$ or $f(t) = \pm t^k g(t^{-1})$ for some integer k. When we consider a surface link up to mirror images, its Alexander polynomial is well-defined up to equivalence in the weak sense.

PROPOSITION 30.19. *Fox's example 10 is not a 3-braid 2-knot. Thus the braid index is 4.*

Proof. The Alexander polynomial of the 2-knot is $2 - t$. Since the span is 1, the 2-knot is in the image of the surjection $\Phi_2^* : G_2^* \to H_2^*$. However the polynomial $2 - t$ (or any equivalent polynomial in the weak sense $\pm t^*(2 - t)$ or $\pm t^*(2t - 1)$) is not on the list for $\alpha = 2$. In fact G_2^* has only one element, which is the empty set, and $\Phi_2^*([\emptyset])$ is not a 2-knot. ∎

PROPOSITION 30.20. *A spun 2-knot of a figure-eight is not a 3-braid 2-knot.*

Proof. The Alexander polynomial of the 2-knot is $t^2 - 3t + 1$. This polynomial or any equivalent polynomial in the weak sense is not on the list for $\alpha = 3$. ∎

	g		Alexander Poly	
0_1	—	$T_{0,1}$	0	A
1_1	—	$S_{1,1}$	1	A
2_1	$\{\}$	$T_{2,1}$	1, -1	A
3_1	$\{\}$	$S_{3,1}$	1, -1, 1	A
4_1	$\{\}$	$T_{4,1}$	1, -1, 1, -1	A
4_2	$\{1\}$	$S_{4,1}$	1, -1, 2, -1	N
5_1	$\{\}$	$S_{5,1}$	1, -1, 1, -1, 1	A
5_2	$\{1\}$	$S_{5,2}$	1, -1, 2, -2, 1	N
5_3	$\{2\}$	$T_{5,1}$	1, -2, 2, -2, 1	A
6_1	$\{\}$	$T_{6,1}$	1, -1, 1, -1, 1, -1	A
6_2	$\{1\}$	$S_{6,1}$	1, -1, 2, -2, 2, -1	N
6_3	$\{2\}$	$S_{6,2}$	1, -2, 2, -3, 2, -1	N
6_4	$\{1,2\}$	$T_{6,2}$	1, -1, 2, -3, 2, -1	N
6_5	$\{1,3\}$	$S_{6,3}$	1, -2, 3, -3, 3, -1	N
6_6	$\{1,4\}$	$T_{6,3}$	1, -2, 3, -3, 2, -1	A
7_1	$\{\}$	$S_{7,1}$	1, -1, 1, -1, 1, -1, 1	A
7_2	$\{1\}$	$S_{7,2}$	1, -1, 2, -2, 2, -2, 1	N
7_3	$\{2\}$	$T_{7,1}$	1, -2, 2, -3, 3, -2, 1	N
7_4	$\{3\}$	$S_{7,3}$	1, -2, 3, -3, 3, -2, 1	A
7_5	$\{1,2\}$	$S_{7,4}$	1, -1, 2, -3, 3, -2, 1	N
7_6	$\{1,3\}$	$T_{7,2}$	1, -2, 3, -4, 4, -3, 1	N
7_7	$\{1,4\}$	$S_{7,5}$	1, -2, 4, -4, 4, -3, 1	N
7_8	$\{1,5\}$	$T_{7,3}$	1, -2, 3, -4, 3, -2, 1	A
7_9	$\{2,3\}$	$S_{7,6}$	1, -2, 3, -4, 4, -2, 1	N
7_{10}	$\{2,4\}$	$S_{7,7}$	1, -3, 4, -5, 4, -3, 1	A

TABLE 30.1

	g		Alexander Poly	
8_1	$\{\}$	$T_{8,1}$	1, -1, 1, -1, 1, -1, 1, -1	A
8_2	$\{1\}$	$S_{8,1}$	1, -1, 2, -2, 2, -2, 2, -1	N
8_3	$\{2\}$	$S_{8,2}$	1, -2, 2, -3, 3, -3, 2, -1	N
8_4	$\{3\}$	$S_{8,3}$	1, -2, 3, -3, 4, -3, 2, -1	N
8_5	$\{1,2\}$	$T_{8,2}$	1, -1, 2, -3, 3, -3, 2, -1	N
8_6	$\{1,3\}$	$S_{8,4}$	1, -2, 3, -4, 5, -4, 3, -1	N
8_7	$\{1,4\}$	$T_{8,3}$	1, -2, 4, -5, 5, -5, 3, -1	N
8_8	$\{1,5\}$	$S_{8,5}$	1, -2, 4, -5, 5, -4, 3, -1	N
8_9	$\{1,6\}$	$T_{8,4}$	1, -2, 3, -4, 4, -3, 2, -1	A
8_{10}	$\{2,3\}$	$T_{8,5}$	1, -2, 3, -4, 5, -4, 2, -1	N
8_{11}	$\{2,4\}$	$S_{8,6}$	1, -3, 4, -6, 6, -5, 3, -1	N
8_{12}	$\{2,5\}$	$T_{8,6}$	1, -3, 5, -6, 6, -5, 3, -1	A
8_{13}	$\{3,4\}$	$T_{8,7}$	1, -2, 4, -5, 5, -4, 2, -1	A
8_{14}	$\{1,2,3\}$	$S_{8,7}$	1, -1, 2, -3, 4, -3, 2, -1	N
8_{15}	$\{1,2,4\}$	$S_{8,8}$	1, -2, 3, -5, 5, -5, 3, -1	N
8_{16}	$\{1,2,5\}$	$S_{8,9}$	1, -2, 4, -5, 6, -5, 3, -1	N
8_{17}	$\{1,2,6\}$	$S_{8,10}$	1, -2, 3, -5, 5, -4, 2, -1	N
8_{18}	$\{1,3,5\}$	$T_{8,8}$	1, -3, 5, -7, 7, -6, 4, -1	N
8_{19}	$\{1,3,6\}$	$S_{8,11}$	1, -3, 5, -6, 7, -5, 3, -1	N
8_{20}	$\{1,4,5\}$	$S_{8,12}$	1, -2, 5, -6, 6, -5, 3, -1	N

TABLE 30.2

	g		Alexander Poly	
9_1	$\{\}$	$S_{9,1}$	1, -1, 1, -1, 1, -1, 1, -1, 1	A
9_2	$\{1\}$	$S_{9,2}$	1, -1, 2, -2, 2, -2, 2, -2, 1	N
9_3	$\{2\}$	$T_{9,1}$	1, -2, 2, -3, 3, -3, 3, -2, 1	N
9_4	$\{3\}$	$S_{9,3}$	1, -2, 3, -3, 4, -4, 3, -2, 1	N
9_5	$\{4\}$	$T_{9,2}$	1, -2, 3, -4, 4, -4, 3, -2, 1	A
9_6	$\{1,2\}$	$S_{9,4}$	1, -1, 2, -3, 3, -3, 3, -2, 1	N
9_7	$\{1,3\}$	$T_{9,3}$	1, -2, 3, -4, 5, -5, 4, -3, 1	N
9_8	$\{1,4\}$	$S_{9,5}$	1, -2, 4, -5, 6, -6, 5, -3, 1	N
9_9	$\{1,5\}$	$T_{9,4}$	1, -2, 4, -6, 6, -6, 5, -3, 1	N
9_{10}	$\{1,6\}$	$S_{9,6}$	1, -2, 4, -5, 6, -5, 4, -3, 1	N
9_{11}	$\{1,7\}$	$T_{9,5}$	1, -2, 3, -4, 4, -4, 3, -2, 1	A
9_{12}	$\{2,3\}$	$S_{9,7}$	1, -2, 3, -4, 5, -5, 4, -2, 1	N
9_{13}	$\{2,4\}$	$S_{9,8}$	1, -3, 4, -6, 7, -7, 5, -3, 1	N
9_{14}	$\{2,5\}$	$S_{9,9}$	1, -3, 5, -7, 8, -7, 6, -3, 1	N
9_{15}	$\{2,6\}$	$S_{9,10}$	1, -3, 5, -7, 7, -7, 5, -3, 1	A
9_{16}	$\{3,4\}$	$S_{9,11}$	1, -2, 4, -5, 6, -6, 4, -2, 1	N
9_{17}	$\{3,5\}$	$T_{9,6}$	1, -3, 5, -7, 8, -7, 5, -3, 1	A
9_{18}	$\{1,2,3\}$	$S_{9,12}$	1, -1, 2, -3, 4, -4, 3, -2, 1	N
9_{19}	$\{1,2,4\}$	$T_{9,7}$	1, -2, 3, -5, 6, -6, 5, -3, 1	N
9_{20}	$\{1,2,5\}$	$S_{9,13}$	1, -2, 4, -6, 7, -7, 6, -3, 1	N
9_{21}	$\{1,2,6\}$	$T_{9,8}$	1, -2, 4, -6, 7, -7, 5, -3, 1	N
9_{22}	$\{1,2,7\}$	$S_{9,14}$	1, -2, 3, -5, 6, -5, 4, -2, 1	N
9_{23}	$\{1,3,4\}$	$S_{9,15}$	1, -2, 4, -5, 7, -7, 5, -3, 1	N
9_{24}	$\{1,3,5\}$	$S_{9,16}$	1, -3, 5, -8, 9, -9, 7, -4, 1	N
9_{25}	$\{1,3,6\}$	$S_{9,17}$	1, -3, 6, -8, 10, -9, 7, -4, 1	N
9_{26}	$\{1,3,7\}$	$S_{9,18}$	1, -3, 5, -7, 8, -8, 5, -3, 1	N
9_{27}	$\{1,4,5\}$	$S_{9,19}$	1, -2, 5, -7, 8, -8, 6, -3, 1	N
9_{28}	$\{1,4,6\}$	$T_{9,9}$	1, -3, 6, -9, 10, -9, 7, -4, 1	N
9_{29}	$\{1,4,7\}$	$S_{9,20}$	1, -3, 6, -8, 9, -8, 6, -3, 1	A
9_{30}	$\{1,5,6\}$	$S_{9,21}$	1, -2, 5, -7, 8, -7, 5, -3, 1	N
9_{31}	$\{2,3,4\}$	$T_{9,10}$	1, -2, 3, -5, 6, -6, 4, -2, 1	N
9_{32}	$\{2,3,5\}$	$S_{9,22}$	1, -3, 5, -7, 9, -8, 6, -3, 1	N
9_{33}	$\{2,3,6\}$	$T_{9,11}$	1, -3, 6, -8, 9, -9, 6, -3, 1	N
9_{34}	$\{2,4,5\}$	$T_{9,12}$	1, -3, 5, -8, 9, -8, 6, -3, 1	N
9_{35}	$\{2,4,6\}$	$S_{9,23}$	1, -4, 7, -10, 11, -10, 7, -4, 1	A
9_{36}	$\{3,4,5\}$	$S_{9,24}$	1, -2, 4, -6, 7, -6, 4, -2, 1	A

TABLE 30.3

CHAPTER 31

Unknotting Surface Braids and Surface Links

31.1. Unknotting Surface Braids

We show that a simple surface braid S is transformed into an unknotted surface braid by surgery along nice 1-handles.

LEMMA 31.1. *Let Γ be a chart and let w be the number of white vertices. After inserting w free edges suitably, we can change Γ into a chart without white vertices by chart moves.*

Proof. For each white vertex, insert a free edge as in Fig. 31.1 and apply chart moves. ∎

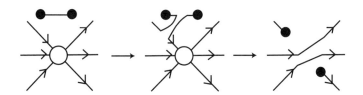

FIGURE 31.1. Deletion of a white vertex

COROLLARY 31.2. *Any simple surface braid is transformed into an ribbon surface braid by surgery along nice 1-handles.*

DEFINITION 31.3. A *nomad* (of degree m) is a union $\nu = \nu_1 \cup \cdots \cup \nu_{m-1}$ of $m - 1$ free edges whose labels are $1, \ldots, m - 1$.

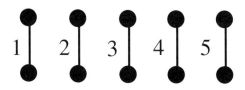

FIGURE 31.2. A nomad of degree 6

LEMMA 31.4. *Let Γ be a ribbon chart (or a chart without white vertices) of degree m. The split union of Γ and a nomad of degree m is transformed into an unknotted chart by chart moves.*

Proof. By Theorem 29.2, we may assume that Γ is a union of some oval nests. The hoops of them are removed by chart moves as in Fig. 31.3 (recall Fig. 28.2). The remainder consists of free edges. ∎

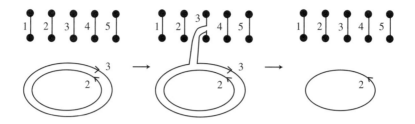

FIGURE 31.3. A nomad absorbing hoops

COROLLARY 31.5. *The product of a ribbon surface braid of degree m and an unknotted surface braid described by a nomad of degree m is an unknotted surface braid.*

Therefore we have the following.

PROPOSITION 31.6. *Any simple surface braid is transformed into an unknotted one by surgery along nice 1-handles. If the surface braid has a chart description with w white vertices, we may assume the number of 1-handles is $w + m$.*

31.2. Wandering Nomads

Let Γ be a chart of degree m in D^2. Put a nomad ν of degree m in a region of $D^2 \setminus \Gamma$. As shown in Fig. 31.4, a nomad can get through any edge of Γ. This is the reason that we call ν a nomad.

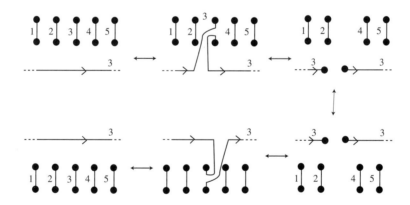

FIGURE 31.4. A nomad getting through edges

Proposition 31.6 is strengthened as follows:

PROPOSITION 31.7. *Let S be a simple surface braid described by a chart Γ with w white vertices. Let S_0 be an unknotted surface braid described by a chart which is a union of a nomad ν and w free edges. Then $S \cdot S_0$ is unknotted.*

Proof. First we notice that the labels of w free edges following ν can change as in Lemma 18.24(D). The nomad ν followed by w free edges can move into any region of $D^2 \setminus \Gamma$. Move them next to a white vertex of Γ and remove the white vertex as in Fig. 31.1. By this process, we lose a free edge but the nomad is intact.

Repeating this procedure, we remove all white vertices. By Lemma 31.4, we have an unknotted chart. ∎

31.3. Unknotting Number

THEOREM 31.8 (Unknotting Theorem). *Any oriented surface link F is transformed into an unknotted surface link by a finite sequence of (oriented) 1-handle surgeries.*

Proof. By 1-handle surgeries, make F connected. F bounds a compact connected oriented 3-manifold in $\mathbb{R}^4$. Along the 1-skeleton of this 3-manifold, apply 1-handle surgeries. The result is a surface link which bounds a handlebody. Thus it is unknotted. ∎

This theorem is true for non-orientable surface links (cf. [**307**]).

The *unknotting number of an oriented surface link F* is the minimum number of oriented 1-handle surgeries that change F into an unknotted surface link. This is denoted by $u(F)$.

The *unknotting number of a simple surface braid S* is the minimum number of 1-handle surgeries along nice 1-handles that change S into an unknotted surface braid. This is denoted by $u(S)$.

Since the closure in $\mathbb{R}^4$ of an unknotted surface braid is an unknotted surface link and since a nice 1-handle surgery is an oriented 1-handle surgery, we have

$$u(F) \leq u(S)$$

for any simple braid presentation S of F.

Thus Proposition 31.6 or 31.7 gives an upper estimation for the unknotting number of an oriented surface link.

EXAMPLE 31.9. Recall that the 2-twist-spun trefoil F is the closure of a surface braid S with chart description as in Fig. 21.15. Figs. 31.5 and 31.6 show that the chart is transformed into a trivial chart by introducing a free edge. The top of Fig. 31.5 is a chart after insertion of a free edge with label 2 ($*$ indicates the inserted free edge). By chart moves, it becomes an unknotted chart. This implies that $u(F) \leq u(S) \leq 1$. Since F is not unknotted, $u(F) = u(S) = 1$.

31.4. Peiffer Transformations

Let G be a group and $P^n(G)$ be the Cartesian product of n copies of G as before.

A *Peiffer transformation of the first kind* is

$$(g_1, \ldots, g_n) \in P^n(G) \mapsto (g_1, \ldots, g_{i-1}, g_i g_{i+1} g_i^{-1}, g_i, g_{i+2}, \ldots, g_n) \in P^n(G)$$

for $i \in \{1, \ldots, n-1\}$. This is the same as the slide action, $\text{slide}(\sigma_i)$, in Sect. 2.7.

A *Peiffer transformation of the second kind* is

$$(g_1, \ldots, g_n) \in P^n(G) \mapsto (g_1, \ldots, g_{i-1}, g_{i+2}, \ldots, g_n) \in P^{n-2}(G)$$

for some i provided $g_i g_{i+1} = 1$ in G.

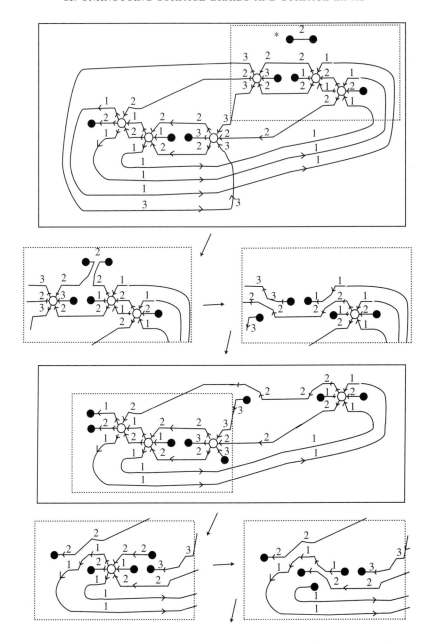

FIGURE 31.5. Unknotting a 2-twist-spun trefoil by a 1-handle surgery (1)

THEOREM 31.10. *Any braid system of a simple surface braid is transformed into the empty braid system by a sequence of Peiffer transformations and their inverses. Moreover Peiffer transformations of the second kind and their inverses are assumed to be deletions or insertions of $(\sigma_i^{-\epsilon}, \sigma_i^{\epsilon})$ for $i \in \{1, \ldots, m-1\}$ and $\epsilon \in \{\pm 1\}$.*

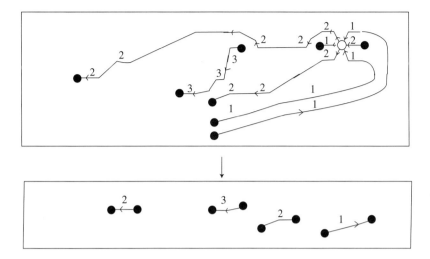

FIGURE 31.6. Unknotting a 2-twist-spun trefoil by a 1-handle surgery (2)

Proof. Consider a chart and a Hurwitz arc system from which we have the braid system. Let w be the number of white vertices. Addition of a chart which is a union of a nomad ν and w free edges with label 1 changes the braid system into the concatenation of the braid system and

$$(\sigma_1^{-1}, \sigma_1, \sigma_2^{-1}, \sigma_2, \ldots, \sigma_{m-1}^{-1}, \sigma_{m-1}, \sigma_1^{-1}, \sigma_1, \ldots, \sigma_1^{-1}, \sigma_1).$$

This is a series of the inverses of Peiffer transformations of the second kind. By Lemma 31.7, the chart becomes an unknotted chart by chart moves. This process is realized by slide actions (Peiffer transformations of the first kind) in the braid system level. The unknotted chart has an unknotted braid system, which is changed into the empty braid system by Peiffer transformations of the second kind. ∎

This theorem was proved by Lee Rudolph by using a relationship between braided surfaces and maps of a disk into a configuration space [**778, 783**]. Our argument of the proof using chart description gives an effective algorithm (recall Sect. 18.10).

31.5. Extended Configuration Space

A geometric braid is regarded as a loop in the configuration space, and a braid is a homotopy class of such a loop. A surface braid is also regarded as a map of a 2-disk into a certain space and the equivalence class is regarded as the homotopy class.

Let E denote the interior of a 2-disk D_1^2.

Let $\widetilde{C}_m^{(1)}(E)$ be the subset of $P^m(E) = E \times \cdots \times E$ consisting of all elements $(w_1, \cdots, w_m)$ satisfying one of the following conditions.

(0) $w_i \neq w_j$ for all distinct $i \neq j$.
(1) $w_i \neq w_j$ for all pairs of distinct $i \neq j$ but a single pair (s, t) with $1 \leq s < t \leq m$ such that $w_s = w_t$.

The subset consisting of elements as in (1) is denoted by $\widetilde{\Sigma}_m^{(1)}$ and is called the *singular locus*. The symmetric group on m letters $\{1, 2, \ldots, m\}$ acts on this space $\widetilde{C}_m^{(1)}(E)$ by permuting the coordinates. We call the quotient space

$$C_m^{(1)}(E) = \widetilde{C}_m^{(1)}(E)/\sim$$

the (*first*) *extended configuration space of unordered m points* of E. The subset

$$\Sigma_m^{(1)}(E) = \widetilde{\Sigma}_m^{(1)}(E)/\sim$$

is the *singular locus*. This locus is regarded as a (real) codimension-two subspace of $C_m^{(1)}(E)$ and the complement $C_m^{(1)}(E) \setminus \Sigma_m^{(1)}(E)$ is the configuration space $C_m = C_m(E)$ as before. Each element of $C_m(E)$ is identified with a set of m distinct points of W, and an element of $\Sigma_m^{(1)}(E)$ is identified with a set of m points which are all distinct except for a pair of points.

For a simple surface braid S in $D_1^2 \times D_2^2$, we define a map

$$f_S : D_2^2 \to C_m^{(1)}(E)$$

by

$$f_s(y) = pr_1(S \cap pr_2^{-1}(y)) \quad \text{for } y \in D_2^2.$$

Then

(1) $f_S(\partial D_2^2) = Q_m$,

(2) f_S intersects $\Sigma_m^{(1)}(E)$ transversely.

Here f_S intersects $\Sigma_m^{(1)}(E)$ transversely at a point $y \in D_2^2$ means that for a regular neighborhood $N(y)$ of y, $f_S^{-1}(f_S(N(y)) \cap \Sigma_m^{(1)}(E)) = \{y\}$ and $f_S|_{\partial N(y)}$ is a meridian loop of $\Sigma_m^{(1)}(E)$ (i.e., a loop in the configuration space $C_m(E)$ represents a closed m-braid which is equivalent to $\widehat{\sigma_1}$ or $\widehat{\sigma_1^{-1}}$). It is obvious that $f_S^{-1}(f_S(D_2^2) \cap \Sigma_m^{(1)}(E))$ is the branch point set $\Sigma(S)$ of S. Conversely if a map $f : D_2^2 \to C_m^{(1)}(E)$ satisfying the above two conditions is given, there is a simple surface braid S_f.

Let $\mathcal{F}$ denote the family of maps $f : D_2^2 \to C_m^{(1)}(E)$ satisfying the above two conditions. This family corresponds to the family of simple surface braids of degree m. Two elements f and f' of $\mathcal{F}$ are *equivalent* if they are homotopic through $\mathcal{F}$. Such a homotopy is called a *transversal homotopy with respect to* $\Sigma_m^{(1)}(E)$. Then corresponding surface braids S_f and $S_{f'}$ are equivalent. (This does not mean that a transversal homotopy $\{f_t\}$ between f and f' gives a (PL) braid isotopy between S_f and $S_{f'}$. However S_f and $S_{f'}$ have equivalent braid monodromies and they are equivalent as surface braids.) This is an interpretation of surface braids in terms of maps into a configuration space.

Let $\pi_2^{\Sigma_m^{(1)}}(C_m^{(1)}(E), Q_m)$ denote the set of transversal homotopy classes (with respect to $\Sigma_m^{(1)}$) of elements of $\mathcal{F}$. It is a commutative monoid by the usual product. This monoid is identified with the simple surface braid monoid $\mathcal{SB}_m$ of degree m.

PROPOSITION 31.11. $\pi_2^{\Sigma_m^{(1)}}(C_m^{(1)}(E), Q_m) \cong \mathcal{SB}_m$.

One of the differences from the classical case is that we have to take a homotopy between f and f' through $\mathcal{F}$. If we give up this condition, what happens? Let f and f' be elements of $\mathcal{F}$ and suppose that there is a homotopy $\{f_t\}$ between them. Replacing the homotopy, we assume that $f_t \in \mathcal{F}$ for all $t \in [0, 1]$ but a finite number of exceptional values such that about each exceptional value a "finger move" or its inverse "Whitney trick" happens (see Fig. 31.7). After a finger move, a pair of intersection points appears. Each of them is a transversal intersection. A finger move introduces a pair of new branch points to the corresponding surface braid S_f, one of whose monodromy is $w^{-1}\sigma_1 w$ and the other is $w^{-1}\sigma_1^{-1} w$ for some $w \in B_m$. A finger move corresponds to a 1-handle surgery along a nice 1-handle attaching to S_f.

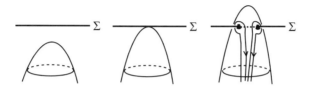

FIGURE 31.7. A finger move in the extended configuration space

Recall that two simple surface braids of degree m are related by a finite sequence of nice 1-handle surgeries and inverse operations (Proposition 31.6). This implies that for any maps f and f' in $\mathcal{F}$, there exists a homotopy $\{f_t\}$ between them such that $f_t \in \mathcal{F}$ for all $t \in [0, 1]$ but a finite number of exceptional values where a finger move or its inverse occurs. Since any map $f : D_2^2 \to C_m^{(1)}(E)$ with $f_S(\partial D_2^2) = Q_m$ is homotopic to a map belonging to $\mathcal{F}$, we have the following.

PROPOSITION 31.12. $\pi_2(C_m^{(1)}(E), Q_m)$ is trivial.

CHAPTER 32

Seifert Algorithm for Surface Braids

32.1. Seifert Algorithm for Classical Braids

Let $a_1 \ldots, a_m$ be mutually disjoint arcs in a 2-disk D^2 which connect $Q_m = \{y_1, \ldots, y_m\}$ to points of ∂D^2 as in Fig. 32.1, and let $R_1, \ldots, R_m$ be rectangles in $D^2 \times [0, 1]$ such that $R_i = a_i \times [0, 1]$ for $i \in \{1, \ldots, m\}$. Put $R = \cup_{j=1}^m R_j$.

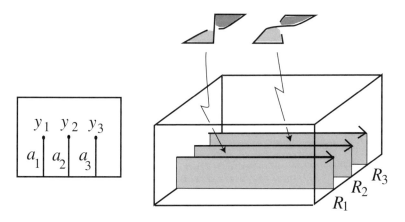

FIGURE 32.1

For a geometric braid b corresponding to a braid word, say $s_1^{\epsilon_1} \cdots s_1^{\epsilon_k}$ for $s_1, \ldots, s_k \in \{\sigma_1, \ldots, \sigma_{m-1}\}$ and $\epsilon_1, \ldots, \epsilon_k \in \{\pm 1\}$, consider a set of simple bands $A_1, \ldots, A_k$ attaching to the trivial braid $Q_m \times [0, 1]$ in this order such that for each $j \in \{1, \ldots, k\}$, the simple band b_j corresponds to $s_j^{\epsilon_j}$. The union $R \cup A_1 \cup \cdots \cup A_k$ is a surface, which can be regarded as a Seifert surface for a closure of b in $\mathbb{R}^3$.

32.2. Seifert Algorithm for Surface Braids

We introduce a method to construct a Seifert manifold for a simple surface braid.

Let S be a simple surface braid of degree m and let Γ be a chart description. Fix a bi-parameterization $D_2^2 \cong I_3 \times I_4$ and consider the motion picture $\{b_t\}$ of S along the I_4-direction. Let R be the union of rectangles in $D_1^2 \times I_3$ as in the previous section and let $M_0 = R \times I_4$, which is a union of 3-disks. The boundary of M_0 is regarded as a Seifert manifold of a closure of the trivial surface braid $S_0 = Q_m \times D_2^2$.

First, we consider the case where Γ has no white vertices. Then Γ is a union of some arcs and loops such that (1) each of them has no self-intersection, (2) the intersection of them consists of degree-4 vertices, and (3) the endpoints of

265

each arc are black vertices of the chart Γ. Let γ be one of such curves, and let $i \in \{1, \ldots, m-1\}$ be the label assigned to γ. Let A be a simple band corresponding to σ_i and consider a 3-manifold $H(\gamma) = A \times \gamma$ in $D_1^2 \times D_2^2$. (Precisely speaking, let $N(\gamma) \cong \gamma \times [-1, 1]$ be a bicollar neighborhood of γ in D_2^2 such that an arc $\{y\} \times [-1, 1] \subset \gamma \times [-1, 1] = N(\gamma)$ intersects γ from the left side of γ. For each $y \in \Gamma$, let A_y be a copy of A in $D_1^2 \times (\{y\} \times [-1, 1])$. The trace A_y for all $y \in \gamma$ is denoted by $A \times \gamma$.)

If γ is an arc, the endpoints of γ are saddle points as in Fig. 14.4. Deform them into simple bands as in Fig. 21.3, which are copies of A. Then $H(\gamma) = A \times \gamma$ is a 3-disk and the intersection of $H(\gamma)$ and M_0 is a pair of disks which is (attaching arcs of A) $\times \gamma$ in $\partial H(\gamma)$. Thus $H(\gamma)$ is a 1-handle attaching to M_0.

If γ is a loop, then $H(\gamma)$ is a solid torus and the intersection of $H(\gamma)$ and M_0 is a pair of parallel annuli in $\partial H(\gamma)$. We call this a *round 1-handle* attaching to M_0. By dividing it into two pieces, it can be regarded as a 1-handle attaching to M_0 followed by a 2-handle.

Consider $H(\gamma)$ for all γ. They are mutually disjoint. Let H_1 be the union of such handles. The union $M_0 \cup H_1$ is a Seifert manifold of the closure of S.

Now we consider the case where Γ has white vertices. Let Γ_1 be a chart obtained from Γ by replacing all white vertices as in Fig. 32.2, and let S_1 be the surface braid described by Γ_1. We assume that saddle points are deformed into saddle bands. Construct a 3-manifold $M_0 \cup H_1$ for S_1 as above. Near a white vertex as in (1) of Fig. 32.2, say W, S_1 looks as in the first row of Fig. 32.3 in a motion picture. Consider a continuous sequence of bands as in the second row. The trace $H(W)$ of the continuous sequence of bands is a 3-disk in $D_1^2 \times D_2^2$. (In Fig. 32.3, we are looking at the braid from the opposite side of M_0.) Thus the intersection of $H(W)$ and M_0 is a pair of disks which are the trace of the attaching arcs. $H(W)$ intersects H_1 at the saddle bands corresponding to the black vertices. Thus the whole intersection of $H(W)$ and $M_0 \cup H_1$ is an annulus in $\partial H(W)$. $H(W)$ is a 2-handles attaching to $M_0 \cup H_1$. The case of (2) in Fig. 32.2 is similar. Let H_2 be the union of such 2-handles $H(W)$ for all white vertices. The union $M_0 \cup H_1 \cup H_2$ is a 3-manifold in $D_1^2 \times D_2^2$ which is a Seifert manifold for a closure of S.

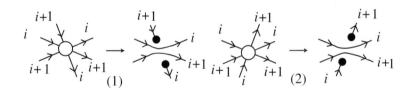

FIGURE 32.2

Let Γ be a chart of degree m with w white vertices. Let Γ_1 be a chart obtained from Γ by removing all white vertices as in Fig. 32.2. The chart Γ_1 is a union of some arcs and loops such that (1) each of them has no self-intersection, (2) the intersection of them consists of degree-4 vertices, and (3) the endpoints of each arc are black vertices of the chart Γ_1. Let u and v be the numbers of the arcs and the loops of Γ_1, respectively.

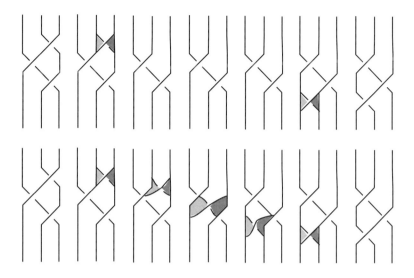

FIGURE 32.3

PROPOSITION 32.1. *In the above situation, a surface link described by Γ has a Seifert manifold which has a handle decomposition with m 0-handles, u 1-handles, v round 1-handles and w 2-handles.*

Since a round 1-handle is a 1-handle followed by a 2-handle, in the theorem, we may say that there is a handle decomposition with m 0-handles, $u + v$ 1-handles and $v + w$ 2-handles.

The algorithm stated here was given in [314]. A similar algorithm was independently introduced by J. S. Carter and M. Saito in [80] (see [87, 89]). Their method does not require a surface link to be in a braid form at all.

CHAPTER 33

Basic Symmetries in Chart Descriptions

33.1. Symmetry Theorem

Recall that we denote by F^* the mirror image of an oriented surface link F and by $-F$ the oriented surface link which is obtained from F by reversing its orientation. We denote by Γ^* a mirror image of a chart Γ and by $-\Gamma$ a chart obtained from Γ by reversing the orientation of all edges (Sect. 29.1).

THEOREM 33.1 (Symmetry Theorem on Chart Description). *Let Γ be a chart and let $F = F(\Gamma)$ be a surface link which is described by Γ. Then the following hold.*

(1) $F(-\Gamma) \cong F(\Gamma)^*$.
(2) $F(\Gamma^*) \cong -F(\Gamma)$.
(3) $F(-\Gamma^*) \cong -F(\Gamma)^*$.

Proof. (1) Let $\{b_t\}$ be the motion picture of a surface braid $S = S(\Gamma)$ associated with Γ. If we change the orientation of edges of the chart Γ, then by definition, all crossings appearing in the braid diagram of each b_t are reversed. This is the mirror image of S. Thus its closure is also a mirror image. (2) Rotate each braid $b_t \subset D_2^2 \times I_3$ and reverse the orientation of strings of b_t as in Fig. 33.1. The surface link is $-F$ and it has a chart description which is a mirror image of Γ. (3) is obtained from (1) and (2). ∎

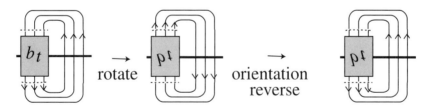

FIGURE 33.1. $F(\Gamma^*) = -F(\Gamma)$

As an easy corollary, if a surface link F has a chart description Γ such that $\Gamma = -\Gamma^*$, then F is negative amphicheiral (i.e., $F \cong -F^*$). However we have proved a stronger result that if $\Gamma = -\Gamma^*$, then F is a ribbon surface link. (Recall that a ribbon surface link is always negative amphicheiral.)

It is interesting that the symmetry theorem on chart description looks similar to the symmetry theorem on twist-spun knots (Theorem 10.4).

CHAPTER 34

Singular Surface Braids and Surface Links

34.1. Singular Surface Links

A classical link is regarded as the image of a generic map of a 1-manifold into 3-space. Then it is natural to consider that a generalization of a link is the image of a generic map of a 2-manifold into 4-space. A generic map of a surface into 4-space is an immersion with transverse double points and not an embedding in general. It is important and necessary, in various situations, to consider singular surface links. We introduce the notion of a singular surface link and a singular surface braid. We work in the PL (or smooth) category.

DEFINITION 34.1. A *singular surface link* is a closed surface F immersed in $\mathbb{R}^4$ locally flatly such that each singular point is a transverse double point. If the underlying surface F_0 of F (i.e., F is the image of an immersion of F_0 in $\mathbb{R}^4$) is connected, it is also called a *singular surface knot*.

For a transverse double point $y \in \mathbb{R}^4$, there is a regular neighborhood N of y in $\mathbb{R}^4$ such that $N \cap F$ is a cone $y * L$ for a Hopf link L in ∂N. When F is oriented (or orientable and connected), we can define the sign for a double point in the usual way.

Two singular surface links are *equivalent* if there is an orientation-preserving homeomorphism of $\mathbb{R}^4$ which carries one to the other. This is equivalent to the condition that they are ambiently isotopic to each other.

34.2. Unknotted Singular Surface Links

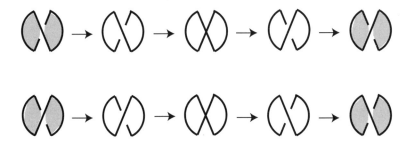

FIGURE 34.1. Standard singular 2-knots with a single double point

The *standard singular 2-knot with one double point* is a singular 2-knot illustrated in Fig. 34.1 in the motion picture method. The first row is a singular 2-knot whose normal Euler number is 2, and the second row is a singular 2-knot whose normal Euler number is -2. A singular surface knot is *unknotted* or *trivial* if it is obtained from some standard surfaces illustrated in Fig. 6.1 and Fig. 34.1 by connected sum. A singular surface link is *unknotted* or *trivial* if it is a split union of unknotted singular surface knots.

PROPOSITION 34.2 (cf. [**329**]). *Let F and F' be unknotted singular surface knots with the same underlying surface. They are equivalent if and only if their normal Euler numbers are the same and the number of double points is the same.*

34.3. Singular Surface Braids

DEFINITION 34.3. A *singular surface braid* is a properly immersed oriented surface S in $D_1^2 \times D_2^2$ satisfying the following conditions.

(1) There is an immersion $f : S_0 \to D_1^2 \times D_2^2$ with $f(S_0) = S$ such that the composition $pr_2 \circ f : S_0 \to D_2^2$ is a branched covering of degree m.
(2) The boundary of S is a trivial closed braid $Q_m \times \partial D_2^2$ in the solid torus $D_1^2 \times \partial D_2^2$.
(3) Each fiber $pr_2^{-1}(y)$ for $y \in D_2^2$ contains at most one double point of S, and if there is a double point in $pr_2^{-1}(y)$, then y is not a branch point of the branched covering.

There are two kinds of singular points of S. One is a singular point of the associated branched covering $pr_2 \circ f : S_0 \to D_2^2$ and the other is a singular point of the immersion f, i.e., a double point of S. The former is referred to as a *branch-point type singular point* and the latter is a *double-point type singular point*. We denote by $\Sigma(S)$ the image of both kinds of singular points of S under $pr_2 : D_1^2 \times D_2^2 \to D_2^2$.

Note that

$$pr_2|_{S \setminus pr_2^{-1}(\Sigma(S))} : S \setminus pr_2^{-1}(\Sigma(S)) \to D_2^2 \setminus \Sigma(S)$$

is a covering map of degree m.

A singular surface braid is *simple* if the associated branched covering is simple. This is equivalent to the condition that $pr_1(S \cap pr_2^{-1}(y))$ consists of m or $m - 1$ points of D_1^2 for each $y \in D_2^2$.

The notions of equivalence, a braid isotopy, a product and the monoid, etc., are defined for singular surface braids in the same way as those for surface braids.

The notion of a singular braided surface is similarly defined [**778**, **783**].

34.4. Braid Monodromy and Charts

For a singular surface braid, the braid monodromy $\rho : \pi_1(D_2^2 \setminus \Sigma(S), y_0) \to B_m$ and a braid system $(b_1, \ldots, b_n)$ are defined.

PROPOSITION 34.4. *An n-tuple of m-braids, $(b_1, \ldots, b_n)$, is a braid system of a singular surface braid (or a simple singular surface braid, resp.) if and only if*

(1) *each b_i is an element of A_m (resp. SA_m) or a conjugate of σ_1^2 or σ_1^{-2}, and*
(2) *the product $b_1 \cdots b_n$ is trivial in B_m.*

Proof. First, we prove the only if part. By the third condition of the definition of a singular surface braid, if a point $y \in \Sigma(S)$ is a branch point, then $pr_2^{-1}(y)$ does not contain a double point of S. The situation is the same as in the non-singular case. So the local monodromy about this point is an element of A_m (resp. SA_m). If $y \in \Sigma(S)$ is not a branch point, then there is a single transverse double point in

$pr_2^{-1}(y)$. Thus the local monodromy is a conjugate of σ_1^2 or σ_1^{-2}. The condition $b_1 \cdots b_n = e$ follows since ∂S is a trivial closed braid.

We prove the if part. If $b_i \in A_m$ (or SA_m, resp.), there is a braided surface (resp. a simple braided surface) with one branch point whose monodromy is b_i. If b_i is a conjugate of σ_1^2 or σ_1^{-2}, there is a singular braided surface with a single singular point whose braid monodromy is b_i. By the same argument as in the proof of Proposition 17.11, we have the desired singular surface braid. ∎

For a simple singular surface braid, we define a chart description. All we have to do is to introduce degree-2 vertices which represent transverse double points of S. In the definition of a chart (Definition 18.7), we change condition (1) and introduce a new condition (5) as follows:

(1′) Every vertex has degree one, *two*, four or six.

(5) For each degree-two vertex (called a *node*), the two edges have the same label, but the orientations are opposite. (See Fig. 34.2.)

We call such a chart a *singular chart*.

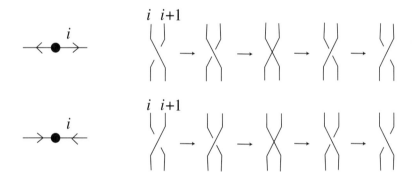

FIGURE 34.2. Degree 2 vertices and motion pictures

For chart moves, the definition of chart move (CI) is modified as follows:

(CI′) There are no black vertices and nodes (degree-one and degree-two vertices) in $\Gamma \cap E$ and $\Gamma' \cap E$.

(CII) and (CIII) are the same as before. New chart moves are introduced as follows:

(CIV) $\Gamma \cap E$ and $\Gamma' \cap E$ are as in Fig. 34.3, where $|i - j| > 1$.

(CV) $\Gamma \cap E$ and $\Gamma' \cap E$ are as in Fig. 34.4, where $|i - j| = 1$.

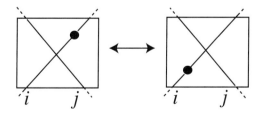

FIGURE 34.3. CIV-move

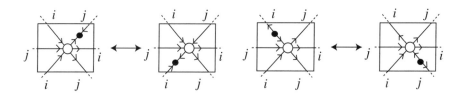

FIGURE 34.4. CV-move

THEOREM 34.5. *Two singular charts present equivalent simple singular surface braids if and only if they are related by ambient isotopies and chart moves.*

This theorem is proved by an argument similar to the non-singular case. Refer to [**323**] for a proof.

34.5. The Braid Description of a Singular Surface Link

THEOREM 34.6 ([**323**]). *Any singular oriented surface link is equivalent to the closure of a simple singular surface braid.*

This is a generalized Alexander's theorem for singular surface links. By this theorem, we can describe a singular surface link by a chart. For example, consider the singular 2-link that is in the first row of Fig. 34.5. We call it the *Fenn-Rolfsen 2-link* [**154**]. Each component is a standard singular 2-knot with one double point. In the second row is a braid form of the middle section. Applying the braiding method in Chapter 23, we have a chart of degree 4 as in Fig. 34.6. This is a chart description of the Fenn-Rolfsen 2-link. Fig. 34.7 is a simplified chart description of the Fenn-Rolfsen 2-link by chart moves.

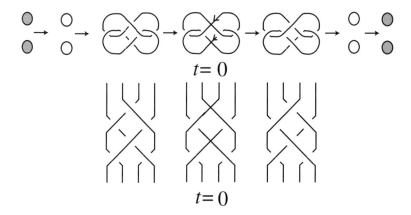

FIGURE 34.5. The Fenn-Rolfsen 2-link

Further results on singular surface braids (for example, finger moves for surface braids) are found in [**322, 323, 325**].

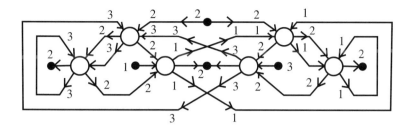

FIGURE 34.6. A chart for the Fenn-Rolfsen 2-link (before chart moves)

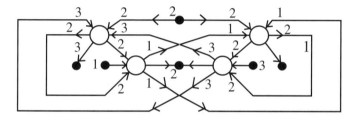

FIGURE 34.7. A chart for the Fenn-Rolfsen 2-link (after chart moves)

Bibliography

[1] C.C. Adams, *The knot book, An elementary introduction to the mathematical theory of knots*, W. H. Freeman and Company, New York(1994)

[2] I.R. Aitchison and D.S. Silver, *On certain fibred ribbon disc pairs*, Trans. Amer. Math. Soc. **306**(1988) 529–551

[3] S. Akbulut and R. Kirby, *An exotic involution of S^4*, Topology **18**(1979) 75–81

[4] S. Akbulut and R. Kirby, *Branched covers of surfaces in 4-manifolds*, Math. Ann. **252**(1980) 111–131

[5] J. W. Alexander, *A lemma on systems of knotted curves*, Proc. Nat. Acad. Sci. USA **9**(1923) 93–95

[6] J. W. Alexander, *Topological invariants of knots and links*, Trans. Amer. Math. Soc **30**(1928) 275–306

[7] J. W. Alexander and G.B. Briggs, *On types of knotted curves*, Ann. of Math. **28**(1927) 562–586

[8] J. J. Andrews and M. L. Curtis, *Knotted 2-spheres in 4-space*, Ann. of Math. **70**(1959) 565–571

[9] J. J. Andrews and S.J. Lomonaco, *The second homotopy group of spun 2-spheres in 4-space*, Bull. Amer. Math. Soc. **75**(1969) 169–171

[10] J. J. Andrews and S.J. Lomonaco, *The second homotopy group of spun 2-spheres in 4-space*, Ann. of Math. **90**(1969) 199–204

[11] J. J. Andrews and D.W. Sumners, *On higher-dimensional fibered knots*, Trans. Amer. Math. Soc. **153**(1971) 415–426

[12] D. Armand-Ugon, R. Gambini and P. Mora, *Intersecting braids and intersecting knot theory*, J. Knot Theory Ramifications **4**(1995) 1–12

[13] M. A. Armstrong and E. C. Zeeman, *Transversality for piecewise linear manifolds*, Topology **6**(1967) 433–466

[14] V. I. Arnold, *A branched covering of $\mathbb{C}P^2 \to S^4$, hyperbolicity and projectivity topology*, Siberian Math. J. **29**(1988) 717–726

[15] E. Artin, *Theorie der Zopfe*, Hamburg Abh. **4**(1925) 47–72

[16] E. Artin, *Zur Isotopie zweidimensionalen Flächen im R_4*, Abh. Math. Sem. Univ. Hamburg **4**(1926) 174–177

[17] E. Artin, *Theory of braids*, Ann. of Math. **48**(1947) 101–126

[18] E. Artin, *Braids and permutations*, Ann. of Math. **48**(1947) 643–649

[19] W. A. Arvola, *Complexified real arrangements of hyperplanes*, Manuscripta math. **71**(1991) 295–306

[20] W. A. Arvola, *The fundamental group of the complement of an arrangement of complex hyperplanes*, Topology **31**(1992) 757–765

[21] K. Asano, *A note on surfaces in 4-spheres*, Math Sem. Notes Kobe Univ. **4**(1976) 195–198

[22] K. Asano, *The embedding of non-orientable surfaces in 4-space*, preprint

[23] K. Asano, Y. Marumoto and T. Yanagawa, *Ribbon knots and ribbon disks*, Osaka J. Math. **18**(1981) 161–174

[24] K. Asano and K. Yoshikawa, *On polynomial invariants of fibred 2-knots*, Pacific J. Math. **97**(1981) 267–269

[25] N. Askitas, *Embedding of 2-spheres in 4-manifolds*, Manuscripta math. **90**(1996) 137–138

[26] M. F. Atiyah, *The geometry and physics of knots*, Cambridge Univ. Press (1990)

[27] M. F. Atiyah, *Representations of braid groups*, in "Geometry of low-dimensional manifolds: 2", London Math. Soc. Lect. Note Series 151, Cambridge Univ. Press (1991) 115–122

[28] M. F. Atiyah, N.J. Hitchin and I.M. Singer, *Self-duality in four-dimensional Riemannian geometry*, Proc. R. Soc. Lond. A **362**(1978) 425–461

[29] M. F. Atiyah and I.M. Singer, *The index of elliptic operators, III*, Ann. of Math. **87**(1968) 546–604

[30] T. F. Banchoff, *Triple points and singularities of projections of smoothly immersed surfaces*, Proc. Amer. Math. Soc. **46**(1974) 402–406

[31] T. F. Banchoff, *Triple points and surgery of immersed surfaces*, Proc. Amer. Math. Soc. **46**(1974) 407–413

[32] D. Bar-Natan, *On the Vassiliev knot invariants*, Topology **34**(1995) 423–472

[33] D. Bar-Natan, S. Garoufalidis, *On the Melvin-Morton-Rozansky conjecture*, Invent. Math. **125**(1996) 103–133

[34] D. Bennequin, *Entrelacements et équations de Pfaff*, Astérisque **107–108**(1983) 87–161

[35] D. Bernardete, Z. Nitecki and M. Gutierrez, *Braids and the Nielsen-Thurston classification*, J. Knot. Theory Ramifications **4**(1995) 549–618

[36] I. Berstein and A. L. Edmonds, *On the construction of branched coverings of low-dimensional manifolds*, Trans. Amer. Math. Soc. **247**(1979) 87–124

[37] I. Berstein and A. L. Edmonds, *On the classification of generic branched coverings of surfaces*, Illinois J. Math. Soc. **28**(1984) 64–82

[38] J. S. Birman, *On braid groups*, Commun. Pure Appl. Math. **22**(1969) 41–72

[39] J. S. Birman, *Mapping class groups*, Commun. Pure Appl. Math. **22**(1969) 213–238

[40] J. S. Birman, *Non-conjugate braids can define isotopic knots*, Commun. Pure Appl. Math. **22**(1969) 239–242

[41] J. S. Birman, *Plat presentations for link groups*, Commun. Pure Appl. Math. **26**(1973) 673–678

[42] J. S. Birman, *Braids, links, and mapping class groups*, Ann. Math. Studies 82 (1974) Princeton Univ. Press, Princeton, N.J.

[43] J. S. Birman, *On the Jones polynomial of closed 3-braids*, Invent. Math. **81**(1985) 287–294

[44] J. S. Birman, *Mapping class groups of surfaces*, in "Braids" (Santa Cruz, CA, 1986), Contemp. Math. 78 (1988) 13–43

[45] J. S. Birman, *Recent developments in braid and link theory*, Math.Intelligencer **13**(1991) 52–60

[46] J. S. Birman, *New points of view in knot theory*, Bull. Amer. Math. Soc. **28**(1993) 253–287

[47] J. S. Birman, *Studying links via closed braids*, Lecture Notes on the Ninth KAIST Mathematical Workshop **1**(1994) 1–67

[48] J. S. Birman, T. Kanenobu, *Jones' braid-plat formula and a new surgery triple*, Proc. Amer. Math. Soc. **102**(1988) 687–695

[49] J. S. Birman and X.-S. Lin, *Knot polynomials and Vassiliev's invariant*, Invent. Math. **11**(1993) 225–270

[50] J. S. Birman and W. Menasco, *Studying links via closed braids IV: Composite links and split links*, Invent. Math. **102**(1990) 115–139

[51] J. S. Birman and W. Menasco, *Studying links via closed braids II: On a theorem of Bennequin*, Topology Appl. **40**(1991) 71–82

[52] J. S. Birman and W. Menasco, *Studying links via closed braids I: A finiteness theorem*, Pacific J. Math. **154**(1992) 17–36

[53] J. S. Birman and W. Menasco, *Studying links via closed braids V: The unlink*, Trans. Amer. Math. Soc. **329**(1992) 585–606

[54] J. S. Birman and W. Menasco, *Studying links via closed braids VI: A non-finiteness theorem*, Pacific J. Math. **156**(1992) 265–285

[55] J. S. Birman and W. Menasco, *Studying links via closed braids III: Classifying links which are closed 3-braids*, Pacific J. Math. **161**(1993) 25–113

[56] S. Bleiler and M. Scharlemann, *A projective plane in R^4 with three critical points is standard. Strongly invertible knots have property P*, Topology **27**(1988) 519–540

[57] F. Bohnenblust, *The algebraical braid group*, Ann. of Math. **48**(1947) 127–136

[58] M. Boileau, B. Zimmermann, *The π-orbifold group of a link*, Math. Z. **200**(1989) 187–208

[59] J. Boyle, *Classifying 1-handles attached to knotted surfaces*, Trans. Amer. Math. Soc. **306**(1988) 475–487

[60] J. Boyle, *The turned torus knot in S^4*, J. Knot Theory Ramifications **2**(1993) 239–249

[61] E. Brieskorn, *Automorphic sets and braids and singularities*, Contemp. Math. **78**(1988) 45–115

[62] E. Brieskorn and K. Saito, *Artin Gruppen und Coxeter Gruppen*, Invent. Math. **17**(1972) 245–271

[63] E. H. Brown Jr., *Generalizations of the Kervaire invariant*, Ann. of Math. **95**(1972) 368–383

[64] M. Brown, *A proof of the generalized Schönflies theorem*, Bull. Amer.Math. Soc. **66**(1960) 74–76

[65] A. M. Brunner, E. J. Mayland Jr., and J. Simon, *Knot groups in S^4 with nontrivial homology*, Pacific J. Math. **103**(1982) 315–324

[66] R. Brussee, *Some remarks on the Kronheimer-Mrowka classes of algebraic surfaces*, J. Differential Geom. **41**(1995) 269–275

[67] S. Bullett, *Braid orientations and Stiefel-Whitney classes*, Quart. J. Math. Oxford **32**(1981) 267–285

[68] G. Burde and H. Zieschang, *Knots*, Studies in Math. 5, Walter de Gruyter (1985)

[69] S. E. Cappell, J. L. Shaneson, *There exist inequivalent knots with the same complement*, Ann. of Math. **103**(1976) 349–353

[70] S. E. Cappell, J. L. Shaneson, *An introduction to embeddings, immersions and singularities in codimension two*, Proc. Sympos. Pure Math. **32**(1978) 129–149

[71] J. S. Carter, D. Jelsovsky, S. Kamada, L. Langford and M. Saito, *State-sum invariants of knotted curves and surfaces from quandle cohomology*, Electron. Res. Announc. Amer. Math. Soc. **5**(1999) 146–156

[72] J. S. Carter, D. Jelsovsky, S. Kamada, L. Langford and M. Saito, *Quandle cohomology and state-sum invariants of knotted curves and surfaces*, preprint (math.GT/9903135)(1998)

[73] J. S. Carter, D. Jelsovsky, S. Kamada and M. Saito, *Computations of quandle cocycle invariants of knotted curves and surfaces*, Adv. in Math. **157**(2001) 36–94

[74] J. S. Carter, D. Jelsovsky, S. Kamada and M. Saito, *Quandle homology groups, their Betti numbers, and virtual knots*, J. Pure Appl. Algebra **157**(2001) 135–155

[75] J. S. Carter, D. Jelsovsky, S. Kamada and M. Saito, *Shifting homomorphisms in quandle cohomology and skeins of cocycle knot invariants*, J. Knot Theory Ramifications, to appear

[76] J. S. Carter, S. Kamada and M. Saito, *Alexander numbering of knotted surface diagrams*, Proc. Amer. Math. Soc. **128**(2000) 3761–3771

[77] J. S. Carter, S. Kamada and M. Saito, *Geometric interpretations of quandle homology*, J. Knot Theory Ramifications **10**(2001) 345–386

[78] J. S. Carter, S. Kamada, M. Saito and S. Satoh, *A theorem of Sanderson on link bordisms in dimension* 4, Algebraic and Geometric Topology **1**(2001) 299–310

[79] J. S. Carter, J. H. Rieger and M. Saito, *A combinatorial description of knotted surfaces and their isotopies*, Adv. in Math. **127**(1997) 1–51

[80] J. S. Carter and M. Saito, *A Seifert algorithm for knotted surfaces*, preprint (previous version to 1997) (1991)

[81] J. S. Carter and M. Saito, *Canceling branch points on the projections of surfaces in 4-space*, Proc. Amer. Math. Soc. **116**(1992) 229–237

[82] J. S. Carter and M. Saito, *Reidemeister moves for surface isotopies and their interpretations as moves to movies*, J. Knot Theory Ramifications **2**(1993) 251–284

[83] J. S. Carter and M. Saito, *A diagrammatic theory of knotted surfaces*, in "Quantum Topology", eds. R. A. Baadhio and L. H. Kauffman, Series on knots and everything, vol.3, World Scientific (1993) 91–115

[84] J. S. Carter and M. Saito, *Knotted surfaces, braid movies and beyond*, in "Knots and Quantum Gravity", Oxford Science Publishing (1994) 191–229

[85] J. S. Carter and M. Saito, *Knot diagrams and braid theories in dimension 4*, in "Real and complex singularities", Papers from 3rd international workshop held in São Carlos, 1994, Pitman Res. Notes Math. Ser. 333 (1995) 112–147

[86] J. S. Carter and M. Saito, *Braids and movies*, J. Knot Theory Ramifications **5**(1996) 589–608

[87] J. S. Carter and M. Saito, *A Seifert algorithm for knotted surfaces*, Topology **36**(1997) 179–201

[88] J. S. Carter and M. Saito, *Normal Euler classes of knotted surfaces and triple points on projections*, Proc. Amer. Math. Soc. **125**(1997) 617–623

[89] J. S. Carter and M. Saito, *Knotted surfaces and their diagrams*, Surveys and monographs, Amer. Math. Soc. **55**(1998)

[90] A. J. Casson, *Three lectures on new-infinite constructions in 4-manifolds*, in "A la Recherche de la Topologie Perdue", Progress in Math., 62, Birkhäuser, Boston (1986) 201–214

[91] A. J. Casson, C. McA. Gordon, *On slice knots in dimension three*, in "Algebraic and Geometric Topology" (Stanford, 1976), Proc. Sympos. Pure Math. 32-II, Amer. Math. Soc. (1978) 39–53

[92] A. J. Casson, C. McA. Gordon, *Cobordism of classical knots*, in "A la recherche de la topologie perdue", Progress in Math., 62, Birkhäuser, Boston (1986) 181–199

[93] L. Cervantes and R. A. Fenn, *Boundary links are homotopy trivial*, Quart. J. Math. Oxford **39**(1988) 151–158

[94] T. A. Chapman, *Locally homotopically unknotted embeddings of manifolds in codimension two are locally flat*, Topology **18**(1979) 339–348

[95] R. Charney, *Artin groups of finite type are biautomatic*, Math. Ann. **292**(1992) 671–683

[96] W.-L. Chow, *On the algebraic braid group*, Ann. of Math. **49**(1948) 654–658

[97] T. D. Cochran, *Ribbon knots in S^4*, J. London Math. Soc. **28**(1983) 563–576

[98] T. D. Cochran, *Slice links in S^4*, Trans. Amer. Math. Soc. **285**(1984) 389–400

[99] T. D. Cochran, *On an invariant of link cobordism in dimension four*, Topology Appl. **18**(1984) 97–108

[100] T. D. Cochran, *Geometric invariants of link cobordism*, Comment. Math. Helv. **60**(1985) 291–311

[101] T. D. Cochran, *Link concordance invariants and homotopy theory*, Invent. Math. **90**(1987) 635–645

[102] T. D. Cochran, *Localization and finiteness in link concordance*, Topology Appl. **32**(1989) 121–133

[103] T. D. Cochran, *Derivatives of links: Milnor's concordance invariants and Massey's products*, Memoirs Amer. Math. Soc., 84, No. 427 (1990)

[104] T. D. Cochran, *Links with trivial Alexander's module but nonvanishing Massey products*, Topology **29**(1990) 189–204

[105] T. D. Cochran, *k-cobordism for links in S^3*, Trans. Amer. Math. Soc. **327**(1991) 641–654

[106] T. D. Cochran, J. P. Levine, *Homology boundary links and the Andrews-Curtis conjecture*, Topology **30**(1991) 231–239

[107] T. D. Cochran, W. B. R. Lickorish, *Unknotting information from 4-manifolds*, Trans. Amer. Math. Soc. **297**(1986) 125–142

[108] T. D. Cochran, K. E. Orr, *Not all links are concordant to boundary links*, Bull. Amer. Math. Soc. **23**(1990) 99–106

[109] T. D. Cochran, K. E. Orr, *Not all links are concordant to boundary links*, Ann. of Math. **138**(1993) 519–554

[110] T. D. Cochran, K. E. Orr, *Homology boundary links and Blanchfield forms: concordance classification and new tangle-theoretic constructions*, Topology **33**(1994) 397–427

[111] F. R. Cohen, *Artin's braid groups and classical homotopy theory*, Contemp. Math. **44**(1985) 207–220

[112] F. R. Cohen, *Artin's braid groups, classical homotopy theory, and sundry other curiosities*, Contemp. Math. **78**(1988) 167–206

[113] J. H. Conway, *An enumeration of knots and links and some of their related properties*, in "Computational Problems in Abstract Algebra" (Oxford, 1967), Pergamon Press (1970) 329–358

[114] D. Cooper, D. D. Long , *Derivative varieties and the pure braid group*, Amer. J. Math. **115**(1993) 137–160

[115] R. Craggs, *On the algebra of handle operations in 4-manifolds*, Topology Appl. **30**(1988) 237–252

[116] R. Craggs, *Freely reducing group readings for 2-complexes in 4-manifolds*, Topology **28**(1989) 247–271

[117] P. R. Cromwell, *Positive braids are visually prime*, Proc. London Math. Soc. **67**(1993) 384–424

[118] R. H. Crowell, *The group G'/G'' of a knot group G*, Duke Math. J. **30**(1963) 349–354

[119] R. H. Crowell, *On the annihilator of a knot module*, Proc. Amer. Math. Soc. **15**(1964) 696–700

[120] R. H. Crowell, *Torsion in link modules*, J. Math. Mech. **14**(1965) 289–298

[121] R. H. Crowell, *The derived module of a homomorphism*, Adv. Math. **6**(1971) 210–238

[122] R. H. Crowell and R. H. Fox, *Introduction to Knot Theory*, Graduate Texts in Mathematics, 57, Springer-Verlag, New York-Heidelberg (1977)

[123] M. Culler, C.M. Gordon, J. Luecke and P.B. Shalen, *Dehn surgery on knots*, Ann. of Math. [Correction (1988), ibid., 127:663] **125**(1987) 237–300

[124] D. Dahm, *A generalization of braid theory*, Princeton Ph.D. thesis (1962)

[125] E. Date, M. Jimbo, K. Miki, T. Miwa, *Braid group representations arising from the generalized chiral Potts models*, Pacific J. Math. **154**(1992) 37–66

[126] T. Deguchi, *Braid group representations and link polynomials derived from generalized $SU(n)$ vertex models*, J. Phys. Soc. Japan **58**(1989) 3441–3444

[127] T. Deguchi, Y. Akutsu, *A general formula for colored Z_n graded braid matrices and the fusion braid matrices*, J. Phys. Soc. Japan. **60**(1991) 2559–2570

[128] P. Dehoronoy, *From large cardinals to braids via distributive algebra*, J. Knot Theory Ramifications **4**(1995) 33–79

[129] R. Dijkgraaf, E. Witten, *Topological gauge theories and group cohomology*, Commun. Math. Phys. **129**(1990) 393–429

[130] S. K. Donaldson, *An application of gauge theory to four-dimensional topology*, J. Diff. Geom. **18**(1983) 279–315

[131] S. K. Donaldson, *Polynomial invariants for smooth four-manifolds*, Topology **29**(1990) 257–315

[132] M. J. Dunwoody, R. A. Fenn, *On the finiteness of higher knot sums*, Topology **26**(1987) 337–343

[133] B. Eckmann, *Aspherical manifolds and higher-dimensional knots*, Comment. Math. Helv. **51**(1976) 93–98

[134] A. L. Edmonds, R. S. Kulkarni, R. E. Stong, *Realizability of branched coverings of surfaces*, Trans. Amer. Math. Soc. **282**(1984) 773–790

[135] Y. Eliashberg, *Topology of 2-knots in R^4 and symplectic geometry*, in "The Floer Memorial Volume", Progress in Math. 133, Birkhäuser (1995) 335–353

[136] E. A. Elrifai, H. R. Morton, *Algorithms for positive braids*, Quart. J. Math. Oxford (2) **45**(1994) 479–497

[137] M. Eudave-Muñoz, *Primeness and sums of tangles*, Trans. Amer. Math. Soc. **306**(1988) 773–790

[138] M. Eudave-Muñoz, *Prime knots obtained by band sums*, Pacific J. Math. **139**(1989) 53–57

[139] M. Eudave-Muñoz, *Band sums of links which yield composite links. The cabling conjecture for strongly invertible knots*, Trans. Amer. Math. Soc. **330**(1992) 463–501

[140] E. Fadell, L. Neuwirth, *Configuration spaces*, Math. Scand. **10**(1962) 111–118

[141] E. Fadell, J. Van Buskirk, *The braid groups of E^2 and S^2*, Duke Math. J. **29**(1962) 243–258

[142] M. S. Farber, *Linking coefficients and two-dimensional knots*, Soviet. Math. Doklady **16**(1975) 647–650

[143] M. S. Farber, *Duality in an infinite cyclic covering and even-dimensional knots*, Math. USSR-Izv. **11**(1977) 749–781

[144] M. S. Farber, *Classification of some knots of codimension two*, Soviet Math. Dokl. **19**(1978) 555–558

[145] M. S. Farber, *Isotopy types of knots of codimension two*, Trans. Amer. Math. Soc. **261**(1980) 185–209

[146] M. S. Farber, *Stable classification of knots*, Soviet. Math. Doklady **23**(1981) 685–688

[147] M. S. Farber, *An algebraic classification of some even-dimensional spherical knots. I*, Trans. Amer. Math. Soc. **281**(1984) 507–527

[148] M. S. Farber, *An algebraic classification of some even-dimensional spherical knots. II*, Trans. Amer. Math. Soc. **281**(1984) 529–570

[149] M. S. Farber, *Hermitian forms on link modules*, Comment. Math. Helv. **66**(1991) 189–236

[150] T. Feng, X.-S. Lin and Z. Wang, *Burau representation and random walks on string links*, Pacific J. Math. **182**(1998) 289–302

[151] R. Fenn, *On Dehn's lemma in 4 dimensions*, Bull. London Math. Soc. **3**(1971) 79–89

[152] R. Fenn, G. T. Jin, R. Rimányi, *Laces*, in "The 3rd Korea-Japan School of knots and links" (Taejon, 1994), Proc. Applied Math. Workshop, 4, KAIST Korea (1994) 21–31

[153] R. Fenn, R. Rimányi, C. Rourke, *The braid-permutation group*, Topology **36**(1997) 123–135

[154] R. Fenn and D. Rolfsen, *Spheres may link homotopically in 4-space*, J. London Math. Soc. **34**(1986) 177–184

[155] R. Fenn, C. Rourke, *On Kirby's caluculus of links*, Topology **18**(1979) 1–15

[156] R. Fenn, C. Rourke, *Racks and links in codimension two*, J. Knot Theory Ramifications **1**(1992) 343–406

[157] R. Fenn, C. Rourke, B. Sanderson, *Trunks and classifying spaces*, Appl. Categ. Structures **3**(1995) 321–356

[158] R. Fenn, C. Rourke, B. Sanderson, *James bundles and applications*, preprint

[159] T. Fiedler, *A small state sum for knots*, Topology **32**(1993) 281–294

[160] S. M. Finashin, M. Kreck, O. Ya Viro, *Exotic knottings of surfaces in the 4-sphere*, Bull. Amer. Math. Soc. **17**(1987) 287–290

[161] S. M. Finashin, M. Kreck, O. Ya Viro, *Non-diffeomorphic but homeomorphic knottings of surfaces in the 4-sphere*, in "Topology and geometry-Rohlin seminar", Lecture Notes in Math., 1346, Springer Verlag (1988) 157–198

[162] R. H. Fox, *Free differential calculus. I. Derivation in the free group ring*, Ann. of Math. **57**(1953) 547–560

[163] R. H. Fox, *Free differential calculus. II. The isomorphism problem of groups*, Ann. of Math. **59**(1954) 196–210

[164] R. H. Fox, *Free differential calculus. III. Subgroups*, Ann. of Math. **64**(1956) 407–419

[165] R. H. Fox, *Congruence classes of knots*, Osaka Math. J. **10**(1958) 37–41

[166] R. H. Fox, *The homology characters of the cyclic coverings of the knots of genus one*, Ann. of Math. **71**(1960) 187–196

[167] R. H. Fox, *Free differential calculus, V. The Alexander matrices re-examined*, Ann. of Math. **71**(1960) 408–422

[168] R. H. Fox, *A quick trip through knot theory*, in "Topology of 3-Manifolds and Related Topics" (Georgia, 1961), Prentice-Hall (1962) 120–167

[169] R. H. Fox, *Some problems in knot theory*, in "Topology of 3-manifolds and related topics" (Georgia, 1961), Prentice-Hall (1962) 168–176

[170] R. H. Fox, *Rolling*, Bull. Amer. Math. Soc. **72**(1966) 162–164

[171] R. H. Fox, *Characterization of slices and ribbons*, Osaka J. Math. **10**(1973) 69–76

[172] R.H. Fox and J. W. Milnor, *Singularities of 2-spheres in 4-space and equivalences of knots*, preprint (1957)

[173] R.H. Fox and J. W. Milnor, *Singularities of 2-spheres in 4-space and cobordism of knots*, Osaka J. Math. **3**(1966) 257–267

[174] R. H. Fox and L. Neuwirth, *The braid groups*, Math. Scand. **10**(1962) 119–126

[175] J. Franks, R. H. Williams, *Braids and the Jones polynomial*, Trans. Amer. Math. Soc. **303**(1987) 97–108

[176] M. H. Freedman, *The disk theorem for four-dimensional manifolds*, in "Proc. Internat. Congr. Math." (Warsaw, 1983) (1984) 647–663

[177] M. H. Freedman, F. Quinn, *The topology of 4-manifolds*, Princeton Univ. Press(1990)

[178] C. D. Frohman, A. Nicas, *The Alexander polynomial via topological quantum field theory*, in "Differential geometry, global analysis, and topology" (Halifax, NS, 1990), CMS Conf. Proc., 12, Amer. Math. Soc., Providence, RI(1991) 27–40

[179] H. Fujii, *Geometric indices and Alexander polynomial of a knot*, Proc. Amer. Math. Soc. **124**(1996) 2923–2933

[180] S. Fukuhara, *A note on equivalence classes of plats*, Kodai Math. J. **17**(1994) 505–510

[181] W. Fulton, R. MacPherson, *A compactification of configuration spaces*, Ann. of Math. (2) **139**(1994) 183–225

[182] D. Gabai, *Genus is superadditive under band connected sum*, Topology **26**(1987) 209–210

[183] H. Z. Gao, *Normal Euler numbers of embeddings of nonorientable surfaces into four-manifolds*, J. Systems Sci. Math. Sci. **9**(1989) 244–250

[184] H. Z. Gao, *On normal Euler numbers of embedding surfaces into 4-manifolds*, J. Systems Sci. Math. Sci. **3**(1990) 166–171

[185] F. A. Garside, *The braid group and other groups*, Quart. J. Math. Oxford (2) **20**(1969) 235–254

[186] C. H. Giffen, *Link concordance implies link homotopy*, Math. Scand. **45**(1979) 243–254

[187] C. A. Giller, *Towards a classical knot theory for surfaces in R^4*, Illinois J. Math. **26**(1982) 591–631

[188] R. Gillette, J. Van Buskirk, *The word problem and consequences for the braid groups and mapping class groups of the 2-sphere*, Trans. Amer. Math. Soc. **131**(1968) 277–296

[189] P. M. Gilmer, *Configurations of surfaces in 4-manifolds*, Trans. Amer. Math. Soc. **264**(1981) 353–380

[190] P. M. Gilmer, *Topological proof of the G-signature theorem for G finite*, Pacific J. Math. **97**(1981) 105–114

[191] P. M. Gilmer, *On the slice genus of knots*, Invent. Math. **66**(1982) 191–197

[192] P. M. Gilmer, *Slice knots in S^3*, Quart. J. Math. Oxford(2) **34**(1983) 305–322

[193] P. M. Gilmer, *Ribbon concordance and a partial order on S-equivalence classes*, Topology Appl. **18**(1984) 313–324

[194] P. M. Gilmer, *Real algebraic curves and link cobordism*, Pacific J. Math. **153**(1992) 31–69

[195] P. M. Gilmer, *Classical knot and link concordance*, Comment. Math. Helv. **68**(1993) 1–19

[196] P. M. Gilmer, *Signatures of singular branched covers*, Math. Ann. **295**(1993) 643–659

[197] P. M. Gilmer, C. Livingston, *On embedding 3-manifolds in 4-space*, Topology **22**(1983) 241–252

[198] P. M. Gilmer, C. Livingston, *The Casson-Gordon invariant and link concordance*, Topology **31**(1992) 475–492

[199] H. Gluck, *The embedding of two-spheres in the four-sphere*, Trans. Amer. Math. Soc. **104**(1962) 308-333

[200] D. L. Goldsmith, *Homotopy of braids– In answer to a question of E. Artin*, Lecture Notes in Math., Vol. 375, Springer, Berlin (1974) 91–96

[201] D. L. Goldsmith, *The theory of motion groups*, Michigan Math. J. **28**(1981) 3–17

[202] D. L. Goldsmith, *Motion of links in the 3-sphere*, Math. Scand. **50**(1982) 167–205

[203] D. L. Goldsmith, L. H. Kauffman, *Twist spinning revisited*, Trans. Amer. Math. Soc. **239**(1978) 229–251

[204] R. E. Gompf, *Smooth concordance of topologically slice knots*, Topology **25**(1986) 353–373

[205] F. González-Acuña, *Homomorphs of knot groups*, Ann. of Math. **102**(1975) 373–377

[206] F. González-Acuña, *A characterization of 2-knot groups*, Rev. Mat. Iberoamericana **10**(1994) 221–228

[207] F. González-Acuña, J. M. Montesinos, *Non-amphicheiral codimension 2 knots*, Canad. J. Math. **32**(1980) 185–194

[208] C. McA. Gordon, *Twist-spun torus knots*, Proc. Amer. Math. Soc. **32**(1972) 319–322

[209] C. McA. Gordon, *Some higher-dimensional knots with the same homotopy groups*, Quart. J. Math. Oxford **24**(1973) 411–422

[210] C. McA. Gordon, *On the higher-dimensional Smith conjecture*, Proc. London Math. Soc. **29**(1974) 98–110

[211] C. McA. Gordon, *Knots in the 4-sphere*, Comment. Math. Helv. **39**(1976) 585–596

[212] C. McA. Gordon, *A note on spun knots*, Proc. Amer. Math. Soc. **58**(1976) 361–362

[213] C. McA. Gordon, *Uncountably many stably trivial strings in codimension two*, Quart. J. Math. Oxford. **28**(1977) 369–379

[214] C. McA. Gordon, *Some aspects of classical knot theory*, in "Knot theory", Lect. Notes in Math., 685, Springer Verlag (1978) 1–60

[215] C. McA. Gordon, *Homology of groups of surfaces in the 4-sphere*, Math. Proc. Cambridge Phil. Soc. **89**(1981) 113–117

[216] C. McA. Gordon, *Ribbon concordance of knots in the 3-sphere*, Math. Ann. **257**(1981) 157–170

[217] C. McA. Gordon, *On the G-signature theorem in dimension four*, in "A la Recherche de la Topologie Perdue" Progress in Math., 62, Birkhäuser, Boston (1986) 159–180

[218] C. McA. Gordon, R. A. Litherland, *On the signature of a link*, Invent. Math. **47**(1978) 53–69

[219] C. McA. Gordon, R. A. Litherland, K. Murasugi, *Signatures of covering links*, Canad. J. Math. **33**(1981) 381–394

[220] C. McA. Gordon, J. Luecke, *Knots are determined by their complements*, Bull. Amer. Math. Soc. (N.S.) **20**(1989) 83–87

[221] C. McA. Gordon, J. Luecke, *Knots are determined by their complements*, J. Amer. Math. Soc. **2**(1989) 371–415

[222] M. A. Gutiérrez, *Homology of knot groups: I. Groups with deficiency one*, Bol. Soc. Mat. Mex. **16**(1971) 58–63

[223] M. A. Gutiérrez, *An exact sequence calculation for the second homotopy of a knot*, Proc. Amer. Math. Soc. **32**(1972) 571–577

[224] M. A. Gutiérrez, *On knot modules*, Invent. math. **17**(1972) 329–335

[225] M. A. Gutiérrez, *Secondary invariants for links*, Rev. Columbina Mat. **6**(1972) 106–115

[226] M. A. Gutiérrez, *Boundary links and an unlinking theorem*, Trans. Amer. Math. Soc. **171**(1972) 491–499

[227] M. A. Gutiérrez, *Unlinking up to cobordism*, Bull. Amer. Math. Soc. **79**(1973) 1299–1302

[228] M. A. Gutiérrez, *An exact sequence calculation for the second homotopy of a knot. II*, Proc. Amer. Math. Soc. **40**(1973) 327–330

[229] M. A. Gutiérrez, *On the Seifert manifold of a 2-knot*, Trans. Amer. Math. Soc. **240**(1978) 287–294

[230] M. A. Gutiérrez, *Homology of knot groups. III. Knots in S^4*, Proc. London Math. Soc. **39**(1979) 469–487

[231] N. Habegger, X. S. Lin, *On link concordance and Milnor's invariants*, Bull. London Math. Soc. **30** (1998) 419–428

[232] N. Habegger, X. S. Lin, *The classification of links up to link-homotopy*, J. Amer. Math. Soc. **3**(1990) 389–419

[233] K. Habiro, T. Kanenobu and A. Shima, *Finite type invariants of ribbon 2-knots*, in "Low-dimensional topology" (Funchal, 1998), Contemp. Math., 233, Amer. Math. Soc. (1999) 187–196

[234] R. Hartley, *On the classification of three-braid links*, Abh. Math. Sem. Univ. Hamburg **50**(1980) 108–117

[235] R. Hartley, *Knots and involutions*, Math. Z. **171**(1980) 175–185

[236] R. Hartley, *Invertible amphicheiral knots*, Math. Ann. **252**(1980) 103–109

[237] R. Hartley, *Identifying non-invertible knots*, Topology **22**(1983) 137–145

[238] R. Hartley, A. Kawauchi, *Polynomials of amphicheiral knots*, Math. Ann. **243**(1979) 63–70

[239] R. Hartley, K. Murasugi, *Covering linkage invariants*, Canad. J. Math. **29**(1977) 1312–1339

[240] Y. Hashizume, *On the uniqueness of the decomposition of a link*, Osaka Math. J. **10**(1958) 283–300

[241] J. Hempel, *3-manifolds*, Ann. of Math. Studies, 86, Princeton Univ. Press (1976)

[242] J. A. Hillman, *A non-homology boundary link with zero Alexander polynomial*, Bull. Austr. Math. Soc. **16**(1977) 229-236

[243] J. A. Hillman, *High dimensional knot groups which are not two-knot groups*, Bull. Austr. Math. Soc. **16**(1977) 449-462

[244] J. A. Hillman, *Longitudes of a link and principality of an Alexander ideal*, Proc. Amer. Math. Soc. **72**(1978) 370-374

[245] J. A. Hillman, *Alexander ideals and Chen groups*, Bull. London Math. Soc. **10**(1978) 105-110

[246] J. A. Hillman, *Orientability, asphericity and two-knots*, Houston J. Math. **6**(1980) 67-76

[247] J. A. Hillman, *Spanning links by non-orientable surfaces*, Quart. J. Math. Oxford **31**(1980) 169-179

[248] J. A. Hillman, *Trivializing ribbon links by Kirby moves*, Bull. Austr. Math. Soc. **21**(1980) 21-28

[249] J. A. Hillman, *Alexander ideals of links*, Lect. Notes in Math., 895, Springer Verlag (1981)

[250] J. A. Hillman, *The Torres conditions are insufficient*, Math. Proc. Cambridge Phil. Soc. **89**(1981) 19-22

[251] J. A. Hillman, *Finite knot modules and the factorization of certain simple knots*, Math. Ann. **257**(1981) 261-274

[252] J. A. Hillman, *A link with Alexander module free which is not a homology boundary link*, J. Pure Appl. Algebra **20**(1981) 1-5

[253] J. A. Hillman, *Aspherical four-manifolds and the centres of two-knot groups*, Comment. Math. Helv. **56**(1981) 465-473

[254] J. A. Hillman, *Alexander polynomials, annihilator ideals, and the Steinitz-Fox-Smythe invariant*, Proc. London Mth. Soc. **45**(1982) 31-48

[255] J. A. Hillman, *Aspherical four-manifolds and the centres of two-knot groups*, Comment. Math. Helv. **58**(1983) 166

[256] J. A. Hillman, *Factorization of Kojima knots and hyperbolic concordance of Levine pairings*, Houston Math. J. **10**(1984) 187-194

[257] J. A. Hillman, *Seifert fibre spaces and Poincaré duality groups*, Math. Z. **190**(1985) 365-369

[258] J. A. Hillman, *Knot modules and the elementary divisor theorem*, J. Pure Appl. Algebra **40**(1986) 115-124

[259] J. A. Hillman, *Finite simple even dimensional knots*, J. London Math. Soc. **34**(1986) 369-374

[260] J. A. Hillman, *On metabelian two-knot groups*, Proc. Amer. Math. Soc. **96**(1986) 372-375

[261] J. A. Hillman, *Two-knot groups with torsion free abelian normal subgroups of rank two*, Comment. Math. Helv. **63**(1988) 664-671

[262] J. A. Hillman, *2-knots and their groups*, Austr. Math. Soc. Lect. Ser., 5, Cambridge Univ. Press (1989)

[263] J. A. Hillman, *A homotopy fibration theorem in dimension four*, Topology Appl. **33**(1989) 151-161

[264] J. A. Hillman, *The algebraic characterization of the exteriors of certain 2-knots*, Invent. Math. **97**(1989) 195-207

[265] J. A. Hillman, *A remark on branched cyclic covers*, J. Pure Appl. Algebra **87**(1993) 237-240

[266] J. A. Hillman, *On 3-dimensional Poincaré duality complexes and 2-knot groups*, Math. Proc. Cambridge Philos. Soc. **114**(1993) 215-218

[267] J. A. Hillman, *Free products and 4-dimensional connected sums*, Bull. London Math. Soc. **27**(1995) 387–391

[268] J. A. Hillman, *Embedding homology equivalent 3-manifolds in 4-space*, Math. Z. **223**(1996) 473–481

[269] J. A. Hillman, *Optimal presentations for solvable 2-knot groups*, Bull. Austral. Math. Soc. **57**(1998) 129–133

[270] J. A. Hillman and A. Kawauchi, *Unknotting orientable surfaces in the 4-sphere*, J. Knot Theory Ramifications **4**(1995) 213–224

[271] S. Hirose, *On diffeomorphisms over T^2-knots*, Proc. Amer. Math. Soc. **119**(1993) 1009–1018

[272] M. W. Hirsch, *Immersions of manifolds*, Trans. Amer. Math. Soc. **93**(1959) 242–276

[273] L. R. Hitt, D. S. Silver, *Ribbon knot families via Stallings' twists*, J. Austral. Math. Soc. Ser. A **50**(1991) 356–372

[274] M. Horibe, *On trivial 2-spheres in 4-space* (in Japanese), Master Thesis, Kobe Univ., Kobe (1974)

[275] F. Hosokawa, *A concept of cobordism between links*, Ann. of Math. **86**(1967) 362-373

[276] F. Hosokawa, *On trivial 2-spheres in 4-space*, Quart. J. Math. Oxford (2) **19** (1968) 249–256

[277] F. Hosokawa and A. Kawauchi, *Proposals for unknotted surfaces in four-spaces*, Osaka J. Math. **16** (1979) 233–248

[278] F. Hosokawa, A. Kawauchi, Y. Nakanishi, M. Sakuma, *Note on critical points of surfaces in 4-spaces*, Kobe J. Math. **1**(1984) 151-152

[279] F. Hosokawa, T. Maeda, S. Suzuki, *Numerical invariants of surfaces in 4-space*, Math. Sem. Notes Kobe Univ. **7**(1979) 409–420

[280] F. Hosokawa, S. Suzuki, *Linking 2-spheres in the 4-sphere*, Kobe J. Math. **4**(1987) 193–208

[281] F. Hosokawa, S. Suzuki, *On singular cut-and-pastes in the 3-space with applications to link theory*, Rev. Mat. Univ. Complut. Madrid **8**(1995) 155–168

[282] F. Hosokawa, T. Yanagawa, *Is every slice knot a ribbon knot?*, Osaka J. Math. **2**(1965) 373–384

[283] J. Hoste, Y. Nakanishi, K. Taniyama, *Unknotting operations involving trivial tangles*, Osaka J. Math. **27**(1990) 555–566

[284] J. Hoste, J. H. Przytycki, *A survey of skein modules of 3-manifolds*, in "Knots 90" (Osaka, 1990), Walter de Gruyter (1992) 363–379

[285] J. Howie, *On the asphericity of ribbon disk complements*, Trans. Amer. Math. Soc. **289**(1985) 285-302

[286] J. F. P. Hudson, *Piecewise linear topology*, Benjamin, N.Y. (1969)

[287] J. F. P. Hudson, *On spanning surfaces of links*, Bull. Austral. Math. Soc. **48**(1993) 337-345

[288] N. V. Ivanov, *Permutation representations of braid groups of surfaces*, Math. USSR-Sb. **71**(1992) 309-318

[289] Z. Iwase, *Good torus fibrations with twin singular fibers*, Japan J. Math. **10**(1984) 321-352

[290] Z. Iwase, *Dehn-surgery along a torus T^2-knot*, Pacific J. Math. **133**(1988) 289-299

[291] Z. Iwase, *Dehn surgery along a torus T^2-knot II*, Japan. J. Math. **16**(1990) 171-196

[292] W. Jaco, *Lectures on three-manifold topology*, Conference Board of Math., 43, Amer. Math. Soc. (1980)

[293] F. Jaeger, *Circuit partitions and the homfly polynomial of closed braids*, Trans. Amer. Math. Soc. **323**(1991) 449-463

[294] B. Jiang, *A simple proof that the concordance group of algebraic slice knots is infinitely generated*, Proc. Amer. Math. J. **83**(1981) 181-192

[295] B. Jiang, *Fixed points and braids*, Invent. math. **75**(1984) 69-74

[296] G. T. Jin, *On Kojima's η-function of links*, in "Differential topology", Lecture Notes in Math., 1350, Springer Verlag (1988) 14–30

[297] G. T. Jin, *The Cochran sequences of semi-boundary links*, Pacific J. Math. **149**(1991) 293-302

[298] V. F. R. Jones, *A polynomial invariant for knots via von Neumann algebras*, Bull. Amer. Math. Soc. (N.S.) **12** (1985) 103–111

[299] V. F. R. Jones, *Hecke algebra representations of braid groups and link polynomials*, Ann. of Math. (2) **126** (1987) 335–388

[300] V. F. R. Jones, *Knots, braids and statistical mechanics*, in "Advances in differential geometry and topology" World Sci. Publ.(1990) 149–184

[301] V. F. R. Jones, *Knots, braids, statistical mechanics and von Neumann algebras*, in "New Zealand mathematics colloquium" (Dunedin, 1991), New Zealand J. Math., 21 (1992) 1–16

[302] A. Joyal, *Braided tensor categories*, Adv. in Math. **102**(1993) 20–78

[303] D. Joyce, *A classical invariant of knots, the knot quandle*, J. Pure Appl. Algebra **23** (1982) 37–65

[304] O. Kakimizu, *Finding disjoint incompressible spanning surfaces for a link*, Hiroshima Math. J. **22**(1992) 225-236

[305] O. Kakimizu, *Incompressible spanning surfaces and maximal fibred submanifolds for a knot*, Math. Z. **210**(1992) 207-223

[306] N. Kamada and S. Kamada, *Abstract link diagrams and virtual knots*, J. Knot Theory Ramifications **9** (2000) 93–106

[307] S. Kamada, *Non-orientable surfaces in 4-space*, Osaka J. Math. **26** (1989) 367–385

[308] S. Kamada, *Orientable surfaces in the 4-sphere associated with non-orientable knotted surfaces*, Math. Proc. Camb. Phil. Soc. **108** (1990) 299-306

[309] S. Kamada, *On doubled surfaces of non-orientable surfaces in the 4-sphere*, Kobe J. Math. **7** (1990) 19-24

[310] S. Kamada, *On deform-spun projective planes in 4-sphere obtained from peripheral inverting deformations*, Proc. KAIST Math. Workshop **5** (1990) 197-203

[311] S. Kamada, *Projective planes in 4-sphere obtained by deform-spinnings*, in "Knots 90", ed. A. Kawauchi (Osaka 1990), Walter de Gruyter (1992) 125–132

[312] S. Kamada, *Surfaces in R^4 of braid index three are ribbon*, J. Knot Theory Ramifications **1** (1992) 137–160

[313] S. Kamada, *2-dimensional braids and chart descriptions*, in "Topics in Knot Theory", Proceedings of the NATO Advanced Study Institute on Topics in Knot Theory held in Turkey (1992) 277–287

[314] S. Kamada, *Seifert circles for surface braids* (in Japanese), Suurikaisekikenkyusho Koukyuroku, (Seminar note at RIMS, Kyoto) **813** (1992) 144–154

[315] S. Kamada, *Generalized Alexander's and Markov's theorems in dimension four*, preprint (1992)

[316] S. Kamada, *A characterization of groups of closed orientable surfaces in 4-space*, Topology (1994) **33** 113–122

[317] S. Kamada, *Alexander's and Markov's theorems in dimension four*, Bull. Amer. Math. Soc. (1994) **31** 64–67

[318] S. Kamada, *Survey on 2-dimensional braids*, Proceedings of the 41st Topology Symposium, held in Ehime, Japan (1994) 162–178

[319] S. Kamada, *On 2-dimensional braids and 2-links*, in "The 3rd Korea-Japan School of Knots and Links", Proc. Applied Math. Workshop, 4, KAIST (Taejon, 1994) (1994) 33–39

[320] S. Kamada, *On braid monodromies of non-simple braided surfaces*, Math. Proc. Camb. Phil. Soc. (1996) **120** 237–245

[321] S. Kamada, *An observation of surface braids via chart description*, J. Knot Theory Ramifications (1996) **4** 517–529

[322] S. Kamada, *Crossing changes for singular 2-dimensional braids without branch points*, Kobe J. Math. (1996) **13** 177–182

[323] S. Kamada, *Surfaces in 4-space: A view of normal forms and braidings*, in "Lectures at Knots 96" (ed. S. Suzuki), World Scientific Publishing Co. (1997) 39–71

[324] S. Kamada, *Standard forms of 3-braid 2-knots and their Alexander polynomials*, Michigan Math. J. **45** (1998) 189–205

[325] S. Kamada, *Unknotting immersed surface-links and singular 2-dimensional braids by 1-handle surgeries*, Osaka J. Math. **36** (1999) 33–49

[326] S. Kamada, *Vanishing of a certain kind of Vassiliev invariants of 2-knots*, Proc. Amer. Math. Soc. **127** (1999) 3421-3426

[327] S. Kamada, *Arrangement of Markov moves for 2-dimensional braids*, in "Low Dimensional Topology" (Madeira, Portugal, 1998), Contemp. Math., 233, Amer. Math. Soc. (1999) 197–213

[328] S. Kamada, *Wirtinger presentations for higher dimensional manifold knots obtained from diagrams*, Fund. Math. **168**(2001) 105–112

[329] S. Kamada, *On 1-handle surgery and finite type invariants of surface knots*, Topology Appl., to appear

[330] S. Kamada, A. Kawauchi, T. Matumoto, *Combinatorial moves on ambient isotopic submanifolds in a manifold*, J. Math. Soc. Japan **53** (2001) 321–331

[331] S. Kamada and Y. Matsumoto, *Certain racks associated with the braid groups*, in "Knots in Hellas, 98", The Proceedings of the International Conference on Knot Theory (Greece, 1998), 2000 118–130

[332] T. Kanenobu, *2-knot groups with elements of finite order*, Math. Sem. Notes Kobe Univ. **8**(1980) 557-560

[333] T. Kanenobu, *Groups of higher-dimensional satellite knots*, J. Pure Appl. Algebra **28**(1983) 179-188

[334] T. Kanenobu, *Fox's 2-spheres are twist spun knots*, Math. Fac. Sci. Kyushu Univ., Ser. A. **37**(1983) 81-86

[335] T. Kanenobu, *Higher dimensional cable knots and their finite cyclic covering spaces*, Topology Appl. **19**(1985) 123-127

[336] T. Kanenobu, *Deforming twist spun 2-bridge knots of genus one*, Proc. Japan Acad. Ser. A **64**(1988) 98-101

[337] T. Kanenobu, *Unions of knots as cross sections of 2-knots*, Kobe J. Math. **4** (1988) 147–162

[338] T. Kanenobu, *Untwisted deform-spun knots: examples of symmetry spun 2-knots*, in "Transformation groups" (Osaka, 1987), Lecture Notes in Math., 1375, Springer Verlag (1989) 145–167

[339] T. Kanenobu, *Weak unknotting number of a composite 2-knot*, J. Knot Theory Ramifications **5** (1996) 161–166

[340] T. Kanenobu and K. Kazama, *The peripheral subgroup and the second homology of the group of a knotted torus in S^4*, Osaka J. Math. **31**(1994) 907-921

[341] T. Kanenobu and Y. Marumoto, *Unknotting and fusion numbers of ribbon 2-knots*, Osaka J. Math. **34** (1997) 525–540

[342] S. J. Kaplan, *Constructing 4-manifolds with given almost framed boundaries*, Trans. Amer. Math. Soc. **254**(1979) 237–263

[343] S. J. Kaplan, *Twisting to algebraically slice knots*, Pacific J. Math. **102**(1982) 55–59

[344] M. M. Kapranov, V. A. Voevodskiĭ, *Braided monoidal 2-categories and Manin-Schechtman higher braid groups*, J. Pure Appl. Algebra **92**(1994) 241–267

[345] A. Katanaga, O. Saeki, M. Teragaito and Y. Yamada, *Gluck surgery along a 2-sphere in a 4-manifold is realized by surgery along a projective plane*, Michigan Math. J. **46** (1999) 555–571

[346] M. Kato, *Higher-dimensional PL knots and knot manifolds*, J. Math. Soc. Japan **21**(1969) 458–480

[347] L. H. Kauffman, *Signature of branched fibrations*, in "Knot theory" (Proc. Sem., Plans-sur-Bex, 1977), Lecture Notes in Math., 685, Springer, Berlin (1978) 203–217

[348] L. H. Kauffman, *On knots*, Ann. of Math. Studies, 115, Princeton Univ. Press (1987)

[349] L. H. Kauffman, *State models and the Jones polynomial*, Topology **26**(1987) 395–407

[350] L. H. Kauffman , *Knots and physics*, Series on Knots and Everything, 1, World Scientific Publ.(1991)

[351] L. H. Kauffman, *Gauss codes, quantum groups and ribbon Hopf algebras*, Rev. Math. Phys. **5**(1993) 735-773

[352] L. H. Kauffman, L. R. Taylor, *Signature of links*, Trans. Amer. Math. Soc. **216**(1976) 351–365

[353] R. K. Kaul, *Chern-Simons theory, coloured-oriented braids and link invariants*, Comm. Math. Phys. **162**(1994) 289–319

[354] A. Kawauchi, *A partial Poincaré duality theorem for infinite cyclic coverings*, Quart. J. Math **26**(1975) 437–458

[355] A. Kawauchi, *On the Alexander polynomials of cobordant links*, Osaka J. Math. **15**(1978) 151–159

[356] A. Kawauchi, *The invertibility problem on amphicheiral excellent knots*, Proc. Japan Acad., Ser. A, Math. Sci. **55**(1979) 399-402

[357] A. Kawauchi, *On links not cobordant to split links*, Topology **19**(1980) 321–334

[358] A. Kawauchi, *On the Robertello invariants of proper links*, Osaka J. Math. **21**(1984) 81–90

[359] A. Kawauchi, *Rochlin invariant and α-invariant*, in "Four-Manifold Theory" (Durham, 1982), Contemp. Math., 35, Amer. Math. Soc. (1984) 315–326

[360] A. Kawauchi, *The signature invariants of infinite cyclic coverings of closed odd dimensional manifolds*, in "Algebraic and topological theories-to the memory of Dr. T. Miyata", Kinokuniya Co. Ltd.(1985) 52–85

[361] A. Kawauchi, *On the signature invariants of infinite cyclic coverings of even dimensional manifolds*, in "Homotopy theory and related topics" (Kyoto, 1984), Adv. Stud. Pure Math., 9, North-Holland (1986) 177–188

[362] A. Kawauchi, *Three dualities on the integral homology of infinite cyclic coverings of manifolds*, Osaka J. Math. **23**(1986) 633–651

[363] A. Kawauchi, *On the integral homology of infinite cyclic coverings of links*, Kobe J. Math. **4**(1987) 31–41

[364] A. Kawauchi, *The imbedding problem of 3-manifolds into 4-manifolds*, Osaka J. Math. **25**(1988) 171–183

[365] A. Kawauchi, *Knots in the stable 4-space; an overview*, in "A fête of topology", Academic Press (1988) 453–470

[366] A. Kawauchi, *The first Alexander modules of surfaces in 4-sphere*, in "Algebra and topology" (Taejon, 1990), Proc. KAIST Math. Workshop, 5, KAIST, Taejon, Korea (1990) 81–89

[367] A. Kawauchi, *Knots 90*, (Osaka, 1990), Walter de Gruyter (1992)

[368] A. Kawauchi, *Splitting a 4-manifold with infinite cyclic fundamental group*, Osaka J. Math. **31**(1994) 489–495

[369] A. Kawauchi, *A survey of knot theory*, Birkhäuser Verlag, Basel·Boston·Berlin (1996)

[370] A. Kawauchi, T. Matumoto, *An estimate of infinite cyclic coverings and knot theory*, Pacific J. Math. **90**(1980) 99–103

[371] A. Kawauchi, H. Murakami, K. Sugishita, *On the T-genus of knot cobordism*, Proc. Japan Acad. Sci., Ser. A Math. Sci. **59**(1983) 91–93

[372] A. Kawauchi, T. Shibuya and S. Suzuki, *Descriptions on surfaces in four-space, I. Normal forms*, Math. Sem. Notes Kobe Univ. **10** (1982) 75–125

[373] A. Kawauchi, T. Shibuya and S. Suzuki, *Descriptions on surfaces in four-space, II. Singularities and cross-sectional links*, Math. Sem. Notes Kobe Univ. **11** (1983) 31–69

[374] C. Kearton, *Classification of simple knots by Blanchfield duality*, Bull. Math. Soc. **79**(1973) 962–955

[375] C. Kearton, *Noninvertible knots of codimension 2*, Proc. Amer. Math. Soc. **40**(1973) 274–276

[376] C. Kearton, *Presentations of n-knots*, Trans. Amer. Math. Soc. **202**(1975) 123–140

[377] C. Kearton, *Blanchfield duality and simple knots*, Trans. Amer. Math. Soc. **202**(1975) 141–160

[378] C. Kearton, *Simple knots which are doubly-null cobordant*, Proc. Amer. Math. Soc. **52**(1975) 471–472

[379] C. Kearton, *Cobordism of knots and Blanchfield duality*, J. London Math. Soc. **10**(1975) 406–408

[380] C. Kearton, *Attempting to classify knot modules and their hermitean pairings*, in "Knot theory", Lect. Notes in Math., 685, Springer Verlag (1978) 227–242

[381] C. Kearton, *Signatures of knots and the free differential calculus*, Quart. J. Math. Oxford **30**(1979) 157–182

[382] C. Kearton, *Factorization is not unique for 3-knots*, Indiana Univ. Math. J. **28**(1979) 451–452

[383] C. Kearton, *A remarkable 3-knot*, Bull. London Math. Soc. **14**(1982) 387–398

[384] C. Kearton, *Spinning, factorization of knots, and cyclic group actions on spheres*, Archiv Math. **40**(1983) 361–363

[385] C. Kearton, *Some non-fibred 3-knots*, Bull. London Math. Soc. **15**(1983) 365–367

[386] C. Kearton, *An algebraic classification of certain simple even-dimensional knots*, Trans. Amer. Math. Soc. **276**(1983) 1–53

[387] C. Kearton, *Simple spun knots*, Topology **23**(1984) 91–95

[388] C. Kearton, *Integer invariants of certain even-dimensional knots*, Proc. Amer. Math. Soc. **93**(1985) 747–750

[389] C. Kearton, *Knots, groups, and spinning*, Glasgow Math. J. **33**(1991) 99–100

[390] C. Kearton and S. M. J.Wilson, *Spinning and branched cyclic covers of knots*, R. Soc. Lond. Proc. Ser. A Math. Phys. Eng. Sci. **455** (1999) 2235–2244

[391] M. A. Kervaire, *Les nœuds de dimensions supérieures*, Bull. Soc. Math. France **93** (1965) 225–271

[392] M. A. Kervaire, *On higher dimensional knots*, in "Differential and combinatorial topology", Princeton Math. Ser., 27, Princeton Univ. Press. (1965) 105–119

[393] M. A. Kervaire, *Knot cobordism in codimension 2*, in "Manifolds" (Amsterdam 1970), Lect. Notes in Math., 197, Springer Verlag (1971) 83–105

[394] M. A. Kervaire, J. Milnor, *On 2-spheres in 4-manifolds*, Proc. Nat. Acad. USA **47**(1961) 1651–1657

[395] M. A. Kervaire, C. Weber, *A survey of multidimensional knots*, in "Knot theory", Lect. Notes in Math., 685, Springer Verlag (1978) 61–138

[396] M. E. Kidwell, *Relations between the Alexander polynomial and summit power of a closed braid*, Math. Sem. Notes Kobe Univ. **10**(1982) 387-409

[397] S. Kinoshita, *On Wendt's theorem of knots*, Osaka Math. J. **9** (1957) 61–66

[398] S. Kinoshita, *On Went's theorem of knots II*, Osaka Math. J. **10** (1958) 259–261

[399] S. Kinoshita, *Alexander polynomials as isotopy invariants I*, Osaka Math. J. **10**(1958) 263–271

[400] S. Kinoshita, *Alexander polynomials as isotopy invariants II*, Osaka Math. J. **11**(1959) 91–94

[401] S. Kinoshita, *On diffeomorphic approximations of polyhedral surfaces in 4-space*, Osaka J. Math. **12**(1960) 191–194

[402] S. Kinoshita, *On the Alexander polynomials of 2-spheres in a 4-sphere*, Ann. of Math. **74**(1961) 518–531

[403] S. Kinoshita, *On elementary ideals of θ-curves in the 3-sphere and 2-links in the 4-sphere*, Pacific J. Math. **49**(1973) 127–134

[404] S. Kinoshita, *On elementary ideals of projective planes in the 4-sphere and oriented θ-curves in the 3-sphere*, Pacific J. Math. **57**(1975) 217–221

[405] S. Kinoshita, *On the distribution of Alexander polynomials of alternating knots and links*, Proc. Amer. Math. Soc. **79**(1980) 664–648

[406] S. Kinoshita, *Branched coverings of knots and links*, Lecture Notes, Kwansei Gakuin Univ. (1980)

[407] S. Kinoshita, *On the branch points in the branched coverings of links*, Canad. Math. Bull. **28** (1985) 165–173

[408] S. Kinoshita, *Elementary ideals in knot theory*, Kwansei Gakuin Annual Studies **35** (1986) 183–208

[409] S. Kinoshita and H. Terasaka, *On unions of knots*, Osaka Math. J. **9** (1957) 131–153

[410] R. Kirby, *A calculus for framed links in S^3*, Invent. Math. **45**(1978) 35–56

[411] R. Kirby, *Problems in low-dimensional manifold theory*, Proc. Symp. Pure Math. **32** (1978) 273–312

[412] R. Kirby, *The topology of 4-manifolds*, Lecture Notes in Mathematics, 1374, Springer Verlag (1989)

[413] R. Kirby, W. B. R. Lickorish, *Prime knots and concordance*, Math. Proc. Cambridge Philos. Soc. **86**(1979) 437–441

[414] R. Kirby, P. Melvin, *Slice knots and property R*, Invent. Math. **45**(1978) 57–59

[415] P. Kirk, *Link maps in the four sphere*, in "Differential Topology" Lect. Notes Math., 1350, Springer Verlag (1988) 31–43

[416] P. Kirk, U. Koschorke, *Generalized Seifert surfaces and linking numbers*, Topology Appl. **42**(1991) 247–262

[417] K. H. Ko, *Seifert matrices and boundary link cobordisms*, Trans. Amer. Math. Soc. **299**(1987) 657–681

[418] K. H. Ko, *A survey on link concordance*, in "The 3rd Korea-Japan School of knots and links" (Taejon, 1994), Proc. Applied Math. Workshop, 4, KAIST, Taejon Korea (1994) 53–68

[419] K. H. Ko and J. S. Carter, *Triple points of immersed surfaces in three dimensional manifolds*, Topology Appl. **32**(1989) 149–159

[420] K. H. Ko and L. Smolinsky, *The framed braid group and 3-manifolds*, Proc. Amer. Math. Soc. **115**(1992) 541–551

[421] K. H. Ko and L. Smolinsky, *The framed braid group and representations*, in "Knots 90" (Osaka, 1990), Walter de Gruyter (1992) 289–297

[422] K. Kobayashi, *On a homotopy version of 4-dimensional Whitney's lemma*, Math. Sem. Notes, Kobe Univ. **5**(1977) 109–116

[423] M. Kobayashi, T. Kobayashi, *On canonical genus and free genus of knot*, J. Knot Theory Ramifications **5** (1996) 77–85

[424] T. Kobayashi, *Uniqueness of minimal genus Seifert surfaces for links*, Topology Appl. **33**(1989) 265–279

[425] T. Kobayashi, *Fibered links which are band connected sum of two links*, in "Knots 90" (Osaka, 1990), Walter de Gruyter (1992) 9–23

[426] T. Kobayashi, *Minimal genus Seifert surfaces for a knot which is a non-trivial band sum*, in "The 3rd Korea-Japan School of knots and links" (Taejon, 1994), Proc. Applied Math. Workshop, 4, KAIST, Korea (1994) 79–89

[427] T. Kobayashi, H. Nishi, *A necessary and sufficient condition for a 3-manifold to have genus g Heegaard splitting (a proof of Hass-Thompson conjecture)*, Osaka J. Math. **31**(1994) 109–136

[428] K. Kodama, M. Sakuma, *Symmetry groups of prime knots up to 10 crossings*, in "Knots 90" (Osaka, 1990), Walter de Gruyter (1992) 323–340

[429] T. Kohno, *Monodromy representations of braid groups* (in Japanese), Sugaku **41**(1989) 305–319

[430] T. Kohno, *New developments in the theory of knots*, Advanced Series in Mathematical Physics, 11, World Scientific Publ. (1990)

[431] S. Kojima, *Classification of simple knots by Levine pairings*, Comment. Math. Helv. **54**(1979) 356–367 [Erratum, ibid., **55**(1980), 652-653]

[432] U. Koschorke, *Higher order homotopy invariants for higher-dimensional link maps*, in "Algebraic topology" (Gottingen, 1984), Lecture Notes in Math., 1172, Springer Verlag (1985) 116–129

[433] U. Koschorke, *Multiple point invariants of link maps*, in "Differential topology" (Siegen, 1987), Lecture Notes in Math., 1350, Springer Verlag (1988) 44–86

[434] U. Koschorke, *On link maps and their homotopy classification*, Math. Ann. **286**(1990) 753–782

[435] U. Koschorke, *Link homotopy with many components*, Topology **30**(1991) 267–281

[436] U. Koschorke, D. Rolfsen, *Higher dimensional link operations and stable homotopy*, Pacific J. Math. **139**(1989) 87–106

[437] M. Kranjc, *Embedding 2-complexes in $\mathbf{R}^4$*, Pacific J. Math. **133**(1988) 301–313

[438] M. Kreck, *On the homeomorphism classification of smooth knotted surfaces in the 4-sphere*, in "Geometry of low-dimensional manifolds 1" (Durham, 1989), London Math. Soc. Lecture Note Ser., 150, Cambridge Univ. Press (1990) 63–72

[439] P. B. Kronheimer, *Embedded surfaces in 4-manifolds*, in "Proc. ICM" (Kyoto, 1990), Math. Soc. Japan (1991) 529–539

[440] P. B. Kronheimer, *The genus-minimaizing property of algebraic curves*, Bull. Amer. Math. Soc. **29**(1993) 63–69

[441] P. B. Kronheimer, T. S. Mrowka, *Gauge theory for embedded surfaces I*, Topology **32**(1993) 773–826

[442] P. B. Kronheimer, T. S. Mrowka, *The genus of embedded surfaces in the projective plane*, Math. Res. Lett. **1**(1994) 797–808

[443] S. Kwasik, *On invariant knots*, Math. Proc. Cambridge Philos. Soc. **94**(1984) 473–475

[444] S. Kwasik, *Low-dimensional concordances, Whitney towers and isotopies*, Math. Proc. Cambridge Philos. Soc. **102**(1987) 103–119

[445] S. Kwasik, R. Schultz, *Pseudofree group actions on S^4*, Amer. J. Math. **112**(1990) 47–70

[446] S. Kwasik, R. Schultz, *Vanishing of Whitehead torsion in dimension four*, Topology **31**(1992) 735–756

[447] R. H. Kyle, *Branched covering spaces and the quadratic forms of links*, Ann. of Math. **59**(1954) 539–548

[448] R. H. Kyle, *Branched covering spaces and the quadratic forms of links II*, Ann. of Math. **69**(1959) 686–699

[449] H. Lambert, *Mapping cubes with holes onto cubes with handles*, Illinois. J. Math. **13**(1969) 606–615

[450] H. Lambert, *A 1-linked link whose longitudes lie in the second commutator subgroup*, Trans. Amer. Math. Soc. **147**(1970) 261–269

[451] S. Lambropoulou, *Knot theory related to generalized and cyclotomic Hecke algebra of type B*, J. Knot Theory Ramifications **8**(1999) 621–658

[452] S. Lambropoulou and C. P. Rourke, *Markov's theorem in 3-manifolds*, Topology Appl. **78** (1997) 95–122

[453] T. Lawson, *Splitting S^4 on $\mathbf{R}P^2$ via the branched cover of $\mathbf{C}P^2$ over S^4*, Proc. Amer. Math. Soc. **86** (1982) 328–330

[454] T. Lawson, *Detecting the standard embedding of $\mathbf{R}P^2$ in S^4*, Math. Ann. **267** (1984) 439–448

[455] T. Lawson, *Representing homology classes of almost definite 4-manifolds*, Michigan Math. J. **34** (1987) 85–91

[456] T. Lawson, *h-cobordisms between simply connected 4-manifolds*, Topology Appl. **28** (1988) 75–82

[457] T. Lawson, *Compactness results for orbifold instantons*, Math. Z. **200** (1988) 123–140

[458] T. Lawson, *Smooth embeddings of 2-spheres in 4-manifolds*, Exposition. Math. **10** (1992) 289–309

[459] T. Q. T. Le, H. Murakami, J. Murakami, T. Ohtsuki, *A three-manifold invariant via the Kontsevich integral*, Osaka J. Math. **36** (1999) 365–395

[460] R. Lee and D. M. Wilczyński, *Locally flat 2-spheres in simply connected 4-manifolds*, Comment. Math. Helv. **65**(1990) 388–412

[461] R. Lee and D. M. Wilczyński, *Locally flat 2-spheres in simply connected 4-manifolds*, Comment. Math. Helv. **67**(1992) 334–335

[462] Y. W. Lee, *Contractibly embedded 2-spheres in $S^2 \times S^2$*, Proc. Amer. Math. Soc. **85**(1982) 280–282

[463] G. I. Lehrer, *A survey of Hecke algebras and the Artin braid groups*, Amer. Math. Soc. Contemp. Math. **78**(1988) 365–385

[464] J. Levine, *A characterization of knot polynomials*, Topology **4**(1965) 135–141

[465] J. Levine, *A classification of differentiable knots*, Ann. of Math. **82**(1965) 15–51

[466] J. Levine, *Unknotting spheres in codimension two*, Topology **4**(1965) 9–16

[467] J. Levine, *Polynomial invariants of knots of codimension two*, Ann. of Math. **84**(1966) 537–554

[468] J. Levine, *A method for generating link polynomials*, Amer. J. Math. **89**(1967) 69–84

[469] J. Levine, *Knot cobordism groups in codimension two*, Comment. Math. Helv. **44**(1969) 229–224

[470] J. Levine, *Invariants of knot cobordism*, Invent. Math. **8**(1969) 98–110

[471] J. Levine, *An algebraic classification of some knots of codimension two*, Comment. Math. Helv. **45**(1970) 185–198

[472] J. Levine, *The role of the Seifert matrix in knot theory*, in "Acta Congr. Intern. Math., 2" (Paris, 1970), Gauthier-Villars (1971) 95–98

[473] J. Levine, *Knot modules*, in "Knots, groups and 3-manifolds", Ann. Math. Studies, 84, Princeton Univ. Press (1975) 25–34

[474] J. Levine, *Knot modules I*, Trans. Amer. Math. Soc. **229**(1977) 1–50

[475] J. Levine, *Some results on higher dimensional knot groups*, in "Knot Theory" (Switzerland, 1977), Lect. Notes in Math., 685, Springer Verlag (1978) 243–269

[476] J. Levine, *Algebraic structure of knot modules*, Lect. Notes in Math. 772, Springer Verlag (1980)

[477] J. Levine, *The module of a 2-component link*, Comment. Math. Helv. **57**(1982) 377–399

[478] J. Levine, *Doubly slice knots and doubled disk knots*, Michigan J. Math. **30**(1983) 249–256

[479] J. Levine, *Localization of link modules*, Amer. Math. Soc. Contemp. Math. **20**(1983) 213–229

[480] J. Levine, *Links with Alexander polynomial zero*, Indiana Univ. Math. J. **36**(1987) 91–108

[481] J. Levine, *Surgery on links and the $\bar{\mu}$-invariants*, Topology **26**(1987) 45–61

[482] J. Levine, *An approach to homotopy classification of links*, Trans. Amer. Math. Soc. **306**(1988) 361–387

[483] J. Levine, *Symmetric presentation of link modules*, Topology Appl. **30**(1988) 183–198

[484] J. Levine, *The $\bar{\mu}$-invariants of based links*, in "Differential topology" (Siegen, 1987), Lecture Notes in Math., 1350, Springer Verlag (1988) 87–103

[485] J. Levine, *Link concordance and algebraic closure of groups*, Comment. Math. Helv. **64**(1989) 236–255

[486] J. Levine, *Link concordance and algebraic closure of groups II*, Invent Math. **96**(1989) 571–592

[487] J. Levine, *Signature invariants of homology bordism with application to links*, in "Knots 90" (Osaka, 1990), Walter de Gruyter (1992) 395–406

[488] J. Levine, *Link invariants via the eta invariant*, Comment. Math. Helv. **69**(1994) 82–119

[489] J. P. Levine, W. Mio, K. E. Orr, *Links with vanishing homotopy invariant*, Comm. Pure Appl. Math. **46**(1993) 213–220

[490] H. Levinson, *Decomposable braids and linkages*, Trans. Amer. Math. Soc. **178**(1973) 111–126

[491] H. Levinson, *Decomposable braids as subgroups of braid groups*, Trans. Amer. Math. Soc. **202**(1975) 51–55

[492] B. H. Li, *Embeddings of surfaces in 4-manifolds I*, Chinese Sci. Bull. **36**(1991) 2025–2029

[493] B. H. Li, *Embeddings of surfaces in 4-manifolds II*, Chinese Sci. Bull. **36**(1991) 2030–2033

[494] G.-S. Li, *An invariant of link homotopy in dimension four*, Topology **36**(1997) 881–897

[495] Y. Q. Li , *Representations of a braid group with transpose symmetry and the related link invariants*, J. Phys. A **25**(1992) 6713–6721

[496] Y. Q. Li , *Multiparameter solutions of Yang-Baxter equation from braid group representations*, J. Math. Phys. **34**(1993) 768–774

[497] Y. Q. Li and M. L. Ge, *Polynomials from nonstandard braid group representations*, Phys. Lett. A **152**(1991) 273–275

[498] Y. Q. Li and M. L. Ge, *Link polynomials related to the new braid group representations*, J. Phys. A **24**(1991) 4241–4247

[499] Y. Q. Li, M. L. Ge, K. Xue, L. Y. Wang, *Weight conservation condition and structure of the braid group representation*, J. Phys. A **24**(1991) 3443–3453

[500] C.-C. Liang, *An algebraic classification of some links of codimension two*, Proc. Amer. Math. Soc. **67**(1977) 147–151

[501] L. Liao, X. C. Song, *Quantum Lie superalgebras and "nonstandard" braid group representations*, Modern Phys. Lett. A **6**(1991) 959–968

[502] A. Libgober, *Levine's formula in knot theory and quadratic reciprocity law*, Enseign. Math. **26**(1980) 323–331

[503] A. Libgober, *Alexander modules of plane algebraic curves*, Amer. Math. Soc. Contemp. Math. **20**(1983) 231–247

[504] A. Libgober, *On divisibility properties of braids associated with algebraic curves*, in "Braids" (Santa Cruz, CA, 1986), Contemp. Math., 78, Amer. Math. Soc. (1988) 387–398

[505] W. B. R. Lickorish, *A representation of orientable combinatorial 3-manifolds*, Ann. of Math. **76**(1962) 531–540

[506] W. B. R. Lickorish, *A finite set of generators for the homeotopy group of a 2-manifold*, Proc. Cambridge Phil. Soc. **60**(1964) 769–778 [Corrigendum, ibid., 62(1966), 679–681]

[507] W. B. R. Lickorish, *Surgery on knots*, Proc. Amer. Math. Soc. **60**(1977) 296–298

[508] W. B. R. Lickorish, *Shake-slice knots*, in "Topology of low-dimensional manifolds" (Sussex, 1977), Lect. Notes in Math., 722, Springer Verlag (1979) 67–70

[509] W. B. R. Lickorish, *The unknotting number of a classical knot*, Contemp. Math. **44**(1985) 117–121

[510] W. B. R. Lickorish, *Unknotting by adding a twisted band*, Bull. London Math. Soc. **18**(1986) 613–615

[511] W. B. R. Lickorish, *Three-manifolds and the Temperley-Lieb algebra*, Math. Ann. **290**(1991) 657–670

[512] W. B. R. Lickorish, *Invariants for 3-manifolds from the combinatorics of the Jones polynomial*, Pacific J. Math. **149**(1991) 337–347

[513] W. B. R. Lickorish and K. C. Millett, *A polynomial invariant of oriented links*, Topology **26**(1987) 107–141

[514] V. T. Liem, G. A. Venema, *Characterization of knot complements in the 4-sphere*, Topology Appl. **42**(1991) 231–245

[515] V. T. Liem, G. A. Venema, *On the asphericity of knot complements*, Canad J. Math. **45**(1993) 340–356

[516] M. Lien, *Construction of high dimensional knot groups from classical knot groups*, Trans. Amer. Math. Soc. **298**(1986) 713–722

[517] V. Ya. Lin, *Artin braids and the groups and spaces connected with them*, J. Soviet. Math. **18**(1979) 736–788

[518] X. S. Lin, *A knot invariant via representation spaces*, J. Diff. Geom. **35**(1992) 337–357

[519] X. S. Lin, *Alexander-Artin-Markov theory for 2-links in R^4*, preprint

[520] D. Lines, *On odd-dimensional fibred knots obtained by plumbing and twisting*, J. London Math. Soc. **32**(1985) 557–571

[521] D. Lines, *On even-dimensional fibred knots obtained by plumbing*, Math. Proc. Cambridge Philos. Soc. **100**(1986) 117–131

[522] S. Lipschutz, *On a finite matrix representation of the braid group*, Arch. Math. **12**(1961) 7–12

[523] P. Lisca, *On tori embedded in four-manifolds*, J. Diff. Geom. **38**(1993) 13–37

[524] P. Lisca, *Smoothly embedded 2-spheres and exotic 4-manifolds*, Enseign. Math. (2) **39**(1993) 225–231

[525] R. A. Litherland, *Deforming twist-spun knots*, Trans. Amer. math. Soc. **250** (1979) 311–331

[526] R. A. Litherland, *Slicing doubles of knots in homology 3-spheres*, Invent. Math. **54**(1979) 69–74

[527] R. A. Litherland, *The second homology of the group of a knotted surface*, Quart. J. Math. Oxford **32** (1981) 425–434

[528] R. A. Litherland, *Cobordism of satellite knots*, in "Four-manifold theory" (Durham, 1982), Contemp. Math., 35, Amer. Math. Soc. (1984) 327–362

[529] R. A. Litherland, *Symmetries of twist-spun knots*, in "Knot theory and manifolds", Lecture Notes in Math., 1144, Springer Verlag (1985) 97–107

[530] R. A. Litherland, *The Alexander module of a knotted theta-curve*, Math. Proc. Cambridge Philos. Soc. **106** (1989) 95–106

[531] C. Livingston, *Homology cobordisms of 3-manifolds, knot concordance, and prime knots*, Pacific J. Math. **94**(1981) 193–206

[532] C. Livingston, *Knots which are not concordant to their inverses*, Quart. J. Math. Oxford **34**(1983) 323–328

[533] C. Livingston, *Stably irreducible surfaces in S^4*, Pacific J. Math. **116**(1985) 77–84

[534] C. Livingston, *Knots with finite weight commutator subgroups*, Proc. Amer. Math. Soc. **101**(1987) 195–198

[535] C. Livingston, *Companionship and knot group representations*, Topology Appl. **25**(1987) 241–244

[536] C. Livingston, *Indecomposable surfaces in the 4-sphere*, Pacific J. Math. **132**(1988) 371–378

[537] C. Livingston, *Links not concordant to boundary links*, Proc. Amer. Math. Soc. **110**(1990) 1129–1131

[538] C. Livingston, *Knot theory*, Carus Mathematical Monographs, 24, Mathematical Association of America (1993)

[539] S. J. Lomonaco, *The second homotopy group of a spun knot*, Topology **8**(1969) 95–98

[540] S. J. Lomonaco, *The fundamental ideal and π_2 of higher dimensional knots*, Proc. Amer. Math. Soc. **38**(1973) 431–433

[541] S. J. Lomonaco, *The third homotopy group of some higher dimensional knots*, in "Knots, groups and 3-manifolds" Ann. Math. Studies, 84, Princton Univ. Press (1975) 35–45

[542] S. J. Lomonaco, *The homotopy groups of knots I, how to compute the algebraic 2-type*, Pacific J. Math. **95**(1981) 349–390

[543] S. J. Lomonaco, *Five dimensional knot theory*, Amer. Math. Soc. Contemp. Math. **20**(1983) 249–270

[544] D. D. Long, *Strongly plus-amphicheiral knots are algebraically slice*, Math. Proc. Cambridge Philos. Soc. **95**(1984) 309–312

[545] D. D. Long, *On the linear representation of braid groups*, Trans. Amer. Math. Soc. **311**(1989) 535–560

[546] D. D. Long and M. Paton, *The Burau representation is not faithful for $n \geq 6$*, Topology **32**(1993) 439–447

[547] J. E. Los, *Knots, braid index and dynamical type*, Topology **33**(1994) 257–270

[548] M. T. Lozano, J. H. Przytycki, *Incompressible surfaces in the exterior of a closed 3-braid I, surfaces with horizontal boundary components*, Math. Proc. Cambridge Philos. Soc. **98**(1985) 275–299

[549] J. Luecke, *Finite covers of 3-manifolds containing essential tori*, Trans. Amer. Math. Soc. **310**(1988) 381–391

[550] R. C. Lyndon and P. E. Schupp, *Combinatorial group theory*, Springer-Verlag, Berlin, Heidelberg, New York (1977)

[551] Z. Q. Ma and B. H. Zhao, *2^n-dimensional representations of braid group and link polynomials*, J. Phys. A **22**(1989) 49–52

[552] C. Maclachlan, *On representations of Artin's braid group*, Michigan Math. J. **25**(1978) 235–244

[553] S. MacLane, *Homology*, Springer Verlag (1963)

[554] T. Maeda, *On the groups with Wirtinger presentations*, Math. Sem. Notes Kobe Univ. **5**(1977) 345–358

[555] T. Maeda, *Knotted surface in the 4-sphere with no minimal Seifert manifold*, in "Combinatorial and geometric group theory" (Edinburgh 1993), London Math. Soc. Lect. Note Ser., 204, Cambridge Univ. Press (1994) 239–246

[556] T. Maeda, K. Murasugi, *Covering linkage invariants and Fox's problem 13*, Amer. Math. Soc. Contemp. Math. **20**(1983) 271–283

[557] W. Magnus, *Über Automorphismen von Fundamentalgruppen berandeter Flächen*, Math. Ann. **109**(1934) 617–646

[558] W. Magnus, *Braids and Riemann surfaces*, Commun. Pure Appl. Math. **25**(1972) 151–161

[559] W. Magnus, *Braid groups: a survey*, in "Proc. 2nd Int. Conf. of Groups" (Canberra, 1973), Lect. Notes in Math., 372, Springer Verlag (1974) 463–487

[560] W. Magnus, A. Karrass and D. Solitar, *Combinatorial group theory*, Wiley New York(1966)

[561] W. Magnus, A. Peluso, *On knot groups*, Commun. Pure Appl. Math. **20**(1967) 749–770

[562] S. Majid, *Braided groups and duals of monoidal categories*, in "Category theory 1991" (Montreal, 1991), CMS Conf. Proc., 13, Amer. Math. Soc. (1992) 329–343

[563] G. S. Makanin, *Separable closed braids*, Math. USSR-Sbornik **60**(1988) 521–531

[564] G. S. Makanin, *An analogue of the Alexander Markov theorem*, Izv. Akad. Nauk SSSR-Ser. Mat. **53**(1989) 200–210

[565] R. Mandelbaum, *Four-dimensional topology: an introduction*, Bull. Amer. Math. Soc. **2**(1980) 1–159

[566] Yu. I. Manin and V. V. Schechtman, *Arrangements of hyperplanes, higher braid groups and higher Bruhat*, Advanced Studies in Pure Mathematics **17** (1986) 289–308

[567] A. A. Markov, *Über die freie Aquivalenz der geschlossner Zopfe*, Rec. Soc. Math. Moscou **1** (1935) 73–78

[568] Y. Marumoto, *On ribbon 2-knots of 1-fusion*, Math. Sem. Notes Kobe Univ. **5**(1977) 59–68

[569] Y. Marumoto, *A class of higher dimensional knots*, J. Fac. Educ. Saga Univ. **31**(1984) 177–185

[570] Y. Marumoto, *On higher dimensional light bulb theorem*, Kobe J. Math. **3**(1986) 71–75

[571] Y. Marumoto, *Some higher dimensional knots*, Osaka J. Math. **24**(1987) 759–783

[572] Y. Marumoto, *Stable equivalence of ribbon presentations*, J. Knot Theory Ramifications **1**(1992) 241–251

[573] Y. Marumoto, *Infinitely many prime equatorial knots of a ribbon 2-knot*, Jour. Osaka Sangyo Univ., Natural Science **94**(1993) 21–26

[574] Y. Marumoto and Y. Nakanishi, *A note on the Zeeman theorem*, Kobe J. Math. **8** (1991) 67–71

[575] Y. Marumoto, Y. Uchida, T. Yasuda, *Motions of trivial links and its ribbon knots*, Michigan Math. J. **42**(1995) 463–477

[576] W. S. Massey, *Algebraic topology: an introduction*, Harbrace College Math. Ser., Harcourt, Brace & World, Inc. (1967)

[577] W. S. Massey, *Proof of a conjecture of Whitney*, Pacific J. Math. **31**(1969) 143–156

[578] W. S. Massey, *The quotient space of the complex projective plane under conjugation is a 4-space*, Geom. Dedicata **2**(1973) 371–374

[579] W. S. Massey, *Imbeddings of projective planes and related manifolds in spheres*, Indiana Univ. Math. J. **23**(1974) 791–812

[580] W. S. Massey, *Singular homology theory*, Springer Verlag(1980)

[581] W. S. Massey, *Completion of link modules*, Duke Math. J. **47**(1980) 399–420

[582] W. S. Massey, *A basic course in algebraic topology*, Springer Verlag(1991)

[583] W. S. Massey, D. Rolfsen, *Homotopy classification of higher-dimensional links*, Indiana Univ. Math. J. **34**(1985) 375–391

[584] W. S. Massey, L. Traldi, *Links with free groups are trivial*, Proc. Amer. Math. Soc. **82**(1981) 155–156

[585] M. Masuda, M. Sakuma, *Knotting codimension 2 submanifolds locally*, Enseign. Math. **35**(1989) 21–40

[586] Y. Matsumoto, *Secondary intersectional properties of 4-manifolds and Whitney's trick*, in "Algebraic and geometric topology" (Stanford, 1976), Proc. Sympos. Pure Math. 32-II, Amer. Math. Soc. (1978) 99–107

[587] Y. Matsumoto, *An elementary proof of Rochlin's signature theorem and its extension by Guillou and Marin*, in "A la recherche de la topologie perdue", Progress in Math., 62, Birkhäuser, Boston (1986) 119–139

[588] Y. Matsumoto, *Lefschetz fibrations of genus two –A topological approach–*, in "Proc. the 37th Taniguchi Sympo.", ed. by S. Kojima et al., World Sci. Publ. (1996) 123–148

[589] Y. Matsumoto, G. A. Venema, *Failure of the Dehn Lemma on contractible 4-manifolds*, Invent. Math. **51**(1979) 205–218

[590] T. Matsuoka, *The Burau representation of the braid group and the Nielsen-Thurston classification*, in "Nielsen theory and dynamical systems" (South Hadley, 1992), Contemp. Math., 152, Amer. Math. Soc. (1993) 229–248

[591] T. Matumoto, *On a weakly unknotted 2-sphere in a simply-connected 4-manifold*, Osaka J. Math. **21**(1984) 489–492

[592] S. V. Matveev, *Distributive groupoids in knot theory*, Math. USSR-Sbornik **47**(1982) 73–83

[593] S. V. Matveev, *Transformations of special spines and the Zeeman conjecture*, Math. USSR Izvestia **31**(1988) 423–434

[594] J. P. Mayberry, K. Murasugi, *Torsion-groups of abelian coverings of links*, Trans. Amer. Math. Soc. **271**(1982) 143–173

[595] E. J. Mayland, *On residually finite knot groups*, Trans. Amer. Math. Soc. **168**(1972) 221–232

[596] E. J. Mayland, *The residual finiteness of the classical knot groups*, Candian J. Math. **17**(1975) 1092–1099

[597] W. A. McCallum, *The higher homotopy groups of the p-spun trefoil knot*, Glasgow Math. J. **17**(1976) 44–46

[598] D. McCullough, A. Miller and B. Zimmermann, *Group actions on handlebodies*, Proc. London Math. Soc. (3) **59**(1989) 373–416

[599] M. McIntyre and G. Cairns, *A new formula for winding number*, Geom. Dedicata **46**(1993) 149–159

[600] F. A. McRobie and J. M. T. Thompson, *Braids and knots in driven oscillators*, Internat. J. Bifur. Chaos Appl. Sci. Engrg. **3**(1993) 1343–1361

[601] W. H. Meeks III, P. Scott, *Finite group actions on 3-manifolds*, Invent. Math. **86**(1986) 287–346

[602] W. H. Meeks III, S. -T. Yau, *Topology of three dimensional manifolds and the embedding problem in minimal surface theory*, Ann. of Math. **112**(1980) 441–484

[603] W. H. Meeks III, S. -T. Yau, *The topological uniqueness of complete minimal surfaces of finite topological type*, Topology **31**(1992) 305–316

[604] M. L. Mehta, *On a relation between torsion numbers and Alexander matrix of a knot*, Bull. Soc. Math. France **108**(1980) 81–94

[605] P. Melvin, W. Kazez, *3-dimensional bordism*, Michigan Math. J. **36**(1989) 251–260

[606] P. Melvin, N. B. Tufillaro, *Templates and framed braids*, Phys. Rev. A **44**(1991) R3419–R3422

[607] W. W. Menasco, *The Bennequin-Milnor unknotting conjectures*, C. R. Acad. Sci. Paris Sér. I Math., 318 (1994) 831–836

[608] W. W. Menasco, A. Thompson, *Compressing handlebodies with holes*, Topology **28**(1989) 485–494

[609] W. W. Menasco, M. B. Thistlethwaite, *The Tait flyping conjecture*, Bull. Amer. Math. Soc. (N.S.) **25**(1991) 403–412

[610] W. W. Menasco, M. B. Thistlethwaite, *The classification of alternating links*, Ann. of Math. **138**(1993) 113–171

[611] K. C. Millett, *Knot theory, Jones' polynomials, invariants of 3-manifolds, and the topological theory of fluid dynamics*, in "Topological aspects of the dynamics of fluids and plasmas" (Santa Barbara, 1991), NATO Adv. Sci. Inst. Ser. E Appl. Sci., 218, Kluwer Acad. Publ. (1992) 29–64

[612] J. W. Milnor, *Link groups*, Ann. of Math. **59**(1954) 177–195

[613] J. W. Milnor, *Isotopy of links*, in "Lefschety symposium" Princeton Math. Ser., 12, Princeton Univ. Press (1957) 280–306

[614] J. W. Milnor, *A duality theorem for Reidemeister torsion*, Ann. of Math. **76**(1962) 137–147

[615] J. W. Milnor, *A survey of cobordism theory*, Enseign. Math. **8**(1962) 16–23

[616] J. W. Milnor, *Spin structures on manifolds*, Enseign. Math. **9**(1963) 198–203

[617] J. W. Milnor, *On the Stiefel-Whitney numbers of complex manifolds and spin manifolds*, Topology **3**(1965) 223–230

[618] J. W. Milnor, *Singular points of complex hypersurfaces*, Ann. of Math. Studies, 61, Princeton Univ. Press (1968)

[619] J. W. Milnor, *Infinite cyclic coverings*, in "Conf. topology of manifolds", Prindle, Weber and Schmdit (1968) 115–133

[620] W. Mio, *On boundary-link cobordism*, Math. Proc. Cambridge Philos. Soc. **101**(1987) 259–266

[621] K. Miyazaki, *On the relationship among unknotting numbers, knotting genus and Alexander invariant for 2-knots*, Kobe J. Math. **3**(1986) 77–85

[622] K. Miyazaki, *Conjugation and the prime decomposition of knots in closed, oriented 3-manifolds*, Trans. Amer. Math. Soc. **313**(1989) 785–804

[623] K. Miyazaki, *Ribbon concordance does not imply a degree one map*, Proc. Amer. Math. Soc. **108**(1990) 1055–1058

[624] K. Miyazaki, *Nonsimple, ribbon fibered knots*, Trans. Amer. Math. Soc. **341**(1994) 1–44

[625] K. Miyazaki, A. Yasuhara, *Knots that cannot be obtained from a trivial knot by twisting*, Contemp. Math. **164**(1994) 139–150

[626] E. E. Moise, *Affine structures in 3-manifolds V, the triangulation theorem and Hauptvermutung*, Ann. of Math. **56**(1952) 96–114

[627] E. E. Moise, *Affine structures in 3-manifolds VII, invariance of the knot-types; local tame imbedding*, Ann. of Math. **59**(1954) 159–170

[628] E. E. Moise, *Geometric topology in dimensions 2 and 3*, Graduate Text in Math., 47, Springer Verlag(1977)

[629] B. G. Moishezon, *Stable branch curves and braid monodromies*, in "Algebraic Geometry" Lect. Notes in Math. 862 (1981) 107–192

[630] B. Moishezon, M. Teicher, *Braid group technique in complex geometry I, line arrangements in* $\mathbf{CP}^2$, Contemp. Math. **78**(1988) 425–555

[631] J. M. Montesinos, *4-manifolds, 3-fold covering spaces and ribbons* , Trans. Amer. Math. Soc. **245**(1978) 453–467

[632] J. M. Montesinos, *On twins in the four-sphere I*, Quart. J. Math. Oxford (2) **34**(1983) 171–199

[633] J. M. Montesinos, *On twins in the four-space II*, Quart. J. Math. Oxford (2) **35**(1984) 73–83

[634] J. M. Montesinos, *Lectures on 3-fold simple coverings and 3-manifolds*, in "Combinatorial methods in topology and algebraic geometry", Contemp. Math., 44, Amer. Math. Soc. (1985) 157–177

[635] J. M. Montesinos, *A note on moves and irregular coverings of S^4*, in "Combinatorial methods in topology and algebraic geometry", Contemp. Math., 44, Amer. Math. Soc. (1985) 345–349

[636] J. M. Montesinos, *A note on twist spun knots*, Proc. Amer. Math. Soc. **98**(1986) 180–184

[637] J. M. Montesinos, H. R. Morton, *Fibred links from closed braids*, Proc. London Math. Soc. **62**(1991) 167–201

[638] J. M. Montesinos, W. Whitten, *Constructions of two-fold branched covering spaces*, Pacific J. Math. **125**(1984) 415–446

[639] J. A. Moody, *The Burau representation of the braid group B_n is unfaithful for large n*, Bull. Amer. Math. Soc. (N.S.) **25**(1991) 379–384

[640] J. A. Moody, *The faithfulness question for the Burau representation*, Proc. Amer. Math. Soc. **119**(1993) 671–679

[641] S. Moran, *The mathematical theory of knots and braids, an introduction*, North-Holland Math. Studies, 82, North-Holland Publ. Co.(1983)

[642] S. Moran, *Some free groups of matrices and the Burau representation of B_4*, Math. Proc. Cambridge Philos. Soc. **110**(1991) 225–228

[643] S. Moran, *A wild variation of Artin's braids*, in "Topics in knot theory" (Erzurum, 1992), NATO Adv. Sci. Inst. Ser. C Math. Phys. Sci., 399, Kluwer Acad. Publ. (1993) 85–106

[644] J. W. Morgan, H. Bass, *The Smith conjecture*, Academic Press (1984)

[645] Y. Moriah, *On the free genus of knots*, Proc. Amer. Math. Soc. **99**(1987) 373–379

[646] K. Morimoto, *On the additivity of the clasp singularities*, Kobe J. Math. **3**(1987) 179–185

[647] K. Morimoto, M. Sakuma, *On unknotting tunnels for knots*, Math. Ann. **289**(1991) 143–167

[648] J. Morita, *A combinatorial proof for Artin's presentation of the braid group B_n and some cyclic analogue*, Tsukuba J. Math. **16**(1992) 439–442

[649] S. Morita, *Mapping class groups of surfaces and three-dimensional manifolds*, in "Proc. ICM" (Kyoto, 1990), Math. Soc. Japan (1991) 665–674

[650] S. Morita, *On the structure of the Torelli group and the Casson invariant*, Topology **30**(1991) 603–621

[651] T. Morita, *Orders of knots in the algebraic knot cobordism group*, Osaka J. Math. **25**(1988) 859–864

[652] Y. Moriah, *On the free genus of knots*, Proc. Amer. Math. Soc. **99**(1987) 373–379

[653] Y. Moriah, *Heegaard splittings of Seifert fibered spaces*, Invent. Math. **91**(1988) 465–481

[654] Y. Moriah, *A note on satellites and tunnel number*, Kobe J. Math. **8**(1991) 73–79

[655] Y. Moriah, *Geometric structures and monodromy representations*, in "Knots 90" (Osaka, 1990), Walter de Gruyter (1992) 593–618

[656] Y. Moriah, *Incompressible surfaces and connected sum of knots*, J. Knot Theory Ramifications **7**(1998) 955–965

[657] Y. Moriah, H. Rubinstein, *Heegaard structures of negatively curved 3-manifolds*, Comm. Anal. Geom. **5**(1997) 375–412

[658] Y. Moriah, J. Schultens, *Irreducible Heegaard splittings of Seifert fibered spaces are either vertical or horizontal*, Topology **37**(1998) 1089–1112

[659] H. R. Morton, *Infinitely many fibered knots having the same Alexander polynomial*, Topology **17**(1978) 101–104

[660] H. R. Morton, *A criterion for an embedded surface in $\mathbf{R}^3$ to be unknotted*, in "Topology of low-dimensional manifolds" (Sussex, 1977), Lecture Notes in Math., 722, Springer Verlag (1979) 93–95

[661] H. R. Morton, *Closed braids which are not prime knots*, Math. Proc. Cambridge Philos. Soc. **86**(1979) 422–426

[662] H. R. Morton, *An irreducible 4-string braid with unknotted closure*, Math. Proc. Cambridge Philos Soc. **93**(1983) 259–261

[663] H. R. Morton, *Fibred knots with a given Alexander polynomial*, in "Nœuds, tresses et singularités", Enseign. Math., 31, Univ. Univ. de Genève (1983) 207–222

[664] H. R. Morton, *Alexander polynominals of closed 3-braids*, Math. Proc. Cambridge Philos. Soc. **96**(1984) 295–299

[665] H. R. Morton, *Exchangable braid*, in "Low-dimensional topology", London Math. Soc. Lecture Note Ser., 95, Cambridge Univ. Press (1985) 86–105

[666] H. R. Morton, *Seifert circles and knot polynomials*, Math. Proc. Cambridge Philos. Soc. **99**(1986) 107–109

[667] H. R. Morton, *Threading knot diagrams*, Math. Proc. Cambridge Philos. Soc. **99**(1986) 247–260

[668] H. R. Morton, *Problems*, in "Braids" (Santa Cruz, 1986), Contemp. Math., 78, Amer. Math. Soc. (1988) 557–574

[669] H. R. Morton, *Polynomials from braids*, in "Braids" (Santa Cruz, 1986), Contemp. Math., 78, Amer. Math. Soc. (1988) 575–585

[670] H. R. Morton, H. B. Short, *Calculating the 2-variable polynomial for knots presented as closed braids*, J. Algorithms **11**(1990) 117–131

[671] H. Murakami, *On the Conway polynomial of a knot with T-genus one*, Kobe J. Math. **2**(1985) 117–121

[672] H. Murakami, Y. Nakanishi, *On a certain move generating link-homology*, Math. Ann. **284**(1989) 75–89

[673] H. Murakami, K. Sugishita, *Triple points and knot cobordism*, Kobe J. Math. **1**(1984) 1–16

[674] H. Murakami, A. Yasuhara, *Crosscap number of a knot*, Pacific J. Math. **171** (1995) 261–273

[675] H. Murakami, A. Yasuhara, *Four-genus and four-dimensional clasp number of a knot*, preprint

[676] J. Murakami, *On local relations to determine the multi-variable Alexander polynomial of colored links*, in "Knots 90" (Osaka, 1990), Walter de Gruyter (1992) 455–464

[677] J. Murakami, *A state model for the multivariable Alexander polynomial*, Pacific J. Math. **157**(1993) 109–135

[678] K. Murasugi, *On the definition of the knot matrix*, Proc. Japan Acad. **37**(1961) 220–221

[679] K. Murasugi, *On a certain numerical invariant of link types*, Trans. Amer. Math. Soc. **117**(1965) 387–422

[680] K. Murasugi, *On the center of the group of a link*, Proc. Amer. Math. Soc. **16**(1965) 1052–1057 [Errata, ibid, **18**(1967), 1142]

[681] K. Murasugi, *On the Minkowski unit of slice links*, Trans. Amer. Math. Soc. **114**(1965) 377–383

[682] K. Murasugi, *On Milnor's invariants for links*, Trans. Amer. Math. Soc. **124**(1966) 94–110

[683] K. Murasugi, *The Arf invariant for knot types*, Proc. Amer. Math. Soc. **21**(1969) 69–72

[684] K. Murasugi, *On Milnor's invariant for links II, the Chen group*, Trans. Amer. Math. Soc. **148**(1970) 41–61

[685] K. Murasugi, *On the signature of links*, Topology **9**(1970) 283–298

[686] K. Murasugi, *On closed 3-braids*, Memoirs Amer. Math. Soc. No. 151, Amer. Math. Soc.(1974)

[687] K. Murasugi, *On the divisibility of knot groups*, Pacific J. Math. **52**(1974) 491–503

[688] K. Murasugi, *On a group that cannot be the group of a 2-knot*, Proc. Amer. Math. Soc. **64**(1977) 154–155

[689] K. Murasugi, *Seifert fibre spaces and braid groups*, Proc. London Math. Soc. **44**(1982) 71–84

[690] K. Murasugi, *On the Arf invariant of links*, Math. Proc. Cambridge Philos. Soc. **95**(1984) 61–69

[691] K. Murasugi, *Jones polynomials and classical conjectures in knot theory*, Topology **26**(1987) 187–194

[692] K. Murasugi, *Jones polynomials and classical conjectures in knot theory II*, Math. Proc. Cambridge Philos. Soc. **102**(1987) 317–318

[693] K. Murasugi, *Covering linkage invariants in abelian coverings of links*, Topology Appl. **25**(1987) 25–50

[694] K. Murasugi, *On the braid index of alternating links*, Trans. Amer. Math. Soc. **326**(1991) 237–260

[695] K. Murasugi, R. S. D. Thomas, *Isotopic closed nonconjugate braids*, Proc. Amer. Math. Soc. **33**(1972) 137-138

[696] Y. Nakagawa, *On the Alexander polynomials of slice links*, Osaka J. Math. **15**(1978) 161–182

[697] Y. Nakagawa, Y. Nakanishi, *Prime links, concordance and Alexander invariants II*, Math. Sem. Notes Kobe Univ. **9**(1981) 403–440

[698] Y. Nakanishi, *Prime links, concordance and Alexander invariants*, Math. Sem. Notes Kobe Univ. **8**(1980) 561–568

[699] Y. Nakanishi, *Unknotting numbers and knot diagrams with the minimum crossings*, Math. Sem. Notes Kobe Univ. **11**(1983) 257–258

[700] Y. Nakanishi, *A remark on critical points of link cobordism*, Kobe J. Math. **3**(1987) 209–212

[701] Y. Nakanishi, *On ribbon knots II*, Kobe J. Math. **7**(1990) 199–211

[702] Y. Nakanishi and M. Teragaito, *2-knots from a view of moving picture*, Kobe J. Math. **8** (1991) 161–172

[703] W. D. Neumann, *Signature related invariants of manifolds I, monodromy and γ-invariants*, Topology **18**(1979) 147–172

[704] W. D. Neumann, *Complex algebraic plane curves via their links at infinity*, Invent. Math. **98**(1989) 445-489

[705] W. Neumann, L. Rudolph, *Unfoldings in knot theory*, Math. Ann. **278**(1987) 409–439

[706] W. Neumann, L. Rudolph, *Unfoldings in knot theory*, Math. Ann. **282**(1988) 349–351

[707] W. D. Neumann, J. Wahl, *Casson invariant of links of singularities*, Comment. Math. Helv. **65**(1990) 58–78

[708] L. Neuwirth, *Knot Groups*, Ann. Math. Studies, 56, Princeton Univ. Press(1965)

[709] H. Noguchi, *On regular neighborhoods of 2-manifolds in 4-Euclidean space I*, Osaka Math. J. **8**(1956) 225–242

[710] H. Noguchi, *A classification of orientable surfaces in 4-space*, Proc. Japan Acad. **39**(1963) 422–423

[711] H. Noguchi, *Obstructions to locally flat embeddings of combinatorial manifolds*, Topology **5**(1966) 203–213

[712] R. A. Norman, *Dehn's Lemma for certain 4-manifolds*, Invent. Math. **7**(1969) 143–147

[713] J. O'Hara, *Energy of a knot*, Topology **30**(1991) 241–247.

[714] J. O'Hara, *Energy functionals of knots II*, Topology Appl. **56**(1994) 45–61.

[715] Y. Ohyama, *On the minimal crossing number and the braid index of links*, Canad. J. Math. **45**(1993) 117–131.

[716] M. Okada, *Delta-unknotting operation and the second coefficient of the Conway polynomial*, J. Math. Soc. Japan **42**(1990) 713–717

[717] A. Omae, *Note on ribbon 2-knots*, Proc. Japan Acad. **47**(1971) 850–853

[718] P. Orlik, *Seifert manifolds*, Lect. Notes in Math., 291, Springer Verlag (1972)

[719] P. Orlik, L. Solomon, *Braids and discriminants*, in "Braids" (Santa Cruz, 1986), Contemp. Math., 78, Amer. Math. Soc. (1988) 605–613

[720] K. E. Orr, *New link invariants and applications*, Comment. Math. Helv. **62**(1987) 542–560

[721] K. E. Orr, *Homotopy invariants of links*, Invent. Math. **95**(1989) 379–394

[722] K. E. Orr, *Link concordance invariants and Massey products*, Topology **30**(1991) 699–710

[723] P. Pao, *Non-linear circle actions on the 4-sphere and twisting spun knots*, Topology **17**(1978) 291–296

[724] C. D. Papakyriakopoulos, *On Dehn's lemma and the asphericity of knots*, Ann. of Math. **66**(1957) 1–26

[725] R. Penrose, *Twistors, particles, strings and links*, in "The interface of mathematics and particle physics" (Oxford, 1988), Inst. Math. Appl. Conf. Ser. New Ser., 24, Oxford Univ. Press (1990) 49–58

[726] R. Piergallini, *Covering moves*, Trans. Amer. Math. Soc. **325**(1991) 903–920

[727] R. Piergallini, *Covering homotopy 3-spheres*, Comment. Math. Helv. **67**(1992) 287–292

[728] U. Pinkall, *Regular homotopy classes of immersed surfaces*, Topology **24**(1985) 421–434

[729] A. Pizer, *Matrices over group rings which are Alexander matices*, Osaka J. Math. **21**(1984) 461–472

[730] A. Pizer, *Non reversible knots exist*, Kobe J. Math. **1**(1984) 23–29

[731] A. Pizer, *Hermitian character and the first problem of R. H. Fox*, Math. Proc. Cambridge Philos. Soc. **98**(1985) 447–458

[732] A. Pizer, *Hermitian character and the first problem of R. H. Fox for links*, Math. Proc. Cambridge Philos. Soc. **102**(1987) 77–86

[733] A. Pizer, *Matrices which are knot module matrices*, Kobe J. Math. **5**(1988) 21–28

[734] S. Plotnick, *Infinitely many disk knots with the same exterior*, Math. Proc. Cambridge Phil. Soc. **98**(1983) 67–72

[735] S. Plotnick, *The homotopy type of four dimensional knot complements*, Math. Z. **183**(1983) 447–471

[736] S. Plotnick, *Fibered knots in S^4-twisting, spinning, rolling, surgery and branching*, in "Four manifold theory", Contemp. Math., 35, Amer. Math. Soc. (1984) 437–459

[737] S. Plotnick, *Equivariant intersection forms, knots in S^4, and rotations in 2-spheres*, Trans. Amer. Math. Soc. **296**(1986) 543–575

[738] S. P. Plotnick, A. I. Suciu, *k-invariants of knotted 2-spheres*, Comment. Math. Helv. **60**(1985) 54–84

[739] V. Poénaru, *A note on the generators for the fundamental group of the complement of a submanifold of codimension 2*, Topology **10**(1971) 47–52

[740] T. M. Price, *Homeomorphisms of quaternion space and projective planes in four space*, J. Austral. Math. Soc. **23**(1977) 112–128

[741] T. M. Price and D. M. Roseman, *Embeddings of the projective plane in four space*, preprint

[742] J. H. Przytycki, *Skein modules of 3-manifolds*, Bull. Polish Acad. Sci. Math. **39**(1991) 91–100

[743] F. Quinn, *Topological transversality holds in all dimensions*, Bull. Amer. Math. Soc. (N.S.) **18**(1988) 145–148

[744] M. Rampichini, *Exchangeable fibred links*, preprint

[745] E. S. Rapaport, *On the commutator subgroup of a knot group*, Ann. of Math. **71**(1960) 157–162

[746] E. S. Rapaport, *Knot-like groups*, in "Knots, groups and 3-manifolds", Ann. Math. Studies, 84, Princeton Univ. Press (1975) 119–133

[747] K. Reidemeister, *Knoten und Gruppen*, Abh. Math. Sem. Univ. Hamburg **5**(1926) 7–23

[748] K. Reidemeister, *Knotentheorie*, Ergebn. Math. Grenzgeb., 1, Springer Verlag (1932)

[749] K. Reidemeister, *Zur dreidimensionalen Topologie*, Abh. Math. Sem. Univ. Hamburg **9**(1933) 189–194

[750] N. Yu. Reshetikhin, V. G. Turaev, *Invariants of 3-manifolds via link polynomials and quantum groups*, Invent. Math. **103**(1991) 547–597

[751] P. M. Rice, *Equivalence of Alexander matrices*, Math. Ann. **193**(1971) 65–75

[752] W. Richter, *High-dimensional knots with $\pi_1 \cong \mathbf{Z}$ are determined by their complements in one more dimension than Farber's range*, Proc. Amer. Math. Soc. **120**(1994) 285–294

[753] R. Riley, *Homomorphisms of knot groups on finite groups*, Math. Comput. **25**(1971) 603–619

[754] R. Riley, *Algebra for Heckoid groups*, Trans. Amer. Math. Soc. **334**(1992) 389–409

[755] R. A. Robertello, *An invariant of knot cobordism*, Commun. Pure Appl. Math. **18**(1965) 543–555

[756] V. A. Rokhlin , *Two-dimensional submanifolds of four-dimensional manifolds*, Func. Anal. App. **5**(1971) 39–48

[757] D. Rolfsen, *Isotopy of links in codimension two*, J. Indian Math. Soc. **36**(1972) 263–278

[758] D. Rolfsen, *Some counterexamples in link theory*, Canad. J. Math. **26**(1974) 978–984

[759] D. Rolfsen, *A surgical view of Alexander's polynomial*, in "Proc. Geometric Topology Conf." (Park City, 1974), Lect. Notes in Math., 438, Springer Verlag (1975) 415–423

[760] D. Rolfsen, *Localized Alexander invariants and isotopy of links*, Ann. of Math. **101**(1975) 1–19

[761] D. Rolfsen, *Knots and links*, Publish or Perish, Inc. (1976)

[762] D. Rolfsen, *Piecewise linear I-equivalence of links*, in "Low dimensional topology", London Math. Soc. Lecture Note Ser., 95, Cambridge Univ. Press (1985) 161–178

[763] D. Rolfsen, *PL link isotopy, essential knotting and quotients of polynomials*, Canad. Math. Bull. **34**(1991) 536-541

[764] D. Rolfsen, *The quest for a knot with trivial Jones polynomial: diagram surgery and the Temperly-Lieb algebra*, in "Topics in knot theory" (Erzurum, 1992), NATO ASI Ser. C, 399, Kluwer Academic Publ. (1993) 195–210

[765] D. Roseman, *Woven knots are spun knots*, Osaka J. Math. **11**(1974) 307–312

[766] D. Roseman, *Projections of knots*, Fund. Math. **89**(1975) 99–110

[767] D. Roseman, *The spun square knot is the spun granny knot*, Bol. Soc. Mat. Mexicana (2) **20**(1975) 49–55

[768] D. Roseman, *Spinning knots about submanifolds; spinning knots about projections of knots*, Topology Appl. **31**(1989) 225-241

[769] D. Roseman, *Motions of flexible objects*, in "Modern geometric computing for visualization" (Tokyo, 1992), Springer, Tokyo (1992) 91–120

[770] D. Roseman, *Reidemeister-type moves for surfaces in four-dimensional space*, in "Knot theory" (Warsaw, 1995), Banach Center Publ., 42, Polish Acad. Sci., Warsaw (1998) 347–380

[771] C. P. Rourke and B. J. Sanderson, *Introduction to piecewise-linear topology*, Springer-Verlag (1972)

[772] D. Ruberman, *Doubly slice knots and the Casson-Gordon invariants*, Trans. Amer. Math. Soc. **279**(1983) 569–588

[773] D. Ruberman, *Invariant knots of free involutions of S^4*, Topology Appl. **18**(1984) 217–224

[774] D. Ruberman, *Concordance of links in S^4*, in "Four-manifold theory" (Durham, 1982), Contemp. Math., 35, Amer. Math. Soc. (1984) 481–483

[775] D. Ruberman, *The Casson-Gordon invariants in high dimensional knot theory*, Trans. Amer. Math. Soc. **306**(1988) 579–595

[776] D. Ruberman, *Seifert surfaces of knots in S^4*, Pacific J. Math. **145**(1990) 97–116

[777] L. Rudolph, *Non-trivial positive braids have positive signature*, Topology **21**(1982) 325–327

[778] L. Rudolph, *Braided surfaces and Seifert ribbons for closed braids*, Comment. Math. Helv. **58**(1983) 1–37

[779] L. Rudolph, *Algebraic functions and closed braids*, Topology **22**(1983) 191–202

[780] L. Rudolph, *Construction of quasipositive knots and links I*, in "Knots, braids and singularities" (Plans-sur Bex, 1982), Enseign. Math., 31 (1983) 233–245

[781] L. Rudolph, *Constructions of quasipositive knots and links II*, in "Four manifold theory" (Durham, 1982), Contemp. Math., 35, Amer. Math. Soc. (1984) 481–483

[782] L. Rudolph, *Some topologically locally-flat surfaces in the complex projective plane*, Comment. Math. Helv. **59**(1984) 592–599

[783] L. Rudolph, *Special positions for surfaces bounded by closed braids*, Rev. Mat. Iberoamericana **1**(1985) 93–133

[784] L. Rudolph, *Isolated critical points of mappings from $\mathbf{R}^4$ to $\mathbf{R}^2$ and a natural splitting of the Milnor number of a classical fibered link I, basic theory; examples*, Comment. Math. Helv. **62**(1987) 630–645

[785] L. Rudolph, *Isolated critical points of mappings from $\mathbf{R}^4$ to $\mathbf{R}^2$ and a natural splitting of the Milnor number of a cassical fibered link II*, in "Geometry and topology" (Athens/Ga., 1985), Lect. Notes in Pure and Appl. Math., 105, Marcel Dekker (1987) 251–263

[786] L. Rudolph, *Mutually braided open books and new invariants of fibered links*, in "Braids" (Santa Cruz, 1986), Contemp. Math., 78, Amer. Math. Soc. (1988) 657–673

[787] L. Rudolph, *Quasipositivity and new knot invariants*, Rev. Mat. Univ. Complut. Madrid **2**(1989) 85–109

[788] L. Rudolph, *A congruence between link polynomials*, Math. Proc. Cambridge Philos. Soc. **107**(1990) 319–327

[789] L. Rudolph, *Quasipositive annuli (Constructions of quasipositive knots and links IV)*, J. Knot Theory Ramifications **1**(1991) 451–466

[790] L. Rudolph, *Totally tangential links of intersection of complex plane curves with round spheres*, in "Topology '90" (Ohio, 1990), Walter de Gruyter (1992) 343–349

[791] L. Rudolph, *A characterization of quasipositive Seifert surfaces (Constructions of quasipositive knots and links, III)*, Topology **31**(1992) 231–237

[792] L. Rudolph, *Quasipositivity as an obstruction to sliceness*, Bull. Amer. Math. Soc. **29**(1993) 51–59

[793] O. Saeki, K. Sakuma, *Immersed n-manifolds in R^{2n} and the double points of their generic projections into R^{2n-1}*, Trans. Amer. Math. Soc. **348** (1996) 2585–2606

[794] M. Saito, *Minimal number of saddle points of properly embedded surfaces in the 4-ball*, Math. Sem. Notes Kobe Univ. **11**(1983) 345–348

[795] M. Saito, *A note on cobordism of surface links in S^4*, Proc. Amer. Math. Soc. **111**(1991) 883–887

[796] T. Sakai, *A remark on the Alexander polynomials of knots*, Math. Sem. Notes Kobe Univ. **5**(1977) 451–456

[797] M. Sakuma, *The homology groups of abelian coverings of links*, Math. Sem. Notes Kobe Univ. **7**(1979) 515–530

[798] M. Sakuma, *On strongly invertible knots*, in "Algebraic and topological theories - to the memory of Dr. T. Miyata", Kinokuniya Co. Ltd. (1985) 176–196

[799] M. Sakuma, *Uniqueness of symmetries of knots*, Math. Z. **192**(1986) 225–242

[800] H. Saleur, *The multivariable Alexander polynomial and modern knot theory*, in "Topological and quantum group methods in field theory and condensed matter physics", Internat. J. Modern Phys. B, 6 (1992) 1857–1869

[801] B. J. Sanderson, *Triple links in codimension 2*, in "Topology, theory and applications II" (Pécs, 1989), Colloq. Math. Soc. János Bolyai, 55, North-Holland (1993) 457–471

[802] N. A. Sato, *Alexander modules of sublinks and an invariant of classical link concordance*, Illinois J. Math. **25**(1981) 508–519

[803] N. A. Sato, *Free coverings and modules of boundary links*, Trans. Amer. Math. Soc. **264**(1981) 499–505

[804] N. A. Sato, *Algebraic invariants of boundary links*, Trans. Amer. Math. Soc. **265**(1981) 359–375

[805] N. A. Sato, *Cobordisms of semi-boundary links*, Topology Appl. **18**(1984) 225–234

[806] Y. Sato, *The reflexivity of 2-knots in $S^2 \times S^2$*, J. Knot Theory Ramifications **1**(1992) 21–29

[807] M. Scharlemann, *Smooth spheres in $\mathbf{R}^4$ with four critical points are standard*, Invent. Math. **79**(1985) 125–141

[808] M. Scharlemann, *Unknotting number one knots are prime*, Invent. Math. **82**(1985) 37–55

[809] M. Scharlemann, *Handlebody complements in the 3-sphere: a remark on a theorem of Fox*, Proc. Amer. Math. Soc. **115**(1992) 1115–1117

[810] M. Scharlemann, *Unlinking via simultaneous crossing changes*, Trans. Amer. Math. Soc. **336**(1993) 855–868

[811] M. Scharlemann, A. Thompson, *Unknotting number, genus, and companion tori*, Math. Ann. **280**(1988) 191–205

[812] P. Scott, *Braid groups and the group of homomorphism of a surface*, Math. Proc. Cambridge Philos Soc. **68**(1970) 605–617

[813] P. Scott, *The geometries of 3-manifolds*, Bull. London Math. Soc. **15**(1983) 401–487

[814] M. Sekine, *Kawauchi's second duality and knotted surfaces in 4-sphere*, Hiroshima Math. J. **19**(1989) 641–651

[815] M. Sekine, *On homology of the double covering over the exterior of a surface in 4-sphere*, Hiroshima Math. J. **21**(1991) 419–426

[816] T. Shibuya, *On the homotopy of links*, Kobe J. Math. **5**(1988) 87–95

[817] T. Shibuya, *Self Δ-equivalence of ribbon links*, Osaka J. Math. **33**(1996) 751–760

[818] A. Shima, *An unknotting theorem for tori in S^4. II*, Kobe J. Math. **13** (1996) 9–25

[819] A. Shima, *On simply knotted tori in S^4*, J. Math. Sci. Univ. Tokyo **4** (1997) 279–339

[820] A. Shima, *Immersions from the 2-sphere to the 3-sphere with only two triple points* (Japanese), in "Topology of real singularities and related topics" (Kyoto, 1997), Suurikaisekikenkyusho Koukyuroku **1006** (1997) 146–160

[821] A. Shima, *On simply knotted tori in S^4. II*, in "KNOTS '96" (Tokyo, 1996), World Sci. Publishing (1997) 551–568

[822] A. Shima, *An unknotting theorem for tori in S^4*, Rev. Mat. Complut. **11** (1998) 299–309

[823] Y. Shinohara, *On the signature of knots and links*, Trans. Amer. Math. Soc. **156**(1971) 273–285

[824] T. Shiomi, *On imbedding 3-manifolds into 4-manifolds*, Osaka J. Math. **28**(1991) 649–661

[825] D. S. Silver, *Examples of 3-knots with no minimal Seifert manifolds*, Math. Proc. Cambridge Philos. Soc. **110**(1991) 417–420

[826] D. S. Silver, *On knot-like groups and ribbon concordance*, J. Pure Appl. Algebra **82**(1992) 99–105

[827] D. S. Silver, *Augmented group systems and n-knots*, Math. Ann. **296**(1993) 585–593

[828] D. S. Silver, *On the existence of minimal Seifert manifolds*, Math. Proc. Cambridge Philos. Soc. **114**(1993) 103–109

[829] D. S. Silver, *Free group automorphisms and knotted tori in S^4*, J. Knot Theory Ramifications **6**(1997) 95–103

[830] J. Simon, *An algebraic classification of knots in S^3*, Ann. of Math. **97**(1973) 1–13

[831] J. Simon, *Wirtinger approximations and the knot groups of F^n in S^{n+2}*, Pacific J. Math. **90**(1980) 177–190

[832] R. K. Skora, *Splittings of surfaces*, Bull. Amer. Math. Soc. **23**(1990) 85–90

[833] R. K. Skora, *Knot and link projections in 3-manifolds*, Math. Z. **206**(1991) 345–350

[834] R. K. Skora, *Closed braids in 3-manifolds*, Math. Z. **211**(1992) 173–187

[835] S. Smale, *A classification of immersions of the two-sphere*, Trans. Amer. Math. Soc. **90**(1959) 281–290

[836] S. Smale, *The classification of immersions of spheres in euclidean space*, Ann. of Math. **69**(1959) 327–344

[837] P. A. Smith, *A theorem on fixed points for periodic transformations*, Ann. of Math. **35**(1934) 572–578

[838] L. Smolinsky, *Doubly sliced knots which are not the double of a disk*, Trans. Amer. Math. Soc. **298**(1986) 723–732

[839] L. Smolinsky, *A generalization of the Levine-Tristram link invariant*, Trans. Amer. Math. Soc. **315**(1989) 205–217

[840] L. Smolinsky, *Invariants of link cobordism*, Topology Appl. **32**(1989) 161–168

[841] N. Smythe, *Boundary links*, in "Topology seminar" (Wisconsin, 1965), Ann. of Math. Studies, 60, Princeton Univ. Press (1966) 69–72

[842] N. Smythe, *Isotopic invariants of links and the Alexander matrix*, Amer. J. Math. **89**(1967) 693–704

[843] N. Smythe, *Topological invariants of isotopy of links I*, Amer. J. Math. **92**(1970) 86–98

[844] N. Smythe, *The Burau representation of the braid group is pairwise free*, Arch. Math. **32**(1979) 309–317

[845] T. Soma, *The Gromov invariant of links*, Invent. Math. **64**(1981) 445–454

[846] A. B. Sosinskii, *Multidimensional knots*, Soviet Math. Doklady **6**(1965) 1119–1122

[847] A. B. Sosinskii, *Homotopy of knot complements*, Soviet Math. Doklady **8**(1967) 1324–1328

[848] A. B. Sosinskii, *Decompositions of knots*, Math. USSR-Sbornik **10**(1970) 139–150

[849] J. Stallings, *On topologically unknotted spheres*, Ann. of Math. **77**(1963) 490–503

[850] J. Stallings, *Homology and central series of groups*, J. of Algebra **2**(1965) 170–181

[851] J. Stallings, *Lectures on polyhedral topology*, Tata Institute of Fundamental Research, Bombay (1968)

[852] J. Stallings, *Group theory and three-dimensional manifolds*, Yale Math. Monographs, 4, Yale Univ. Press (1971)

[853] J. Stallings, *Construction of fibered knots and links*, in "Algebraic and geometric topology" (Stanford, 1976), Proc. Symp. Pure Math., 32-II, Amer. Math. Soc. (1978) 55–60

[854] T. Stanford, *The functoriality of Vassiliev-type invariants of links, braids, and knotted graphs*, J. Knot Theory Ramifications **3**(1994) 247–262

[855] T. Stanford, *Braid commutators and Vassiliev invariants*, Pacific J. Math. **174** (1996) 269–276

[856] D. W. J. Stein, *Computing Massey product invariants of links*, Topology Appl. **32**(1989) 169–181

[857] D. W. J. Stein, *Massey products in the cohomology of groups with applications to link theory*, Trans. Amer. Math. Soc. **318**(1990) 301–325

[858] N. Stoltzfus, *Algebraic computations of the integral concordance and double null concordance group of knots*, in "Knot theory" (Plans-sur-Bex,1977), Lecture Notes in Math., 685, Springer Verlag (1978) 274–290

[859] R. Stong, *Uniqueness of connected sum decompositions in dimension 4*, Topology Appl. **56**(1994) 277–291

[860] P. Strickland, *Which finite simple knots are twist-spun?*, Proc. London Math. Soc. **56**(1988) 114–142

[861] A. I. Suciu, *Infinitely many ribbon knots with the same fundamental group*, Math. Proc. Cambridge Philos. Soc. **98**(1985) 481–492

[862] A. I. Suciu, *Iterated spinnig and homology spheres*, Trans. Amer. Math. Soc. **321**(1990) 145–157

[863] A. I. Suciu, *Inequivalent frame-spun knots with the same complement*, Comment. Math. Helv. **67**(1992) 47–63

[864] D. W. Sumners, *Invertible knot cobordisms*, Comment. Math. Helv. **46**(1971) 240–256

[865] D. W. Sumners, *Polynomial invariants and the integral homology of coverings of knots and links*, Invent. Math. **15**(1972) 78–90

[866] D. W. Sumners, *On the homology of finite cyclic coverings of higher-dimensional links*, Proc. Amer. Math. Soc. **46**(1974) 143–149

[867] D. W. Sumners, J. M. Woods, *The monodromy of reducible curves*, Invent. Math. **40**(1977) 107–141

[868] P. A. Sundheim, *Reidemeister's theorem for 3-manifolds*, Math. Proc. Cambridge Philos. Soc. **110**(1991) 281–292

[869] P. A. Sundheim, *The Alexander and Markov theorems via diagrams for links in 3-manifolds*, Trans. Amer. Math. Soc. **337** (1993) 591–607

[870] S. Suzuki, *Local knots of 2-spheres in 4-manifolds*, Proc. Japan Acad. **45**(1969) 34–38

[871] S. Suzuki, *Knotting problems of 2-spheres in the 4-sphere*, Math. Sem. Notes Kobe Univ. **4**(1976) 241–371

[872] S. Suzuki, *Alexander ideals of graphs in the 3-sphere*, Tokyo J. Math. **7**(1984) 233–247

[873] S. Suzuki, *Almost unknotted θ_n-curves in the 3-sphere*, Kobe J. Math. **1**(1984) 19–22

[874] G. A. Swarup, *An unknotting criterion*, J. Pure Appl. Algebra **6**(1975) 291–296

[875] R. Takase, *Note on orientable surfaces in 4-space*, Proc. Japan Acad. **39**(1963) 424

[876] I. Tamura, *Unknotted codimension one spheres in smooth manifolds*, Topology **23**(1984) 127–132

[877] K. Taniyama, *Cobordism, homotopy and homology of graphs in $\mathbf{R}^3$*, Topology **33**(1994) 509–523

[878] K. Taniyama, A. Yasuhara, *On C-distance of knots*, Kobe J. Math. **11**(1994) 117–127

[879] K. Tatsuoka, *An isoperimetric inequality for Artin groups of finite type*, Trans. Amer. Math. Soc. **339**(1993) 537–551

[880] M. Teragaito, *Fibered 2-knots and lens spaces*, Osaka J. Math. **26**(1989) 57–63 and **26**(1989) 953

[881] M. Teragaito, *Twisting symmetry-spins of 2-bridge knots*, Kobe J. Math. **6**(1989) 117–126

[882] M. Teragaito, *Twisting symmetry-spins of pretzel knots*, Proc. Japan Acad. **66**(1990) 179–183

[883] M. Teragaito, *A note on untwisted deform-spun 2-knots*, Proc. Japan Acad. **68**(1992) 75–78

[884] M. Teragaito, *Composite knots trivialized by twisting*, J. Knot Theory Ramifications **1**(1992) 467–470

[885] M. Teragaito, *Symmetry-spun tori in the four-sphere*, in "Knots 90" (Osaka, 1990), Walter de Gruyter (1992) 163–171

[886] M. Teragaito, *Roll-spun knots*, Math. Proc. Camb. Phil. Soc. **113**(1993) 91–96 and **116**(1994) 191

[887] M. Teragaito, *Twist-roll spun knots*, Proc. Amer. Math. Soc. **122** (1994) 597–599

[888] H. Terasaka, F. Hosokawa, *On the unknotted sphere S^2 in E^4*, Osaka J. Math. **13**(1961) 265–270

[889] M. B. Thistlethwaite, *Kauffman's polynomial and alternating links*, Topology **27**(1988) 311–318

[890] R. S. D. Thomas, *The structure of the fundamental braids*, Quart. J. Math. Oxford **26**(1975) 283–288

[891] R. S. D. Thomas, *Partially closed braids*, Canad. Math. Bull. **17**(1975) 99–107

[892] A. Thompson, *A polynomial invariant of graphs in 3-manifolds*, Topology **31**(1992) 657–665

[893] W. P. Thurston, *Three-dimensional geometry and topology*, Edited by Silvio Levy, Princeton Mathematical Series, 35, Princeton University Press, Princeton, NJ (1997)

[894] W. P. Thurston, *Finite state algorithms for the braid groups*, preprint

[895] G. Torres, *On the Alexander polynomial*, Ann. of Math. **57**(1953) 57–89

[896] G. Torres, R. H. Fox, *Dual presentations of the group of a knot*, Ann. of Math. **59**(1954) 211–218

[897] B. Trace, *On the Reidemeister moves of a classical knot*, Proc. Amer. Math. Soc. **89**(1983) 722–724

[898] B. Trace, *A note concerning Seifert manifolds for 2-knots*, Math. Proc. Cambridge Philos. Soc. **100**(1986) 113–116

[899] B. Trace, *Some comments concerning the Levine approach to slicing classical knots*, Topology Appl. **23**(1986) 217–235

[900] B. Trace, *A note concerning the 3-manifolds which span certain surfaces in the 4-ball*, Proc. Amer. Math. Soc. **102**(1988) 177–182

[901] P. Traczyk, *Nontrivial negative links have positive signature*, Manuscripta Math. **61**(1988) 279–284

[902] P. Traczyk, *A new proof of Markov's braid theorem*, in "Knot theory" (Warsaw, 1995), Banach Center Publ., 42, Polish Acad. Sci., Warsaw (1998) 409–419

[903] P. Traczyk, *3-braids with proportional Jones polynomials*, Kobe J. Math. **15** (1998) 187–190

[904] P. Traczyk, *A criterion for signed unknotting number*, in "Low Dimensional Topology" (Madeira, Portugal, 1998), Contemp. Math., 233, Amer. Math. Soc. (1999) 215–220

[905] L. Traldi, *The determinantal ideals of link modules I*, Pacific J. Math. **101**(1982) 215–222

[906] L. Traldi, *A generalization of Torres' second relation*, Trans. Amer. Math. Soc. **269**(1982) 593–610

[907] L. Traldi, *Linking numbers and the elementary ideals of links*, Trans. Amer. Math. Soc. **275**(1983) 309–318

[908] L. Traldi, *The determinantal ideals of link modules II*, Pacific J. Math. **109**(1983) 237–245

[909] L. Traldi, *Some properties of the determinantal ideals of link modules*, Math. Sem. Notes Kobe Univ. **11**(1983) 363–380

[910] L. Traldi, *Milnor's invariants and the completions of link modules*, Trans. Amer. Math. Soc. **284**(1984) 401–429

[911] H. F. Trotter, *Periodic automorphism of groups and knots*, Duke Math. J **28**(1961) 553–557

[912] H. F. Trotter, *Non-invertible knots exist*, Topology **2**(1964) 341–358

[913] H. F. Trotter, *On S-equivalence of Seifert matrices*, Invent. Math. **20**(1973) 173–207

[914] C. M. Tsau, *Algebraic meridians of knot groups*, Trans. Amer. Math. Soc. **294**(1986) 733–747

[915] C. M. Tsau, *Isomorphisms and peripheral structure of knot groups*, Math. Ann. **282**(1988) 343–348

[916] V. G. Turaev, *Reidemeister torsion in knot theory*, Russian Math. Surveys **41-1**(1986) 119–182

[917] V. G. Turaev, *On Torres-type relations for the Alexander polynomials of links*, Enseign. Math. (2) **34**(1988) 69–82

[918] V. G. Turaev, *Quantum invariants of knots and 3-manifolds*, de Gruyter Studies in Math., 18, Walter de Gruyter (1994)

[919] V. G. Turaev, O. Ya. Viro, *State sum invariants of 3-manifolds and quantum 6j-symbols*, Topology **31**(1992) 865–902

[920] J. Van Buskirk, *Braid groups of compact 2-manifolds with elements of finite order*, Trans. Amer. Math. Soc. **122** (1966) 81–97

[921] V. A. Vassiliev, *Cohomology of knot space*, in "Theory of Singularities and Its Applications", Advances in Soviet Math., vol. 1, Amer. Math. Soc. (1990)

[922] G. A. Venema, *A topological disk in a 4-manifold can be approximated by piecewise linear disks*, Bull. Amer. Math. Soc. **83** (1977) 386–387

[923] G. A. Venema, *Approximating topological surfaces in 4-manifolds*, Trans. Amer. Math. Soc. **265** (1981) 35–45

[924] O. Ya. Viro, *Linkings, two-sheeted branched coverings and braids*, Math. USSR-Sb. **16**(1972) 223–236

[925] O. Ya. Viro, *Branched coverings of manifolds with boundary, and invariants of links*, Math. USSR-Izv. **7**(1973) 1239–1356

[926] O. Ya. Viro, *Local knotting of submanifolds*, Math. USSR-Sb. **19**(1973) 166–176

[927] O. Ya. Viro, *Two-fold branched coverings of the three-sphere*, J. Soviet Math. **8-5**(1977) 531–553

[928] O. Ya. Viro, Lecture given at Osaka City University, September, 1990

[929] O. Ya. Viro, *Moves of triangulations of a PL-manifold*, in "Quantum groups" (Leningrad, 1990), Lecture Notes in Math., 1510, Springer Verlag (1992) 367–372

[930] T. L. Vo, G. A. Venema, *Characterization of knot complements in the 4-sphere*, Topology Appl. **42**(1991) 231–245

[931] T. L. Vo, G. A. Venema, *Complements of 2-spheres in 4-manifolds*, in "Topology Hawaii" (Honolulu, 1990), World Sci. Publ. (1992) 157–163

[932] T. L. Vo, G. A. Venema, *On the asphericity of knot complements*, Canad. J. Math. **45**(1993) 340–356

[933] V. A. Voevodskii, M. M. Kapranov, *Free n-category generated by a cube, oriented matroids, and higher Bruhat orders*, Func. Anal. App. **25**(1991) 50–52

[934] P. Vogel, *Representation of links by braids: A new algorithm*, Comment. Math. Helv. **65** (1990) 104–113

[935] M. Wada, *Twisted Alexander polynomial for finitely presentable groups*, Topology **33**(1994) 241–256

[936] B. Wajnryb, *Markov classes in certain finite symplectic representations of braid groups*, Contemp. Math. **78**(1988) 687–695

[937] B. Wajnryb, *A braidlike presentation of $Sp(n,p)$*, Israel J. Math. **76**(1991) 265–288

[938] M. Wakui, *On Dijkgraaf-Witten invariant for 3-manifolds*, Osaka J. Math. **29**(1992) 675–696

[939] F. Waldhausen, *On irreducible 3-manifolds which are sufficiently large*, Ann. of Math. **87**(1968) 56–88

[940] C. T. C. Wall, *Unknotting tori in codimension one and spheres in codimension two*, Math. Proc. Cambridge Philos. Soc. **61**(1965) 659–664

[941] C. T. C. Wall, *Locally flat PL submanifolds with codimension two*, Math. Proc. Cambridge Philos. Soc. **63** (1967) 5–7

[942] C. T. C. Wall, *Surgery on compact manifolds*, Academic Press (1970)

[943] S. C. Wang, *3-manifolds which admit finite group actions*, Trans. Amer. Math. Soc. **339**(1993) 191–203

[944] C. Weber, *Torsion dans les modules d'Alexander*, in "Knot theory" (Plans-sur-Bex, 1977), Lecture Notes in Math., 685, Springer Verlag (1978) 300–308

[945] D. J. A. Welsh, *Knots and braids: some algorithmic questions*, in "Graph structure theory" (Seattle, 1991), Contemp. Math., 147, Amer. Math. Soc. (1993) 109–123

[946] H. Wenzl, *Representations of braid groups and the quantum Yang-Baxter equation*, Pacific J. Math. **145**(1990) 153–180

[947] H. Wenzl, *Unitary braid representations*, in "Infinite analysis, Part A, B" (Kyoto, 1991), Adv. Ser. Math. Phys., 16, World Sci. Publ. (1992) 985–1006

[948] H. Wenzl, *Braids and invariants of 3-manifolds*, Invent. Math. **114**(1993) 235–275

[949] J. H. C. Whitehead, *On doubled knots*, J. London Math. Soc. **12**(1937) 63–71

[950] H. Whitney, *The self-intersections of a smooth n-manifold in 2n-space*, Ann. of Math. **45**(1944) 220–246

[951] W. Whitten, *Symmetries of links*, Trans. Amer. Math. Soc. **135** (1969) 213–222

[952] W. Whitten, *Algebraic and geometric characterizations of knots*, Invent. Math. **26**(1974) 259–270

[953] W. Whitten, *Knot complements and groups*, Topology **26**(1987) 41–44

[954] R. F. Williams, *The braid index of an algebraic link*, in "Braids" (Santa Cruz, 1986), Contemp. Math., 78, Amer. Math. Soc. (1988) 697–703

[955] R. F. Williams, *The braid index of generalized cables*, Pacific J. Math. **155**(1992) 369–375

[956] E. Witten, *Quantum field theory and the Jones polynomial*, Commun. Math. Phys. **121**(1989) 351–399

[957] Y. S. Wu, *Braid groups, anyons and gauge invariance*, in "Physics in $(2+1)$-dimension" (Sorak Mountain Resort, 1991), World Sci. Publ. (1992) 108–132

[958] T. Yajima, *On the fundamental groups of knotted 2-manifolds in the 4-space*, J. Math. Osaka City Univ. **13** (1962) 63–71

[959] T. Yajima, *On simply knotted spheres in $\mathbf{R}^4$*, Osaka J. Math. **1** (1964) 133–152

[960] T. Yajima, *On a characterization of knot groups of some spheres in $\mathbf{R}^4$*, Osaka J. Math. **6** (1969) 435–446

[961] T. Yajima, *Wirtinger presentations of knot groups*, Proc. Japan Acad. **46** (1970) 997–1000

[962] T. Yajima and S. Kinoshita, *On the graphs of knots*, Osaka Math. J. **9** (1957) 155–163

[963] S. Yamada, *The minimal number of Seifert circles equals the braid index of a link*, Invent. Math. **89** (1987) 347–356

[964] S. Yamada, *An operator on regular isotopy invariants of link diagrams*, Topology **28**(1989) 369–377

[965] S. Yamada, *An invariant of spatial graphs*, J. Graph Theory **13**(1989) 537–551

[966] Y. Yamada, *An extension of Whitney's congruence*, Osaka J. Math. **32**(1995) 185–192

[967] Y. Yamada, *Decomposition of S^4 as a twisted double of a certain manifold*, Tokyo J. Math. **20** (1997) 23–33

[968] M. Yamamoto, *Classification of isolated algebraic singularities by their Alexander polynomial*, Topology **23**(1984) 277–287

[969] T. Yanagawa, *Brunnian systems of 2-spheres in 4-space*, Osaka J. Math. **1**(1964) 127–132

[970] T. Yanagawa, *On ribbon 2-knots I, the 3-manifold bounded by the 2-knots*, Osaka J. Math. **6**(1969) 447–464

[971] T. Yanagawa, *On ribbon 2-knots II, the second homotopy group of the complementary second homotopy group of the complementary domain*, Osaka J. Math. **6**(1969) 465–474

[972] T. Yanagawa, *On ribbon 2-knots III, on the unknotting ribbon 2-knots in S^4*, Osaka J. Math. **7**(1970) 165–172

[973] T. Yanagawa, *On cross sections of higher dimensional ribbon knots*, Math. Sem. Notes Kobe Univ. **7**(1977) 609–628

[974] T. Yanagawa, *A note on ribbon n-knots with genus 1*, Kobe J. Math. **2**(1985) 99–102

[975] T. Yasuda, *A presentation and the genus for ribbon n-knots*, Kobe J. Math. **6**(1989) 71–88

[976] T. Yasuda, *Ribbon knots with two ribbon types*, J. Knot Theory Ramifications **1**(1992) 477–482

[977] T. Yasuda, *On ribbon presentations of ribbon knots*, J. Knot Theory Ramifications **3**(1994) 223–231

[978] A. Yasuhara, *On slice knots in the complex projective plane*, Rev. Mat. Univ. Complut. Madrid **5**(1992) 255–276

[979] A. Yasuhara, *Link homology in 4-manifolds*, Bull. London Math. Soc. **28**(1996) 409–412

[980] D. N. Yetter, *Markov algebras*, in "Braids" (Santa Cruz, 1986), Contemp. Math., 78, Amer. Math. Soc. (1988) 705–730

[981] D. N. Yetter, *Framed tangles and a theorem of Deligne on braided deformations of Tannakian categories*, in "Deformation theory and quantum groups with applications to mathematical physics" (Amherst, MA, 1990), Contemp. Math., 134, Amer. Math. Soc. (1992) 325–349

[982] K. Yoshikawa, *On 2-knot groups with the finite commutator subgroup*, Math. Sem. Notes Kobe Univ. **8**(1980) 321–330

[983] K. Yoshikawa, *On a 2-knot with nontrivial center*, Bull. Austr. Math. Soc. **25**(1982) 321–326

[984] K. Yoshikawa, *On 2-knot groups with abelian commutator subgroups*, Proc. Amer. Math. Soc. **92**(1984) 305–310

[985] K. Yoshikawa, *A ribbon knot group which has no free base*, Proc. Amer. Math. Soc. **102**(1988) 1065–1070

[986] K. Yoshikawa, *Certain abelian subgroups of two-knot groups*, in "Knots 90" (Osaka, 1990), Walter de Gruyter (1992) 231–240

[987] K. Yoshikawa, *An enumeration of surfaces in four-space*, Osaka J. Math. **31**(1994) 497–522

[988] E. C. Zeeman, *Unknotting spheres*, Ann. of Math. **72**(1960) 350–361

[989] E. C. Zeeman, *Linking spheres*, Abh. Math. Sem. Univ. Hamburg **24**(1960) 149–153

[990] E. C. Zeeman, *Isotopies and knots in manifolds*, in "Topology of 3-manifolds and related topics" (Georgia,1961), Prentice-Hall (1962) 187–198

[991] E. C. Zeeman, *Unknotting combinatorial balls*, Ann. of Math. **78**(1963) 501–520

[992] E. C. Zeeman, *Twisting spun knots*, Trans. Amer. Math. Soc. **115**(1965) 471–495

[993] R. B. Zhang, *Braid group representations arising from quantum supergroups with arbitrary q and link polynomials*, J. Math. Phys. **33**(1992) 3918–3930

[994] C. J. Zhu and D. M. Tong, *Some important representations of the braid groups B_n*, Acta Sci. Natur. Univ. Jilin. **1991**(1991) 51–55

[995] G. M. Ziegler, *Higher Bruhat orders and cyclic hyperplane arrangements*, Topology **32**(1993) 259–279

[996] B. Zimmermann, *On groups associated to a knot*, Math. Proc. Cambridge Philos. Soc. **109**(1991) 79–82

Index

309

suspension, 53
symmetric braid word, 144
symmetric group, 10
symmetrically equivalent, 144
symmetry, 83
Symmetry theorem on chart description, 269
Symmetry theorem on twist-spun knots, 82

tame, 28
tetrahedral move, 60
tomography, 58
topological link, 27
topological surface link, 53
trace, 63
trace map, 26
transversal homotopy, 262
trefoil, 36
triple point, 57
trivial, 29, 53, 67, 271, 272
trivial braid, 8
trivial pointed braided surface, 119
trivial surface braid, 107
trivialization, 28
tubular neighborhood, 28
twisting, 83
type α, 47, 193
type β, 47, 193
type γ, 193
type-I bubble move, 59
type-I saddle move, 60
type-II bubble move, 59
type-II saddle move, 60
type-III move, 60
type $\mathcal{R}$, 162

unbraiding sequence, 162
under crossing, 32
under edge, 32
unknot, 29
unknotted, 29, 53, 233, 271, 272
unknotted braid system, 235
unknotting conjecture, 55
unknotting number, 259
unknotting theorem, 30
upper level relator, 92
upper presentation, 36, 92
upper relator, 36, 91

vertical product, 113

white vertex, 135
wild, 28
Wirtinger presentation, 230